The Physics and Metaphysics of Transubstantiation

Mark P. Fusco

The Physics and Metaphysics of Transubstantiation

palgrave
macmillan

Mark P. Fusco
Warrenton, VA, USA

ISBN 978-3-031-34639-2 ISBN 978-3-031-34640-8 (eBook)
https://doi.org/10.1007/978-3-031-34640-8

This Palgrave Macmillan imprint is published by the registered company Springer Nature Switzerland AG.
The registered company address is: Gewerbestrasse 11, 6330 Cham, Switzerland

For my parents—Peter and Eleanor, my brother and sister-in law—Francis and Erica, and my nieces and nephew Emilia, Sophia, Chiara, Maria, Peter, and Michaela and grand nieces and nephews Cleo, Daisy, Rocco, and Titus.

Preface

The postmodern world has in many ways parted ways with traditional religious thought and their explanations of reality. In this work, I critique the concept and event of transubstantiation—the Eucharistic doctrine, which teaches that the moment of consecration of bread and wine are transformed into the real body and blood, soul and divinity of the person Jesus Christ—from the perspectives of sacramental and systematic theology, philosophy, and metaphysics. This assessment serves to ground my speculative reading of Ur-Kenotic Trinitarian identity and warrants applying these findings to the act of transubstantiation in the liturgy. While I hold that transubstantiation is a supernatural occurrence, I also hypothesize that it also perfects the principles and physical laws that determine natural reality. Consequently, I place my speculative reading of transubstantiation in conversation with post-Newtonian physics, that is, my reading of the theory of quantum mechanics, information theory, and black hole entropy. The findings of each of these disciplinary approaches have encouraged my understanding of essential identity based on holographic principles. The term I apply to this new understanding of metaphysics is "holo-cryptic," and its perfect embodiment as a spiritual or essential being is referred to as "holo-somatic." Thus, I explore reality centered on what results when one outlines metaphysics based on how the holo-somatic image conveys past philosophical and theological systems articulated in the idea of essential identity's historical and physical existence. Finally, a correlation is drawn between the transformation of the Eucharist and the properties of the quantum object as wave and particle, quantum superposition, and teleportation as a holo-somatic informational structure.

Warrenton, VA, USA
February 2021

Mark P. Fusco

Acknowledgments

While I am responsible for any shortcomings of my speculative holocryptic metaphysical approach and its application to Eucharistic theology, simultaneously, it is also true that any lasting value in this work follows from the thoughts, criticisms, and support of family, friends, and colleagues.

Lacking the support and encouragement of Francis E. Fusco and Dr. Richard L. Fern, this present work would have never reached print. I can see in retrospect that the seeds of this thesis originated in countless conversations I shared with Francis and Richard. At Yale University, I had the opportunity to study under Richard and have been blessed by his friendship and mentoring ever since. It is impossible to overestimate the encouragement, review commentaries, conversations, notes, and editing that Francis and Richard provided in seeing this work to its completion. Without their involvement, what follows would never have moved beyond semi-formed, radically ungrounded musings. This work testifies to the genius of these two remarkable men.

The theological and philosophical principles that this book develops can be traced back many years and took root, in part, from conversations with friends (David C. Schindler, Timothy Fortin, Mark Cole, Noel Sweeney, and George Bodnar), at annual retreat (Timothy Kaiser, Sigurd Thage Peterson, John Conmy, and Mark Hunter), and at my brother's weekly "Garage Meeting" (John Mallory, Sean Blanchette, Matt Pierce, Pete Belleville, James Scrivener, Nicholas Santschi, and Eliot Brubaker). At these symposiums, I was introduced to and challenged by a diverse group of opinions and disciplinary expertise. The accompaniment of all the above individuals has been key voices in my intellectual formation.

I was fortunate that Sr. Gill Goulding, CSJ and Robert Sweetman guided my theological and metaphysical perspectives during my doctoral studies in systematic theory at the University of Toronto. From the former, my understanding of Hans Urs von Balthasar's theology was developed, and from the latter, my grasp of Thomas Aquinas' metaphysics was honed. This work draws upon and develops upon my doctoral work. I am grateful for the support of St. Michael's and Regis Colleges, as well as the Jesuit community at this world-class institution. The support and friendship of Peter Nguyen, S.J., and Elton Fernandes, S.J. helped make possible the foundational work upon which this book rests.

Most recently, the support of the Jesuit Community at Santa Clara University must be recognized. I am especially grateful for the support of fellow Jesuits Paul Mariani, S.J., Paul Soukup, S.J., Joseph Maria Israel, and Salvatore A. Tassone, S.J. I must recognize two further Jesuits—James Blaettler, S.J., who generously shared his expertise regarding important artistic pieces and movements that helped set my project within the larger tapestry of art history, and Fady Chidiac, S.J. for our theological conversations about the mathematics of infinity. Thankfully, the formatting of the submitted draft was the result of the generous work of Maria Joseph Israel, S.J.

I wish to further thank Aleksandar I. Zecevic at Santa Clara University for sharing his time and insights into the nature of quantum mechanical theory, science, and religion. To my colleagues at the Department of Religious Studies—thank you for your support and friendship. I also wish to acknowledge the financial assistance provided by the university's Franzia Foundation.

A special debt is also owed to Kevin C. Spinale, S.J., James R. Blaettler, S.J., David Pinault, Paul Crowley, S.J., and Monica Lavia for their comments on some of the earliest versions of the manuscript. I am especially grateful for the confidence Dr. Phil Getz has put in this work and me. Finally, I would like to thank the Palgrave Macmillan and Springer team, in particular Tikoji Rao Mega Rao, Nirmal Kumar Gnana Prakasam, Susan Westendorf, Eliana Rangel and Rachel Jacobe.

Contents

Acronyms

AE: Eric Przywara, *Analogia Entis: Metaphysics: Original Structure and Universal Rhythm*. Translated and Introduced by John R. Betz and David Bentley Hart.
Being and Event: Alain Badiou, *Being and Event*.
Brevil: *Bonaventure, Breviloquium* in *The Complete Works of Bonaventure: Cardinal Seraphic Doctor and Saint*.
The Bride of the Lamb, Sergius Bulgakov, *The Bride of the Lamb*.
Church Dogmatics, Karl Barth, *Church Dogmatics* (31 vols).
De Anima: Aquinas, *Commentary on Aristotle's De Anima*.
De pot: Aquinas, *Quaestiones Disputate De Potentia Dei: On the Power of God*.
De Trin: Augustine, *On the Holy Trinity*.
De ver: Aquinas, *The Truth*.
Dict. Non-Phil: François Laruelle, *Dictionary of Non-Philosophy*.
Echo Objects: Barbara Maria Stafford, *Echo Objects: The Cognitive Work of Images*.
Emblems of Mind: Edward Rothstein, *Emblems of Mind: The Inner Life of Music and Mathematics*.
Euclid's Window, Leonard Mlodinow, *Euclid's Window: The Story of Geometry from Parallel Lines to Hyperspace*.
GL: Hans Urs von Balthasar, *Glory of the Lord*, 7 volumes.
God Without Being, Jean-Luc Marion, *God Without Being*.
The Grand Design, Stephen Hawking and Leonard Mlodinow, The Grand Design.

Hex:	Bonaventure, *In Hexaemeron*.
	Homo Abyssus, Ferdinand Ulrich, *Homo Abyssus: The Drama of the Question of Being*.
In Metaph:	Aquinas, *Commentary on Aristotle's Metaphysics*.
Key Concepts:	Alain Badiou, *Key Concepts*.
LA:	Balthasar, *Love Alone is Credible*.
LG:	*Lumen Gentium Dogmatic Constitution on the Church*
Logics of Worlds:	Alain Badiou, *Logic of Worlds: Being and Event*, 2.
MP:	Balthasar, *Mysterium Paschale*.
NPNF1:	*The Nicene and Post-Nicene Fathers of the Christian Church*. Series 1.
NPNF2:	*The Nicene and Post-Nicene Fathers of the Christian Church*. Series 2.
ONE:	The Pure Transcendent Being
One:	Finite appropriations of the ONE.
PG:	*Patrologiae Cursus Completus: Series Graeca*. Ed. J-P Migne. Paris, 1857-66.
PL:	*Patrologiae Cursus Completus: Series Latina*. Ed. J.-P Migne. Paris, 1841-94.
QMO:	*Quantum Mechanical Object*.
QMT:	*Quantum Mechanical Theory*.
SC:	*Sources chrétiennes*. Paris, 1941-.
SCG:	Aquinas, *Summa Contra Gentiles* (5 vols.).
Sent:	Bonaventure, *Commentary on the Sentences*.
	Smith, Wolfgang Smith, *The Quantum Enigma: Finding the Hidden Key*.
ST.:	Aquinas, *Summa Theologiae:* Complete English Edition in Five Volumes, Translated by the English Dominican Province.
	The Pillar and the Ground of Truth, Pavel Florensky, *The Pillar and the Ground of Truth: An Essay in Orthodox Theology in Twelve Letters*.
TD:	Balthasar, *Theo-Drama, 5 volumes*.
TL:	Balthasar, *The Theo-logic, 3 volumes*.

CHAPTER 1

Hidden Worlds

1.1 Introduction: The Mystery of the Infinite and Nothingness Under and In All

The contemporary study of theology has found its greatest challenge and most pressing critique along two fronts—that of postmodern thought and the apparent irreconcilability of many faith tenets with a scientifically based worldview. The term postmodern is an ambiguous term. Generally speaking it makes claim that an all-embracing and systematic explanation of truth and being is no longer possible.[1] Ironically echoing what they critique, postmodern thinkers eulogize the end of overarching "grand narratives" and replace them with a pseudo-religious "faith" in truth's unraveling with uncertainty. Given the epistemic act's inevitable favoring of the self-referential over sustained exploration of the supernatural, unknown other, and ultimately the Transcendent, these movements arguably lead to an individual's alienation from communal life and Nature.[2]

[1] See Lee Braver, *A Thing of This World: A History of Continental Anti-Realism: An Outline* (Evanston, IL: Northwestern University Press, 2007) and John Mullarkey, *Post-Continental Philosophy: An Outline* (London, UK: Continuum International Publishing Group, 2006). Hereafter cited as *Post Continental Philosophy.*

[2] Postmodern thought is a complex movement that exceeds what is possible in this work. See Jean-François Lyotard, *The Postmodern Condition: A Report on Knowledge*, Translated by Geo Bennington and Brian Massumi (Minneapolis, MN: University of Minnesota Press, 1983).

M. P. Fusco, *The Physics and Metaphysics of Transubstantiation*,
https://doi.org/10.1007/978-3-031-34640-8_1

The postmodern funeral dirge seems to leave little room for the Roman Catholic Church's centering of its liturgical life on the Eucharist or a traditional metaphysical approach to reality.

In what follows, I argue that current philosophical questions and what I consider to be the repeatable and falsifiable statements of quantum mechanical theory (QMT) can be profitably revisited, translated, and analogously adapted to make the doctrine of transubstantiation accessible to our post-Newtonian world. This move is justified by setting out the conceptual framework and principles of a new way to perform metaphysics. Thomas Aquinas (1225–1274) reasons that one philosophizes to know the truth of a thing and not personal opinions. Although each discipline has its own tools, methods, and languages, they all purport to stipulate some informational truth about being: "contrary to common opinion, method shares with logic its inability to separate itself completely from its context. There is no method valid for every domain, just as there is no logic that can set aside its objects."[3] On this basis, I suggest that no approach is complete; each approach parlays something of the parallax of perspectives that derive from a subjective understanding of a discipline's methods and findings.

It is neither theology's task to verify a scientific view of reality or mimic its methodological quantification nor the task of science to promote a particular view concerning the Transcendent or the world. Every theory about reality benefits from a cross-disciplinary approach, as truth is an interdisciplinary revelation. This cross-fertilization is not only possible but also inevitable if one believes Max Planck (1858–1947), who sees no contradiction between faith and reason.[4] Theologians such as Hans Urs von Balthasar (1905–1988) criticize formal systems for often losing sight of being's transcendental orientation by abstracting its relational capacities. Most mathematical and scientific explanations of physical reality are adamant that traditional metaphysical and philosophical appropriations of being are irrelevant to their own projects. Thinkers who defend and consider both disciplines necessary will profit from this study.

My task in this book is to outline a new approach to metaphysics that I call holo-cryptic metaphysics. The word metaphysics originally combined

[3] Giorgio Agamben, *The Signature of All Things: On Method* trans. Luca di Santo and Kevin Attell (New York, NY: Zone Books, 2009), 7.

[4] Max Planck, *Where is Science Going?* Translated by J. Murphy (New York, NY: W.W. Norton Co., 1932), 214.

"meta" meaning after and "physics" as Aristotle's seminal work on the subject followed his book on natural physics in an earlier collection of his works. I choose the designation holo-cryptic because it synthesizes the word "holo" for image and "cryptic" for the mysterious or secret. My approach makes explicit being's physical and configurationally immaterial properties as a holographic image. This correspondence is not a simply reworking of past so-called word-picture philosophies of language but rather a metaphysics that takes seriously that a primitive physical object or being is symmetrical to the truth of its essential or informational identity. The startling nature of scientific discoveries and current theoretical physics inspires our belief that this correlation of being to information moves beyond simple description.

Studies of thermodynamics and quantum mechanics show that a physical being's identity is convertible to its formal quantification and representation as a phase state or inversely its configuration space. The central theme of this book is that this exchange concludes in a being's physical effulgence as a holographic image. This principle is then applied to the mystery of transubstantiation insofar as it provides an analogy for how the resurrected Jesus Christ images Godself's real presence in the Eucharist. The liturgy is a sacramental concession and celebration of material reality's characteristic property of being completely transformable into ordered or "informational energy". More specifically, a holograph offers a new perspective on a being's essential identity as photic energy (light). This is not seen as a justification for Manichean doctrine but rather revisits how the soul-body relationship finds an analogy in theoretical physics. Christian theologians such as Thomas Aquinas give us a way to understand how a soul embodies or informs the body—a state that begins at conception and remains until an individual's physical death. Furthermore, Christianity holds that with an individual's resurrection, the person's nonmaterial soul continues to exist. Consequently, in the following work, I develop how a traditional definition of "essence" or "form" can be analogously understood as information communicated as an image. When released or liberated from its physical limitations, the soul is able to exist in any number of spatiotemporal configurations or complex dimensions. If this were not the case, it would be impossible for the resurrected person's soul to experience her promised immortality after death. This personal theonification is made possible given that the Trinity's eternal decision exists as a structuring component of the created order. The historical order and the human person participate in this divine decree.

I use the term holo-somatic to refer to the application of holo-cryptic metaphysics to created reality. In this way, the concept holo-somatic introduces how an individual being's identity can be seen to exist as a holographic image in space-time. The question of the essential nature of a being's historical embodiment is addressed in the concept of the holo-somatic. The holo-somatic reality does not demand that we accept crude correlative theories of body-soul dualism or object-oriented philosophies, which pits an irreducible conflict between the physical and spiritual. Through baptism and the Eucharist, the Christian participates in Jesus Christ's resurrected state and this "imprint" I term holo-somatic. The soul cannot eternally form a union with a material body but can perfectly unify with its energetic identity at one's resurrection.

Holo-cryptic metaphysics derives its principles and concepts from interpreting previous scientific, philosophical, and theological interpretations in light of the findings of post-Newtonian physics, postmodern thought, and a revision of Thomistic metaphysics given a new understanding of Trinitarian hypostatic relations as analogously "differing" or Ur-Kenotic. Below, we briefly outline how each of these three areas justify my forwarding a theory of holo-cryptic metaphysics.

Now, Aristotle's teaching on physical reality cannot be read without criticism. Holo-cryptic metaphysics shows how and why Aristotelian hylomorphic thought must be revised in light of quantum mechanical theory (QMT), Einstein's general theory of relativity (GTR) and special theory of relativity (STR), as well as information's entropic and dimensional manifestation in black hole radiation. If a philosophy of metaphysics is to have any validity for today's mathematically and scientifically astute individuals, it must adapt in conversation with these theories. Holo-cryptic metaphysics is a dialog partner in a diverse "Theo-philo-scientific" discussion that centers on how the holographic representation of being advances our understanding of the ontological and noetic properties of being. Holography gives us a way to speak about a being's physical state within three dimensions and its translation into a nonmaterial informational state within complex multidimensional and spatial configurations. In the extreme physical conditions of a black hole, a physical being's translation into its holographic representation is realized. The region that separates and bridges the black hole's event horizon from the rest of the universe becomes for me a physical analogate for the mereological question. From the perspective of the universe's side, the black hole localizes such an intense gravity that no escape velocity is possible when matter is captured by its hold. Situated within the black hole, one can look upward to the underside of the event horizon. The surface gravity here acts as a boundary condition or metaphoric screen upon which the Quantum Mechanical Object's physical properties can be imagined to manifest as a holographic identity.

The QMO's holographic identity translates its physical properties both as an excitation and thermodynamic perturbation and as an inverse complex dimensional indentation. Cosmologically, matter's holo-cryptic identity at the event horizon cyclically displays physically as electromagnetic radiation. These eruptions from within the black hole's core bounce over the black hole's membrane and inwardly in-fold as a field leaving behind a dimensional imprint. The point of contact between the interior and exterior, physical and immaterial forces on the event horizon is a thermal equilibrium point or bulk dimensions that communicate the QMO's physical and informational state. This point stabilizes physical identity in its symmetrical essential or informational identity one I take as an analogy for traditional philosophical configurations of the essence's relationship to existence.

Postmodern thinkers such as Alain Badiou (1937–) and François Laruelle (1937–) allow for genuine insights into the nature of Trinitarian hypostatic relationships and the metaphysical properties of formal entities such as numbers and the infinite. The former theme, postmodern atheism, challenges past theological assumptions about the Trinitarian community. A response to these concerns can be seen in Hans Urs von Balthasar's theological vision. Jesus Christ's questioning of the nature of divine abandonment on the Cross is analogously reread by him into the heart of divine consciousness. Thus, the Transcendent eternally considers the nature of Jesus Christ's thoughts about Godself's own nonexistence. The Divine's thoughts about nonexistence are not proof of negativity's universal application, but rather that God's mysterious eternal positive configuring of Nothingness and relational ordering makes possible a Creation writ large with created freedom and therein, charitable works and faith. Furthermore, Badiou's work on the philosophy and metaphysics of number provides a challenge for my arguments about the formal basis of holo-somatic metaphysics. This correlation is analogously evident in how a particular number formally or "holographically" mirrors in an infinite series. A number's irreducible identity makes it a suitable member of an infinite set. The number's position within a mathematical set metaphorically presents what theologians make explicit when speaking of Nature's plenitude and a being's immortality.

Whereas some postmodern philosophers argue that difference's universal presence in the created order follows when Pure Being or the ONE is defined by its antithesis, the non(One), Christians hold that the ONE or God is an eternally actual community of three Divine Persons. Ur-Kenosis explains monotheistic Transcendent identity realized in the perfect relational unity as three unique Divine Persons. The Divine's essence and hypostatic relations are mutually defined as Ur-Kenotic, that is, as an eternal ordering. The identities of the Father, Son, and Spirit are absolute, as they are inseparable and equal. This speculative move builds upon the previous theological work done by Thomas Aquinas and Hans Urs von Balthasar.

The human person mirrors divine identity in its real and non-subsisting relationships. The most important of these correspondences define a being self-reflexively, in relationship to the other beings in the universe and apophatically in accordance with the Transcendent. The constellation of these alliances ground finite identity in heno-triadic truth.[5] The finite person exists as a tri-relational identity of essence, esse, and *actus essendi* given its holo-somatic modeling of the changeless Ur-Kenotic identity of the Transcendent Divine Persons. This union actualizes in Creation given the Spirit's sharing of Godself's Absolute Freedom throughout reality. Holocryptic metaphysics proposal that finite being's identity resolves as a "one-in-three" relationship develops many theoretical threads including Ferdinand Ulrich's (1931–2020) concept of transnihilation.

In this chapter, *The Mystery of the Infinite and Nothingness Under and In All*, I begin to sketch out how a balance can be struck between a scientific and theological account of reality. More specifically, finite reality is conceptually situated between the two poles of nothingness and infinity. No finite being, Thomas argues, is the total cause of any effect.[6] Postmodern philosophy is a variegated and byzantine project that rarely makes explicit how its hermeneutical methods, presuppositions, or definitive schemes are shared with theology. However, postmodern philosophy's importance to faith traditions should not be discounted, even with its glib rejection of theological tenets. At the very least, one must recognize that postmodern themes articulate and drive many of the assumptions that underlie the common consciousness of the twenty-first-century individual. To con-

[5] The term "heno-triadic" from the Greek word "*henos*" for one and "*tri*" for three from the Latin offers a shorthand way to talk about divine nature and action that remains Trinitarian and simple.

[6] See Thomas Aquinas, *Quaestiones Disputate De Potentia Dei*: *On the Power of God*, Translated by the English Dominican Fathers (Westminster, MD: The Newman Press, 1952), 3, 1–4, accessed at Html. edition by Joseph Kenny, O.P. at. https://isidore.co/aquinas/QDdePotentia3.htm#3:3., Hereafter cited as *De pot.* See also Saint Thomas Aquinas, *On Evil*, Translated by Jean Oesterle (Notre Dame, IN: University of Notre Dame Press, 1995).

vince contemporary persons that hylomorphic theories of physical reality are applicable, one must take seriously postmodern philosophy and post-Newtonian physics.

The QMO's analogous heno-triadic identity is defined physically as a wave (ψ) or particle, in correspondence with other physical particles and forces and the universe as a whole. Theologically we find analogously the modalities of relational difference that finite beings experience as grounded in the logic of an Ur-Kenotic Creator. The hypostatic filiation among the Divine Persons reflects the self-giving prodigality of divine love and simplicity. Through hypostatic union and a shared divine essence, the Trinitarian communion is one perfect identity of three uniquely definable persons. The eternal mutual self-ordering and absolute transparency among the Divine Persons defines their identity as Ur-Kenotic. The event of transubstantiation ritualistically mirrors Trinitarian identity by highlighting the metaphysical principles of being, infinity, and nothingness as part of physical reality's free, emergent, and non-subsistent movements. Nothingness is not an invasive force. Its foremost purpose is to channel the hypostatic and incarnate identity of the Transcendent Son in the physical universe.

The concept of Ur-Kenosis influences the holo-cryptic metaphysics critique of postmodern accounts of being, as this theological perspective enables a better way to understand created reality as represented in a post-Newtonian world. QMT has led to the most precise and repeatable experimental results. No serious scientist rejects the findings of QMT, and few postmodern educated non-specialists doubt that this theory best explains the nature of physical reality at impossibly small scales. However, this advancement has had the unfortunate side-effect of suggesting to some that all truth is horizontal—that is, that the miraculous can be reduced to the measured. QMT presents the doctrine of transubstantiation with one of its most imposing challenges and paradoxically its greatest opportunity. The calculation of the exact point of contact, if any, between a QMT worldview and the created order envisioned by the doctrine of transubstantiation cannot be fully developed in a single work. Indeed, I intend only to establish what a nonscientist can perceive from the periphery that might be promising avenues for a metaphysics that respects science. It is risky even to suggest such a course of action given theological dogma's reluctance to directly embrace mathematical and scientific quantifications and qualifications of reality.

A being is variously described in accordance with one's theoretical presuppositions, but some common frames include form-matter, part-whole, definitional context, top-down (supervening) correlations, and so on. We synthesize and advance these conceptualizations as the basis of our new interpretation of holo-cryptic and holo-somatic metaphysics. Being receptive to the truth's manifestation in physical reality and even in the concept of the void enables the specialist and non-specialist alike to begin to apprehend beauty's multiple expressions in Nature. Pictorial arts are especially hopeful in these investigations. Philippe Vandenberg states, "art is the result of the misfortune, the anxiety and the fear of the void that haunts imperfect man. He attempts to reconcile this with his search for the absolute. This reconciliation, which is admittedly temporary, may pass through some form of absence."[7] Directed by these insights, holo-cryptic metaphysics models reality in a way that comprehends the concepts of multiplicity, nothingness, void, and infinity as means to better understand Ur-Kenosis. Holo-cryptic metaphysics argues that the created order embodies the Nothingness found at Creation. A subject's humble openness to the Transcendent's measuring or revelation means one accept also that negativity in all its forms, serves to order material reality to its perfection in its Theo-aesthetic and ritualistic transmutations.

The second chapter, *Ur-Kenosis*, hinges on the groundwork done by the Swiss theologian Hans Urs von Balthasar. The Ur-Kenotic communication of the hypostatic relations opens the Trinitarian economy to reality's descriptions in an array of non-subsistent and nonlinear transversal and vertical stimuli. Paradoxically, by means of an anabatic reach, our interior desires and conscious operations follow the Transcendent's own annexing of finite reality in history within the context of the divine's providential plan: "[t]he paradox of the spiritual creature that is ordained beyond itself by the innermost reality of its nature to a goal that is unreachable for it and that can only be given as a gift of grace."[8] In its point of origins and subsequent material actualization, Creation and consciousness manifest this spiritual informing—the metaphoric "geometric" principles of finite kenosis. Similarly, human consciousness' receptive and cognitive capacities depend on its ability to incorporate and optimize the region

[7] Philippe Vandenberg, "What Counts is Kamikaze," in *Philippe Vandenberg: Œuvre 2000–2006*: *La Folie Ne Laisse Pas D'oeuvre* (Philippe Vandenberg, 2206), 94.

[8] See also Hans Urs von Balthasar, *Theo-Logic*, vol. 2: *Truth of God*, Translated by Adrian J. Walker (San Francisco, CA: Ignatius Press, 1985), 181. Hereafter cited as *TL* 2., Thomas Aquinas, *Summa Theologiae*, Translated by the English Domican Province, (West Minister, MD; Classic Christian Classics, 1981). Hereafter cited as *ST. ST.* I–II, 5, 5–7 and *ST.* I, 43, 3: "By the gift of sanctifying grace the rational creature is perfected so that it can freely use not only the created gift itself but also enjoy the Divine person Himself; and so, the invisible mission takes place according to the gift of sanctifying grace; and yet the divine person Himself is given."

within the mind ready to receive information. These cognitive places or "no-places" are more properly understood as part of reason's transcendental momentum, as Balthasar comments, "in its origin… an action, an expression, a clearing, [and] a bestowal of participation."[9]

The French Maoist philosopher Badiou's position recalls Balthasar's theory, but as an atheist his conclusions about being and consciousness reject the existence of an Actual Transcendent Being. Where some Christian theologians see consciousness revealed in the Divine's "Ur-Kenotically" defined omnipotence, for Badiou, the One's ordering is a sign of material being's "fragmented" knowledge and reduced self-sufficiency. Badiou writes "the One is that which inscribes no difference in the subject to which it addresses itself. The One is only insofar as it is for all; such is the maximum of universality when it has its root in the event."[10] Badiou calls the human subject to continually ferment negativity, destabilization, or chaos. However, in Badiou's philosophy, a certain sense of co-belonging and mutuality reveals in finite relational ordering. Without a "ladder" to the Transcendent, Rowan Williams rightly notes of the self-enclosed universe, that its "difference, then, is always plural, is always more; it is always opening to more. Difference is always the ordering of difference. Difference is never only two, even when it is 'a' difference between two things, because that difference will make a difference."[11]

Causes are never fully equivocal or univocal. Holo-cryptic metaphysics explains that this state is possible given the continued presence of the Nothingness by which the Transcendent creates. Nothingness is not an invasive force but the medium that guarantees finite freedom and non-subsistent relationships. It is perhaps the same interpretative spirit that influenced Dun Scotus to note that common nature is also "less-than-numerical" unity. Nothingness' foremost purpose is to channel the hypostatic and incarnate identity of the Transcendent Son into the physical universe. Ur-Kenosis inaugurates divine identity as eternally relational and thus in a manner that fully engages the postmodern preference to demarcate human experience by

[9] See Hans Urs von Balthasar, *Theo-logic Theological Logical Theory*, vol. 1 *Truth of the World*, Translated by Adrian J. Walker (San Francisco, CA: Ignatius Press, 2000), 217. Hereafter cited as *TL* 1. 217.

[10] Alain Badiou, *Paul: The Foundations of Universalism*, Translated by Ray Brassier (Stanford: Stanford University Press, 2003), 76.

[11] Rowan Williams, "Afterword: Making Difference," in *Balthasar at the End of Modernity*, Forward by Fergus Kerr and afterword by Rowan Williams (Edinburgh: T&T Clark Ltd, 1999), 127. Hereafter cited as *End of Modernity*.

formulations such as distance, alienation, abyss, separation, difference, and diastasis. Thus, a way to participate in the plenum's ongoing introduction of new beings called to exist freely as integrated parts of a larger whole is referenced in faith and secular systems.[12] At the deepest ontological level, every created being participates in this ordering as part of its interior nature because proportional relationships are necessary for personal identity. In other words, proportions and analogies bring out the similarity between relationships, not between natures. Analogies invite conscious thought and free action. Thomas argues, that this concession supports the vital correspondence of being's identity as essence and esse.[13]

Ontological ordering can be taken to be the primordial fact (*Ur tatsache*) of finite existence because it speaks to the question of "What it is to be?" by asking the essential question, namely, "What is it, to be something?" The limits that actual existence places on a being's essential nature and desires are relative and are perhaps better thought of as permutable veil behind which the Transcendent resides. A physical being's fleeting appearance is part of the immaterial essence's virtual hold on the Transcendent insofar as both witness the causal force of Creation's exemplar source projecting through Nothingness (*creatio ex nihilo*) in the created order.[14] The essential-esse relationship internally qualifies and externally quantifies a being by revealing its substantial and accidental properties in the world.[15] A human person's existence in the world, for example, holo-somatically images that person's interior ontological ordering as part of their appearance and historical actions in the world (e.g., *resipiscence*).

Holo-cryptic metaphysics speculates on how the soul, consciousness, and ontological ordering all manifest in a being's historical existence and the essence's reformulation in terms of informational constructs such as number and interior geometries. The impenetrable barrier separating finite beings

[12] Hans Urs von Balthasar, *Theological Dramatic Approach*, vol. 4: *The Action*, Translated by Graham Harrison (San Francisco, CA: Ignatius Press, 1994), 345. Hereafter cited as *TD* 4. See *ST.* I, 6, 3, resp. 3.

[13] See *ST.* I, 3, 6 and John F. Wippel, "Essence and Existence in Later Medieval Philosophy," in *Cambridge History of Later Medieval Philosophy* (Cambridge, UK: Cambridge University Press, 1982), 385–410.

[14] See Joseph Owens, "The Accidental and Essential Character of Being in the Doctrine of St. Thomas Aquinas," *Medieval Studies* 20, no. 1 (1958): 40.

[15] George P. Klubertanz, *Introduction to Philosophy of Being*, 2nd ed. (New York, NY: Appleton-Century Crofts, 1963), 80–83. See also *ST.* I, 75, 5 and *GL* 5, 650–56.

from the Transcendent testifies to their asymmetric correspondence. The need for finite consciousness to draw boundary conditions around all beings follows from the more fundamental division between the Transcendent and Creation. However, if we take seriously the scriptural account of humanity's identity as imago Dei, then the chasm or Nothingness that separates Divine and created order is bridged in finite being's holographic representation. Divine thought apophatically discloses itself in created reality and infuses the material order with semantic significance.[16] The ability to project one's interior measure as part of one's rational and moral life is made possible by the Divine's *a priori* revelation of the divine essence and hypostatic relations in Jesus Christ's holo-somatic actualization of the Trinity. In the person of Jesus Christ's human, and divine nature, we see an example of how holo-somatic identity materially and historically actualizes a relationship to the Transcendent Father. As theorists and researchers continue to reveal the hidden subatomic world, a disciplinarily relevant view of our own holo-somatic metaphysics obliquely divulges itself in the natural order when read through the interpretative lens of QMT as part of a larger speculative metaphysics. A syntactical difference exists between a QMO's motion and human action. This correspondence is analogous to reality's dualistic preference to exist as either matter or energy. The QMO's inimitability is reflexive of its state as wave or particle and their shared holographic identity within physical reality. Where theologians speak of nothingness, image, and exemplar, quantum mechanics define the QMO in terms of its relational identity and boundary conditions or surfaces. Nicholas J. Teh represents the advent of the QMO's holographic identity in terms of its logical structure. Thus, on a QMO's internal mathematical functions and symmetrical properties, he writes: "[e]mergence is an asymmetric relation between a top theory X and a bottom theory Y such that X displays novel features with respect to Y."[17]

The physical accidents and internal architecture that characterize a being's state of constant change under evolutionary and erosive forces in Nature promote its ongoing identity by means of non-subsistent relationships. Mathematical set theory gives us one way to model a being's theoretical infinite change. The mathematical set allows for the infinite

[16] See Jacob Böhme, *Works of Jakob Böhme: The Teutonic Philosopher*. vol. 4 (Whitefish: Kessinger, 2003), 9–10. See also: William Perkes Swainson, Jacob Bohme: The Teutonic Philosopher, Transcribed by Wayne Kraus for Jacob Bohme online.com.

[17] Nicholas J. Teh, "Holography and emergence," *Studies in History and Philosophy of Modern Physics Science B: Studies in History and Philosophy of Modern Physics* 44 (2013): 303. Hereafter cited as *Holography and Emergence*.

abstraction and synchronization of an individual being or number around a defined relationship. Every number within a set is open to further transformation by its placement within equations, sets, or new geometric placements, and these propositions are determined as valid or invalid given one's interpretation and application of the standards of consistency and inconsistency. A numeric being's reflexive and reflective identity is closely aligned with the question of how a physical being's authenticity is inseparable from its corresponding sequences. Concurrently, without a number's capacity for differentiation the meaning of a number's representation is strained when the mathematical terms themselves become indistinguishable.[18] This mathematical interpretation revisits the scholastic formulation of being that states that the cause is responsible not only for the being of the effect (*esse effectus*) but also for its form (*forma effectus*). If a being's essential identity is realized within the context of infinity's representation within a mathematical set, it becomes possible to discern its physical or conceptual boundaries without sacrificing its openness to potentially unlimited new correspondences, sequential placements, and reconfigurations.

A mathematical set's logical structure guarantees the malleability of each of its members (e.g., numbers or beings) by allotting an evaluation of its enigmatic potential nature. The notions of consistency, inconsistency, multiplicity, and infinity are philosophical properties that Badiou employs to define a being's potency.[19] In comparison to Badiou's conceptualization of universal metamorphism by means of mathematical sets, holo-somatic metaphysics is guided by re-interpreting conceptual ordering as a matter of a being's freedom and relationship to an existing Transcendent Being. More precisely, an individual being's liberty is holo-somatically defined by theology's overarching categorical realization of a being's spatiotemporal existence as an eschatological state that images the resurrected body. Consequently, while both approaches normalize a being or number's analogous biological growth, decay, and conscious intelligibility, my metaphysical approach sees in reality and being's prevalent states of consistency (identity) and inconsistency (emergence) a holo-somatic embodiment and spatiotemporal holo-

[18] See Ignacio Jane, "The Role of the Absolute Infinite in Cantor's Conception of Set," *Erkenntnis* 42, no. 3 (May 1995): 379. Quoted in: Kenneth A. Reynhout, "Alain Badiou: Hidden Theologian of the Void?" *The Heythrop Journal* (2011): 227. Hereafter cited as *Hidden Void*.

[19] See Alain Badiou, *The Subject of Change: Lessons from the European Graduate School*, Edited by Duane Rousselle (New York, NY: Atropos, 2013), 41–51. Hereafter cited as *The Subject of Change*.

graphic reflection of our identity as the Imago Dei. In this way, the human person's identity is sourced in divine immutability and eternity.

The divine substance (ὀμοουςιος) presents in free hypostatic relationships that requires no further additive reflexive or reflective associations. Given the Divine's absolute simplicity, it seems a natural choice to create from Nothing (*creatio ex nihilo*) and know everything: "now not only does God understand many things by his essence...In addition, this means he understands the several types of things; or that many ideas are in his intellect understood by him. ... Such relations, whereby ideas are multiplied, are not caused by the things themselves but by the divine intellect comparing its own essence with these things."[20] Creation's "Edenic layer" is built upon Nothingness' capacity for endless metaphysical and moral qualification. The concept of "Nothingness" is not a fabrication made by febrile minds possessed by mythic misapprehension favored in the prescientific past but rather a means to analogously investigate the mystery of creation in light of such physical realities as black holes and the cosmic void.

The divine exemplars project the Transcendent's thoughts in a non-evasive manner from Nothingness that is in a manner open to the influences of further causes and created freedom. From this barrier between the Divine and Creation, the Transcendent determines four archetypes or transcendentals to emerge that categorically defines being in realationship to "the One, the True, the Good, and the Beautiful". This fundamental correspondence, writes Balthasar, "is the unlimited abundance of reality which is beyond all comprehension, as it, in its emergence from God, attains subsistence and self-possession within the finite entities."[21] Therefore, with reference to creation, Nothingness introduces a way in which one can conceptually ground, compile, and theoretically actualize the eternal and Transcendent as part of rational being's purview of reality's manifestation of the transcendentals (e.g., one, true, good, beautiful). The transcendentals encourages a person to trace a continuum running from Nothingness to the infinite and ultimately the Eternal.[22] Nothingness' diverse re-orientations and re-packaging by being's engagement with the

[20] *ST*. I., 15, 2. See also Robert J. Russell, *Cosmology: From Alpha to Omega, The Creative Mutual Interaction of Theology and Science* (Minneapolis, MN: Fortress Press, 2008), 34.

[21] Hans Urs von Balthasar, *The Glory of the Lord: A Theological Aesthetic*, vol. 5: *The Realm of Metaphysics in the Modern Age*, Translated by Oliver Davis, Andrew Louth, Brian McNeil C.R.V., John Saward and Rowan Williams, Edited by Brian McNeil C.R.V. and John Riches (San Francisco, CA: Ignatius Press, 1991), 12. Hereafter cited as *GL* 5.

[22] See *ST*. III, 76, 5.

potential pervades and makes possible our physical and conceptual inhabiting of the everyday world of relational ordering and non-subsistence.

Badiou argues that difference's ubiquity makes possible love and truth's complimentary structuring as modes of a finite embodiment—something that analogously communicates in the mathematical processes of addition and subtraction. These philosophical concepts and abstract functions furnish Badiou with logical processes that clarify the nature of human love: "it is not for love's sake that there are two counted from the vantage point of the third. In matters of love, the three is not (*n'est pas*), and the Two is subtracted from every count."[23] This conceptual framing of differing is implicated in the subject's own thought processes and influences, Badiou reasons, as all experiences of love "involves a separation or disjuncture based on simple difference between two people and their infinite subjectivies."[24] In his schema, love is a revolutionary choice for "addition" or unity against the infinite potency awaiting the ongoing actualization in the universe. The successive moments or "foliation" of love's revelation are only possible when one presumes some underlying space or distance (e.g., qualified nothingness) that unifies primitive experiences of the divisive. Thus, love exemplifies the truth of being by orchestrating a being's unique differences toward a single purpose. As Sigi Jottkandt explains: "[f]or Badiou, love must be regarded in terms of an exceptional logic that simultaneously confirms the philosophical counting operation, while generating another number that is not a product of ordination."[25]

Postmodern philosophy offers many insights into how the traditional concept of ontological difference or "othering" correspondences products in unifying events. Our holo-cryptic metaphysics argues that human love virtually indicates the logic of difference that is already a unity, as attested to in the Transcendent's Pure Ur-Kenotic identity. The divine love of eternal hypostatic unity and the personal "count" of the Divine Persons' identities supremely materializes in the finite person and historical mission of Jesus Christ. The liturgy dramatizes and prolongs this mystery in the event of transubstantiation, and its explication is a foundational principle of holo-somatic identity.

[23] Alain Badiou, *Conditions*, Translated by Steven Corcoran (New York, NY: Continuum, 2008), 187. Hereafter cited as *Conditions*, and *Logics of Worlds*, 507–14.

[24] Alain Badiou and Nicolas Truong, *In Praise of Love*, Translated by Peter Bush (Paris, Flammarion S.A., 2009), 27.

[25] AJ. Bartlett and Justin Clemens, *Alain Badiou: Key Concepts*, (London, UK: Routledge Press, 2014), 73. Hereafter cited as *Key Concepts*.

Holo-cryptic metaphysics does not succumb to a hackneyed physicalism. In Chap. 3, *Holographic Matter*, I show that even a fallen and limited physical being can be understood as a holo-somatic or virtual exposition of its potential future status as redeemed or resurrected. This truth, I argue, is also at the core of the Eucharistic feast's characteristic engendering of non-subsisting or virtual relationships through prayer and in ecclesial life. Because this chapter offers a more in-depth discussion of holo-cryptic metaphysics, it is subdivided into six sub-sections that develop a holo-cryptic metaphysics in light of theological, philosophical, and mathematical interpretations of infinity and finite identity. Jesus' economic mission makes possible this eventuality, and thus, the holo-somatic state is understood to be more than an ongoing abstract or therapeutic process. Holo-somatic identity is inseparable from a human person's potential reach into the supernatural realm.

Holo-cryptic metaphysics recommends the created order's transfiguration in the Eternal Word by means of the work of the Spirit, who shoulders in absolute freedom the completion of Jesus' salvific work. The human person sacramentally participates in this exchange at the Eucharistic table and whenever transforming forces are adapted to manifest the created order's divinely allotted task and final destination. Thus, this holo-cryptic metaphysics develops insights raised most pointedly in Balthasar's Theo-aesthetics and Theo-logics. From this schema, I interpret the work of a select group of great modern and postmodern artists whose art—from my perspective—points us to the "transfigurative" power of human creativity and reality.

QMT contends that information (energy) is the most fundamental representational unit of physical beings and can be rendered in a binary manner (e.g., 0 or 1). When the essential identity of a material being is analogously equated to its quantifiable equivalent of information, we attempt to capture one of the principles of hylomorphic theory, namely, that truth (essence) and being (existing individual) are convertible. Taking the Theo-philosophical concept of essence as information opens up the possibility of seeing cross-disciplinary perspectives around a being's material and energetic properties as theoretically compatible without subordinating them to a Gnostic interpretation.

The Laws of Conservation of mass and energy, the Laws of Thermodynamics and entropy, and so forth give us a non-morally binding and analytic analysis of being. In general, terms, these concepts and laws imply that being can be understood as an existing physical state or pattern that over time moves from greater to lesser organization.[26] Thus, entropy is seen as analogous but asymmetrical to my advocating for divine thought's motivation and transformation of Nothingness' pure entropic orchestration of information at Creation. The concepts of Nothingness, the virtual image, and informational entropy illumine the precarious balance in which the postmodern and QMT worldview are set into conversation. I argue that holographic information evidences an epistemological and ontological dense estimation of the "physical" or "real" in relation to the two extremes of Nothingness and the Transcendent.

In Chap. 3, Sect. 3.1, *Holo-somatic Identity and the Material Order*, I compare the scientific view of the QMO to my own theologically based interpretation of created being. The semblance between these methodologies follows from their mutual and virtual description by means of symbols and signs. Most postmodern philosophers would likely hold the quixotic position that surface meaning completely supplants a being's interiority, whereas more traditional views forward the essential displays, embodies and is dignified in a physical being. My defense of the latter position depends in part on my belief it is consistent with QMT and informational entropy, that reveal the natural law.

Holo-cryptic metaphysics rejects all philosophies and theologies that eschew freedom, probability, and the essential to denote a deterministic meaning to "thing-in-itself" based solely on its physical quantification. The liminal power of a being's superficial differentiation in horizontal relationships threatens to see a material being as the countless repetition of the limited and soulless.[27] I am not here forwarding a naive view of material or supernatural realities. It is no more possible for our physical senses to move beyond the phenomenal accidents of the bread and wine and see "face to face" the "sacramentally incarnate" Transcendent Son, than it is for one to intellectually take hold of the infinite interior numeric identity of a physical being that resides in and through its "informational" contours. Faith permeates and guides the human condition.The symbol depicts a being's complex landscape in a way that allows consciousness apprehension of the

[26] See John Gribbin, *Deep Simplicity: Bringing Order to Chaos and Complexity* (New York, NY: Random House, 2004), 26–27.

[27] See François Laruelle, *Dictionary of Non-Philosophy*, Translated by Taylor Adkins Collaborators Tony Brachet, Gilbert Kieffer, Laurent Leroy, Daniel Nicolet, Anne-Francoise Schmid, Serge Valdinoci (Minneapolis, MN: Univocal, 2013), 55–56, 158–60. Hereafter cited as *Dict. Non-Phil.* See also Ray Brassier, "Axiomatic Heresy: The Non-Philosophy of François Laruelle," *Radical Philosophy* 121 (Sept/Oct 2003): 32.

essential core of a being's truncated phenomenal appearance. The holo-somatic impression gives us a way to understand a being's virtual informational matrix as a sign of its essential or holistic meaning in its material characteristics. I understand the Eucharist's properties metaphorically as a holographic instant that plots invisible, supernatural multi-dimensions in the spatiotemporal properties of matter (bread and wine).

In Chap. 3, Sect. 3.2, *Holo-somatic Metaphysics and the Infinite*, I critique past formulations of mathematical infinity by thinkers such as Aristotle, Thomas Aquinas, Georg Cantor, and Alain Badiou. A being's infinite aptitude, the postmodern Badiou argues, can be mathematically portrayed by means of sets. Set theory can correlate the finite, infinite, and various types of infinity. This is achieved by defining the relationships (mappings) that result given each number's intrinsic identity: "[m]aping is an essential technique of mathematics. In every case of a mapping, we first have to define certain characteristics of the sets we are mapping and determine the internal relationships between their elements; only then can we proceed to make external mappings."[28] The interior properties of the members of the Set establish its external correspondences. These external properties introduce a way to explain inconsistency, multiplicity, and polysemous meaning.

In three parts, I outline some modifications made by secular and atheistic views of infinity. One can, for example, explore the properties of a mathematical set without explicitly taking into account how these results suggest further external findings. In Chap. 3, Sect. 3.2.1, *Identity as Closure and Identity as Closure*, I scrutinize how a being may isolate itself around its own subjective becoming. The Promethean gesture conflates ego and will and projects them as a self-enclosed manifestation of the infinite. Once external correspondences are rejected, the inestimable potential of Creation reduces the individual's experiences to a self-imposed recursive loop. In contrast to a being's self-reflexive reduction, in Sect. 3.2.2, *Holo-somatic Infinities*, I turn to the contrary problems associated with being's unfettering that is highlighted in the work of postmodern thinkers such as Badiou and Derrida. The infinite multiplication of beings along horizontal lines does not, I argue, represent a perfect being or lead to a Utopian community. Rather, if relationships remain lateral, that is, confined to the subjective and materially given, the individual risks being dispossessed of its supernatural identity against the forces of implicit and

[28] Edward Rothstein, *Emblems of Mind: The Inner Life of Music and Mathematics* (New York, NY: Random House/Times Books, 1995), 47. Hereafter cited as *Emblems of Mind.*

explicit surface ordering. Cantor's insight into the existence of various hierarchies of infinity counters this tendency at the conceptual and abstract levels of mathematics. Art and theology give us a way to experience this truth. Through the interpretation and creation of images, these two disciplines have a medium that allows individuals to sensually and consciously experience the meaning of the infinite. These movements make immediately relevant how it is possible to center on the act of perception.

In Chap. 3, Sect. 3.2.3, *Infinity and Holo-somatic Non-subsistence*, I once again find proof that the theory of hylomorphism can be revised by my holo-somatic engagement of the infinite. The infinite and non-subsistent relation exist as a mean between the two philosophical termini of Nothingness and Being. The unlimited and non-subsistent relation are inevitable aspects of our holo-somatic becoming and therefore, how we exist in the world.[29] Unlike most postmodern approaches, the Christian can see in holo-somatic metaphysics a way to understand the person of Jesus Christ as a theophany of the Eternal Logos. Non-subsistence and relational ordering opportune one's free and rational life and contemplation of the Transcendent.

My survey of the infinite's intensive and extensive properties drives my position on the nature of the rational and spiritual life. Chap. 3, Sect. 3.2.4, *Holo-somatic Self-Consciousness*. Human consciousness has the capacity to discover theorems that abstract from our immediate and three-dimensional realm to include a multitude of potential and infinite manifestations. The theoretical manipulation of numbers is key for how mathematicians understand complex or hyper-dimensional topologies and spaces. Interior conscious operations are able to intuit these trans-spatiotemporal dimensions. The ability of consciousness to consider the concept of infinity is taken up again in greater detail in Chap. 4 with allusion to the non-rational manner in which the QMO's own strange existence is represented.

Following an analysis of infinity's extensive metaphysical, geometrical, and spatiotemporal movements within human consciousness, I turn in Chap. 3, Sect. 3.3, *Infinity and Providential Design*, to show how my interpretation of the unbounded serves a theological interpretation of history as a holographic categorizing of the Divine's providential plan that is commensurate to an individual's holographic informational structure. Whereas many modern and postmodern philosophies propose that history

[29] The realist nature of atomic matter is fundamental in this position, see Alain Badiou, *Logics of Worlds: Being and Event II*,Translated by Alberto Toscano, (New York, New York: Continuum Press, 2009), 452. Hereafter cited as Logics of Worlds. See also Alain Badiou, *Being and Event*, Translated by Oliver Feltham (New York, NY: Continuum, 2010), xiii–xxii. Hereafter cited as *Being and Event*.

is ultimately meaningless, the Judo-Christian perspective sees divine purposes at work. If Creation's conversion is unidirectional, that is, ordered to its eschatological fulfillment, then there are divine exemplary and anti-entropic holographic forces at play. Stated otherwise, the essential or informational status of each being is never finally lost in space or in time.

The second section of the chapter concludes with a discussion of the relationship between angelic nature and infinity. Chapter 3, Sect. 3.4, *Angelic Nature and Holo-somatic Embodiment*, mutually endorses Thomas' doctrine on the angels and Cantor's theory of infinity. The angelic identity is indispensable to our own holo-cryptic metaphysics, as this nature shows that a purely spiritual being is not necessarily the same as Hollywood's depictions of ghosts or ghouls. The angelic body, Thomas informs us, is unique and nonmaterial. Indeed, in angelic nature, we see a hint of holo-somatic metaphysics of the resurrected body made manifest in space-time. Furthermore, the angelic nature gives us a spiritual analogy for what will be materially quantifiable in the QMO, its entanglement, and non-locality. Quantum non-locality, for instance, pertains to the formal structure and relational properties of the QMO. Counter to our common sense intuitions, QMT unwittingly explores under the directive force of its own principles what theologians explain in terms of saintly bilocation and immaterially embodied divine messengers. From their own respective perspectives, theology and science attempt to connect a physical entity's formal description and prescribed spatiotemporal place in the created order in the most expansive terms.[30] The QMO wave mimics the unconstrained nature of angelic existence and supports the fact that a physical being's freedom is writ large in the universe. Angelic nature acts as a stand-in for my investigation of the QMO's movement between the states of wave and particle or Nothingness mediation of infinity and physical reality. The QMO's condensation from wave into particle or the angel's revelation in history presupposes that this determination could be observed beforehand.[31]

The eyewitness' recognition of an object or being is theologically embedded in the Spirit's identity and mission as an eternal Transcendent observer. Having outlined how infinity initiates within the created order, I turn in Chap. 3, Sect. 3.5, *Holo-cryptic Personhood*, to rival postmodern

[30] See Henry M. Stapp, *Mindful Universe: Quantum Mechanics and the Participating Observer* 2nd ed. (Berlin, DE: Springer-Verlag, 2011), 183–185.

[31] See Lee Smolin, *Three Roads to Quantum Gravity* (New York, NY: Basic Books, 2001), 3–5. See *ST.* I, q. 28, a.1. On causal relata as a descriptor, see Michael McDermott, "Redundant Causation," *British Journal of Philosophy* 46 (1995): 423–44.

views to my holo-somatic theory of the human person. The Christian theologian believes that the human person is fundamentally defined by his/her relationship to Jesus Christ. My holo-cryptic metaphysics draws attention to the Spirit's presence in the act of transubstantiation as a means to personally receive Jesus Christ.

The concept of relation for the finite creature does not necessarily indicate its substantial inhering in another being, but it references a being's intended motion toward something or an individual's perception.[32] A relation can be likened to a "disposition" (*dispositio*) that democratically exists between two beings or as the limitless essential movements within a being's consciousness. Through a regime of presentation and re-presentation, Badiou argues, individual existence symbolically exposes its covert desires by actualizing intervals (*intervallum*) by perceiving the overt multiplicities that historical reality is.[33]

The distinction between a being's conceptual delineation and its spatio-temporal conveyance reveals in the hylomorphic and commensurate structuring of truth and being. These complementary structures extend to their symbolic representation in mathematics and theoretical physics. The relationship between two or more beings can be symbolically or holographically represented. Comparative relationships such as similarity or opposition can be used in lieu of logical propositions.[34] This process can continue indefinitely, much as one can, for example, write out and refine an algorithm for a being or number as a data set. By repeating a mathematical fractal's defining equation, a spatial pattern is infinitely generated in an elaborate and organically lithe manner: "[a] fractal is a type of set produced by a rule called recursive—one keeps applying the same transformations to parts of a set that one applies to the whole. This means that any proportion of a fractal curve contains the same types of movements as the whole; any proportion, magnified, will reveal as much information as the whole."[35]

Heidi Ann Russell explains how fractal information also gives us a way to reflect on the strange subatomic dimensions of the QMO: "Fractals

[32] On causal relata as a descriptor, see Michael McDermott, "Redundant Causation," *British Journal of Philosophy* 46 (1995): 423–44.

[33] See *Being and Event*, 24 and *Hidden Void*, 224.

[34] The polar relationship between the Transcendent and (non)Transcendent is thus reflected in the factual-counterfactual relation in immanent beings; see Peter Menzies, "Difference-Making in Context," in *Causation and Counterfactuals,* Edited by J. Collins, N. Hall and L.A. Paul (Boston, MA: M.I.T. Press, 2004), 139–80.

[35] *Emblems of Mind*, 162.

often involve stretching and folding so two points on a line that are very far away from each other may end up right next to each other as the line is stretched and folded. This intricate pattern is the result of a fractal not fitting neatly into our preconceived ideas of dimensions."[36] Russell stresses that these patterns direct our eyes away from Nothingness' clandestine forces, as "To see only in two or three dimensions is to miss the wonder and beauty of the between." Mathematically elongating the specialized two-dimensional fractal pattern hyper-dimensionally suggests what takes place at the event of transubstantiation. At the liturgy the Spirit converts the three-dimensional spatial objects of bread and wine into the supernatural identity of the Eternal Word, Jesus Christ.

Chapter 4, *Toward a Holo-cryptic Metaphysics*, begins with a brief review of some of the questions and developments in the field of QMT that are most significant for my theory of the holo-somatic interpretation of the doctrine of transubstantiation. It is relatively easy to represent a QMO as binary given its orthogonal structure. Its primitive state is depicted in one of two ways as $|0\rangle$ or as $|1\rangle$ (e.g., with Ket and Bra). These two schemes make four combinations possible for the QMO: (0, 0), (0,1) (1,0), or (1,1). Early QMT research has shown that subjective observation influences how a QMO manifests in the experimental setting. I understand the interior geometric structuring of the QMO wave (e.g., vectors, eigenvectors, and eigenvalues) as understood as an analogous mathematical virtualization of the QMO's interior motion or freedom. I reject the traditional relegation of passive potency solely to matter to include energy's imaging ontological identity in a speculative reading of QMT.[37]

The best scientific models can lean on past artworks and aesthetic theories in their own representational and symbolic presentation of reality. Werner Heisenberg points out that both "Science and art form in the course of the centuries a human language by which we can speak about the more remote parts of reality, and the coherent sets of concepts as well as the different styles of art are different words or groups of words in this language." Modern art orders a way to represent both the internal and external, horizontal and vertical dimensions of being because modern art

[36] Heidi Ann Russell, *Quantum Shift: Theological and Pastoral Implications of Contemporary Developments in Science* (Collegeville, MN: Liturgical Press, 2015), 86. Hereafter cited as *Quantum Shift*.

[37] I follow the practice of using vector mathematics to introduce the QMO's geometric structure. See the appendix for a brief introduction to vectors.

"Transcends concrete experience and creates a symbolic construct of time and space, especially in connection with eternal archetypes and spiritual essences which evoke emotive responses. Modern artists have the singular opportunity of presenting the spiritual significance of the totality of human experience in their recognition of the foundational necessity of the religious imagination—a significant turning point in cultural history."[38] The "morphic" and dimensional properties of the QMO allow one move beyond its restatement in mathematical vectors. The conceptual vernacular of Cubism with its homage to the parallax of perspectives, dimensional exuberance, mirror imaging, and so on suggest avenues into something of the world of the observer in GTR and the QMOs superposition within convoluted spatiotemporal dimensions: "In the fourth dimension, the self-orientated sense of up-down and left-right fades away, and objects can be viewed from all sides at once. Even more startling is the discovery of time and motion, as understood in three-dimensions, as having been mere illusions produced by inadequate perception of four-dimensional extension."[39]

As the fourth dimension is commonly taken to introduce spatiotemporal elongation, it typifies how mathematics executes an iteration of complex dimensions. Without the Transcendent as a natural point, the continuation of this sequence, the metaphoric "tower of turtles," suggests in Cantor's insight that a hierarchical ordering of unequal infinities is possible. The actual identity of the QMO as wave and particle gives us a primitive representation of non-subsistence's ontological and relational influence on identity. Because a physical object's dimensional configuration is in part, relative to its conscious apprehension, it can be variously represented. An object's spatiotemporal identity can be portrayed in complex mathematical or eschatological dimensions. The symmetry between an object's physical and eschatological attributes allows us to speak of the QMO's placement in complex space (e.g., Hilbert space) or in terms of its holosomatic identity. Given the predictable orthogonal structure of the QMO, one can also associate a logical proposition to a specific quantum state. The ability to virtually attribute an informational or logical value to a QMO's state is necessary, as it physically exists in an uncertain and "deterministic" universe. Max Born (1882–1970) recognizes a paradox at play here,

[38] *Diane Apostolos-Cappadona, "Beyond Belief: The Artistic Journey," in Beyond Belief: Modern Art and the Religious Imagination*, Ed., Rosemary Crumlin (Melbourne, AU: National Gallery of Victoria, 1998), 22.

[39] Linda Dalrymple Henderson, "Mysticism, Romanticism, and the Fourth Dimension," Edited by Maurice Tuchman and Judi Freeman, et al., in *The Spiritual In Art: Abstract Painting*: 1890–1985 (New York, NY: Abbeville Press, 1986), 221.

noting that for "The motion of the particle obeys laws of probability, while the probability itself develops in accordance with the causal law."[40]

Even though it is impossible to separate truth from being, it is simultaneously impractical to avoid the introduction of statistical or ontogenetic differences when making propositional statements.[41] Here, the physical identity of the QMO and its logical propositional value is likened to its holographic representation and informational structuring.[42] The discordant presence of uncertainty reveals itself, for example, in the fact that one can speak of a particle as a wave and a wave as a particle, but neither simultaneously and never without fully understanding why such a duality of identity exists outside of transcendent purposes, that is, its sensual and perceptual observation.

After all, QMO relationships typify how in the ideal situation, a material object's interior mathematical properties accidentally comport according to being's self-reflexive measure. Analogy articulates the fundamental law of order in terms of a mathematical proportion.[43] An analogy is a "A form of speech which, by means of one set of ideas immediately presented, points to something else, which is hidden, or a form of speech that does not point out the aim in thought directly, but gives its instruction by an indirect signification."[44] Serving to convey the unlikely marriage of identity and difference, John R. Betz and David Bentley Hart explain "*Ana*(-*logia*)" agrees not only with its stem, *logos*, but also with the *dia*(-*logon*) (in "dialectic"). For all of their differences, what *ἀναλογίζεσθαι*, *διαλογίζεσθαι*, and *λογίζεσθαι* have in common is that they indicate a kind of "measured

[40] See Max Born, "Quantenmechanik der Stoßvorgänge," *Z. Physik* 38 (1926): 804, see Robert J. Russell, "Quantum Physics and the Theology of Non-Interventionist Objective Divine Action," in *The Oxford Handbook of Religion and Science*, Edited by Philip Clayton (New York, NY: Oxford University Press, 2008), 579–595 and "The Information Interpretation of Quantum Mechanics," https://www.informationphilosopher.com/introduction/physics/interpretation.

[41] See Robert W. Spekkens, "Evidence for the epistemic view of quantum states: A toy theory," *Physical Review* A 75.3 (2005): 2, arXiv:quant-ph/0401052v2.

[42] On the philosophical foundations of the hierarchical ordering of judgments about identity and their negation through contradiction, on the material and formal aspects of faith and knowing see *ST.* II–II, 1, 1.

[43] See Aristotle, "Nicomaehcean *Ethics*",. in The Basic Works of Aristotle, Edited and Introduced by Richard McKeon (New York, New York: Random House, 1941), Book 5, chapter 3, 1131 a-b 31f. Hereafter cited as *Ethic. Nic.*

[44] Nyssa, *Commentary on Proverbs 1:6, Catenabible.com.*

consideration."[45] Where analogy works mathematically to "add," "synthesize," or "bridge" two or more beings, dialect momentarily amends or "subtracts" some characteristic of a being and hence signifies less a connecting span than a relationship of alterities.[46] The analogous arrangement between proportionate equations of truth and being suggests that one must give credence to the existence of an analogy of difference laboring in created reality. The so-called analogy of being recommends ways in which created reality can be hierarchically ordered, whereas the analogy of difference gives us a systematic way to organize horizontally even the apparent contradictory and unbiased.[47] Physical appearances change over time and with each new presentation past iterations are developed.

As a simple pure spiritually subsisting essence, the Divine eternally and perfectly possesses interior and exterior freedom. In comparison, finite internal and external freedom communicates in the spatiotemporal. Finite relations make plausible a created being's relative freedom in the world. The complete segregation of a finite being's interior and exterior manifestations is impossible given that one always exists in non-subsistent relations. While not equivalent, Niels Bohr propounds his own version of this when stating, "in our description of nature the purpose is not to disclose the real essence of phenomena, but only to track down as far as possible [the] relations between the multi-fold aspects of our experience."[48] The systematic and conceptual placement and re-positioning of being in reality shows how being moves toward perfection even with contingent and chaotic forces in play. Given the subject's ability to objective or distance themselves from the relationships that define them gives us a means to speak about non-subsistent correspondences having a metaphoric real identity. Intentional consciousness is able to manipulate logical and formal designations to reveal unknown truths.[49]

[45] Eric Przywara, *Analogia Entis: Metaphysics: Original Structure and Universal Rhythm*, Translated and introduced., John R. Betz and David Bentley (Grand Rapids, MI: Wm. B. Eerdmans Publishing Company, 2014), 192. Hereafter cited as *AE*.

[46] See Thomas Aquinas, *Commentary on the Metaphysics of Aristotle*, Translated by John Y. Rewan vols. 1–2 (Chicago, IL: Henry Regnery Company, 1961), 5, lecture. 17, nos. 1010 and 1017, pages 385–386. Hereafter cited as *Metaph*.

[47] See Dionysius the Areopagite "The Divine Names," 6, 3 in Works (1897), transcribed by Roger Pearse, http://www.tertullian.org/fathers/areopagite/divinenames.htmc6.

[48] Niels Bohr, *Atomic Physics and the Description of Nature* (Cambridge, UK: Cambridge University Press, 1934), 18.

[49] See *ST*. I, 28, 1.

It is only when the subject exists in relationship to an actual personal Transcendent Being that a "metaphysically thick" sense of human freedom and consciousness is advanced in the latest mathematical and scientific findings about reality that moves beyond its immediate ratio. The QMO's mutability gives us an analogous concept of non-sentient freedom because both rational and non-conscious beings act spontaneously. This principle resonates in being's essential resourcefulness and mathematically resolves in physical laws such as that of probable distribution as a preamble to the study of holography and cosmic forces such as gravity. Boltzmann points out that the laws of probability calculus imply that if a system were large enough, it would include some region of improbable distribution.[50] Most scientists and scholars believe that a complete description of the physical laws governing the universe will one day be achieved. With this universal template in hand, we will finally see the axiomatic reconciliation of the "near-impossibly" small world of the subatomic particles and QMT and the unimaginably large scales where the force of gravity manipulates the very shape of the cosmos.

Previous theories of transubstantiation depend upon a theory of Nature (*Natura naturans*) grounded in an interpretation of the Aristotelian scientific worldview. However, the controlled experiments of QMT continue to validate that reality's uncertainty reveals freedom as holographic—that is, as information. Unfortunately, most explanations of the sacramental order have settled on near purely deterministic interpretations of empirical being or reality as a whole—a stand that barely moves beyond sympathetic magic. In Chap. 5, *Black Hole Entropy and the Holographic Universe*, I pursue my thesis that a probabilistic and entangled view of the QMO's material and holographic exchange of information opens up the doctrine of transubstantiation to a more scientifically sensitive interpretation of nature. The unlikely site of a black hole validates my construal of the virtual as informational, thanks to its display of the hyper-dimensional and probable description of the quanta. Indeed, the QMO, the infinite mathematical series or the Eucharist's analogous presentation of the Eternal Son in material form are all seen as specific phenomenological explanations of a physical object holographically forecasted in an observer's environs. This natural phenomenon gives us insights into how non-locality can inspire theological models.

Hawking codified the radiating thermal quanta at a black hole's event horizon as temperature and entropy. Information's transmission and

[50] See Ludwig Boltzmann, "Über statische Mechanik," in *Populäre Schiften* (Leipzig, DE: Barth, 1905), 362.

entropic dissipation at the black hole's event horizon illumines the path that physical beings undertake when piloting reality's uncertain but efficacious and transforming nature. This view has ramifications not only for the belief that the Transcendent remains scientifically inaccessible and irreducible to subjective experience. Such a claim holds in a general sense relative to the question of design, as "theistic evolutionists" such as Richard L. Fern believe, for instance, or as the traditional Catholic characterization of "miracles" suggests.

I revisit the idea of soul embodiment from the perspective of information conservation within and projecting from a black hole. When matter enters a black hole, it is torn apart by extreme internal gravitational forces. However, as contemporary theorists and physicists have shown, an image of obliterated physical matter remains at the black hole's entry as a holographic protuberance. Many parallels can be drawn between a black hole's perceived and theoretical operation in Nature and religious speculations concerning the Nothingness at the moment of Creation and its attempted rejection in Hell. The manner in which information emanates holographically from within a black hole can also symbolically represent how even Jesus Christ's sojourn in Hell is included in the drama of the divine-human real presence of the Eucharist. The forces of Hell cannot silence Jesus Christ, the Eternal Word made flesh. The soul represents the metaphoric informational identity of the historical person, Jesus Christ. Jesus' physical body and the Eternal Logos can be likened to the virtual embodiment of two entangled particles—one of which enters a black hole while the other escapes its grasp.

The results of introducing a physical being into a black hole provide us with a way to understand how the physical body's death never obliterates its informational or essential properties. The plasticity that exists between a phenomenal being (e.g., *esse*) and its formal or axiomatic representation (e.g., *essence*) together acts as a pedestal for its external and internal metaphysical edifice. This relational differing constitutes the apex of the Aristotelian interpretation of reality.[51] Comparatively, when time and space are pushed to a near infinite actualization at a black hole's event horizon, they display material identity as pure energy. This "purification" of information occurs because here the "surface gravity is the force required at infinity to hold a mass in place just above the horizon."[52] The event

[51] *AE*, 219–220.

[52] Sean A. Hartnoll, Andrew Lucas, and Subir Sachdev, *Holographic Quantum Matter* (Cambridge, MA: The MIT Press, 2018), 4.

horizon of a black hole inspires new insights into how the infinity that is the person of Jesus Christ becomes physically present in bread and wine at transubstantiation. The faithful consume the consecrated species that "holographically" makes present the Eternal Son in the Eucharist.

In Chap. 6, *Transubstantiation and Quantum Mechanical Theory*, the strange world of quantum mechanics introduces a new spatiotemporal rubric through which we can better understand the act of transubstantiation. The mysterious spatiotemporal and dimensional world of the quanta gives us a way to extend our understanding of the Eucharist beyond its historical unfolding in the liturgy to include our own holo-somatic and trans-historical interpretation. Thomas raised a similar metaphorical eyebrow when he was so bold as to apply his readings of Aristotelian physics to his own theological investigations. The holo-somatic delegation of the quanta presents a more sophisticated explanation for the inherent nonlocal and trans-dimensional nature of the Eucharistic species than previous form-matter or classical scientific deterministic constructions. The hyperdimensional Hilbert space that the QMO is often seen to inhabit, for example, is a superior way to identify the Eucharist's three dimensions as a theoretic emblem of the trans-dimensional existence of Jesus Christ and the Eternal Son.[53] Furthermore, even though QMO's teleportation is a nascent field of study, as there is as yet a limited amount of experimental evidence to back its theoretic conjectures, teleportation hypothesizes that it is possible for a QMO's informational identity to be transported at near superluminal speeds. QMT gives us the most scientific and mathematically precise model of reality on which to base our understanding of the event of transubstantiation. Most postmodern schemes reject or replace the Divine with various philosophical, theological, and scientific demarcations of nothingness, as found in the concept of non-dimensional spaces, the void or relational differing. In the non-classical geometries of the QMO, for example, its temperature and entropic identity presents as a specific density in a black hole with discontinuous spaces.[54]

In the concluding Chap. 7, *Ur-Kenosis in a Holo-cryptic Universe*, I explain how various philosophical, mathematical, and scientific

[53] On quantum thought's aiding of contextualizing nonlocal event, see Emmanuel M. Pathos and Jerome R. Busemeyer, "A quantum probability explanation for violations of 'rational' decision theory," *Proc. R. Society* B 276 (2009): 2171–2178.

[54] See Sean A. Hartnoll, Andrew Lucas, and Subir Sachdev, *Holographic Quantum Matter* (Cambridge, MA: The M.I.T Press, 2018), 67 and 114.

interpretations of Nothingness motivate my own spiritual critique of traditional theories of transubstantiation. The Pure Groundlessness of the Spirit, as Divine Sophia, is not an obtrusive force in the created realm (1 Corinthians. 15:28). The Spirit is not an abstract "sophiological principle" but exists in the created order as the Pure Freedom who is a Divine Person. It is the Absolute Nothingness or Freedom of the Spirit that guarantees the divine fullness: "for if we assert that God is everywhere and in all things on the ground, [then] nothing can be empty of God."[55] The Eternal Word commands everything into existence by virtually mediating divine thought—from the void the Divine predicates or "calls into existence the things that do not exist" (Romans. 4:17).

The Trinitarian community creates the material universe by simply thinking about "nothing" (Genesis 1:1-2).[56] Divine knowledge of this type recalls Thomas' knowledge of simple intelligence (*scientia simplicis intelligentia*).[57] Creation's origins in Nothingness serve as an apophatic exemplification of all material beings' ongoing relationship to Prime Matter, nonbeing and the essential's receptivity.[58] As Thomas states, "by matter I mean that which in itself is neither a quiddity nor a quantity nor anything else expressed by any of the other categories."[59] The concept of nonbeing naturally finds itself described in relationship to being; however, it is impossible, Plato surmises, to put a precise limit on nonbeing, as its "existence" is boundless (*Sophist*, 256e). The possible negative consequences of nonbeing's formlessness however can be demarcated: "there are many kinds of being but an infinite amount of nonbeing; this is so

[55] Origen, "*De Principiis,*" *Ante-Nicene Fathers* vol. 4, 3, 6, 2, The Writings of the Fathers down to A.D. 325. Edited by the Rev. Alexander Roberts, D.D., and James Donaldson, LL.D. Revised and chronologically arranged, with brief prefaces and occasional notes by A. Cleveland Coxe, D.D., (Buffalo: The Christian Literature Company, 1885). *The Logos Catholic Edition.* See also *ST.* I, 7, 1.

[56] Nothingness is seen as a necessary premise in Jewish and Christian understandings of creation and is implied in most mathematical symmetries upon which the theoretic quantum world is understood; see James Owen Weatherall, *Void: The Strange Physics of Nothing* (New Haven, CT: Yale University Press/Templeton Press, 2016), 3–5, and 9–11. Hereafter cited as *Void.*

[57] See *ST.* I, 3, 2.

[58] See *AE*, 188, *ST.* I, 45,1, and Ferdinand Ulrich, *Homo Abyssus: The Drama of the Question of Being*, Translated by, D.C. Schindler (Washington, DC: Humanum Academic Press, 2018), 322. Hereafter cited as *Homo Abyssus.*

[59] Aquinas, *Commentary on the Metaphysics*, vols. 1–2, Translated by John P. Rowan (Chicago, IL: Henry Regnery Company, 1961): Book 7, chapter 2, 1285.

because nonbeing is not only the negation of being but also encompasses all ideas that have and can have no reality. Nonbeing is the area of appearance, illusion, delusion, and error; here there are many dark corners where rogues and mountebanks can hide out and ply their Sophistic mischief."[60] The advances of postclassical physics show us that the notion of nonbeing is neither an illusion nor a delusive path but rather a critical concept necessary for a more profound explanation of the fundamental structure of physical reality. The existence of a black hole asks us to re-evaluate our perceptions regarding being and nonbeing. Thomas anticipates this from the metaphysical viewpoint with his interpretation of Prime Matter and the philosophical accident. Scholastic theologians argue that matter is foundational (*principle per se*), whereas all privations are determined to be accidental (*per accidens*) and therefore, to some degree, self-reflexive.[61] According to Thomas, Prime Matter is pure potency that positively defines an existing material being by the concepts of nonbeing and accidental properties. With this turn, it is possible to see previous philosophical conceptions of Prime Matter in Christian interpretations of Nothingness via *creation ex nihilo*.[62] "Prime Matter," Thomas states, "is satisfied when form meets its aptitude for determination."[63] Prime Matter gives us a way to conceive of divine exemplarity's relationship toward an actual existing physical being.[64]

Philosophical and speculative devising of nothingness, nonbeing, potency, void, multiplicity, or the empty mathematical set are taken as analogous expressions of non-subsistence. The exact convertibility of being and nonbeing never occurs within the created order, but each mutually images its oppositional term. On the general parameters of this "holographic ordering", Etienne Gilson states that "just as essence is in potency

[60] Thorleif Boman, *Hebrew Thought Compared with Greek*, Translated by Jules L. Moreau (Philadelphia, PA: The Westminster Press, 1960), 55.

[61] See Aristotle *Metaphysics:*. vols. 1–2, Introduction and Commentary by W.D. Ross (Oxford, UK: Clarendon Press, 1924) 5, 17, 1004. Hereafter cited as *Metaph.*

[62] See Thomas Aquinas, *Commentary on Aristotle's Physics*, Book 1, lecture 15, 191 b35-192 b5, Translated by Richard J. Blackwell, Richard J. Spath, and Edmund Thirlkel, Introduction by Vernon J. Bourke, Forward by Ralph McInerny (Notre Dame, IN: Dumb Ox Books, 1999).

[63] See *ST.* I, 44, 2.

[64] See *De ver.*, 3, 1–4. See Thomas Aquinas, *Truth.* vol. 1: Questions I-IX, Translated by Robert W. Mulligan, S.J. (Chicago, IL: Henry Regnery Company, 1952). Hereafter cited as *De ver.*

to the act of its own existence, so is also the act of existence in potency to the formal act of its own essence."[65]

The Eucharist supremely manifests how the potency invested in material beings achieves its supernatural end by holo-somatically making real the presence of Jesus Christ. In conclusion, I outline some of the pivotal aspects of celebrating the Divine Liturgy given my holo-cryptic investment in quantum theory. It is my hope that revisiting the doctrine of transubstantiation will assist the current generation's spirituality.

1.2 Hidden Universes and New Negations

The following does not utilize metaphysics in a traditional Leibnizian or Kantian sense, nor am I treating the doctrine of transubstantiation as a scientific theory. However, I draw on the wisdom and insights of both, as well as on a range of other thinkers. Leibniz's theory of philosophical and mathematical intelligibility and the monad-logical structure of the phenomenon must be considered, as it was by early quantum physicists. Leibniz's theories helped mathematician arrive at a vision of the complex geometries of space-time that especially influenced Albert Einstein and the other founders of QMT. Kant's defense of the inherent *a priori* logical framework of the subjective mind—a move at least partially proven by QMT—demands consideration, as a theory of reality should secure some position on idealist philosophical interpretations of subjective consciousness and intelligibility. Consequently, I draw upon a variety of disciplines' theories of the real to advance the structures and principles of my own holo-cryptic ontological and epistemic "transcendental" hermeneutic. Thus, I develop past hylomorphic theories of non-subsistence by adjudicating their principles with an Ur-Kenotic interpretation of super-essential and representational dimensions.[66]

Newtonian space is absolutely uniform and inert "ether," whereas time is homogeneous and forward moving. The rigidity of Newton's space-time was transformed by Einstein's proposition that there is no privileged inertial frame of reference and that the speed of light is constant at 186,000 miles per second (c). Taken together, these two hypotheses rewrite how

[65] Etienne Gilson, *Being and Some Philosophers*, 2nd ed. (Toronto, ON: Pontifical Institute of Medieval Studies, 1952), 171.

[66] See D. C. Schindler, *A Companion to Ferdinand Ulrich Homo Abyssus* (Washington, DC: Humanum Academic Press, 2019), 57–58.

we understand space-time. As one reaches the speed of light, Einstein argues, an individual experiences time as slowing down (temporal dilation). Concurrently, given space's intractable relationship to time, one's immediate localized space decreases as one moves closer to the speed of light. The continuum of time and space are most evident to the subject at the speed of light, for at this point the past seems to fold into the present moment. At relativistic time frames, space and time intermingle and can be imagined as circles within circles, rather than as a one-way sign directed only to the future.[67]

Einstein defines how mass and energy are symmetrical concepts that exist in a reversible or bidirectional relationship akin to laws of conservation. Energy is quantifiable and qualitatively symmetrical to mass in relation to the communication of this relation at the speed of light, as noted in Einstein's famous equation: $E = mc^2$. The nature of our spatiotemporal and material existence makes it difficult to sensually perceive and conceptualize the "transitional equilibrium point" between the terminal points of matter and energy. For me, energy and mass serve as a metaphor for the incomprehensible middle ground between light and consecrated matter delineated in ritualistic practice. Furthermore, the spiritualization of the spatiotemporal arena and a material object are fundamental artistic representations.[68] Spatiotemporal distortion promotes productive cognitive dissonance of the viewer, that is, a contemplative mode open to interior movements. The abstract art movement of the early twentieth century is key in its attempt to refocus a being's material identity by advocating its interior and exterior movements over its realistic depiction. This art focused on the spatial and dimensional properties of a particular environment or being. Drawing from past and modern mystical writings and movements, many of these artists deemed the material to be a shamanistic or energetic force rather than a commentary on a local being's material and accidental properties. Abstract artists attempt to release the essential or informational form from within its embodied and surface abodes as line and color. The artist liberates the interior movements of the represented being by aesthetically manifesting its inner truth. The energetic brush-

[67] See Leonard Shlain, *Art and Physics: Parallel Visions in Space, Time and Light* (New York, NY: William Morrow and Company, Inc., 1991), 125. Hereafter cited as *Shlain*.

[68] See Dörte Zbikowski, "Dematerialized Emptiness and Cyclic Transformation," in *Iconoclash*, Edited by Bruno Latour and Peter Weibel (Cambridge, MA: MIT Press, 2002), 433.

stroke and mass of paint, the crust of consecrated bread, and the QMO's spatiotemporal actualization share a single commitment to revealing hidden truths to the receptive observer. These varied perspectives, descriptive languages, and models influence my argument that the holo-somatic body is proleptic as it is related to final spatiotemporal and material truth, that is, the resurrected body.

The resurrected body is understood to be the eschatological perfection of the material corporeal body in its spiritual state and thus images the perfective state of Jesus Christ in the Eucharist. In physical terms, this dynamic is manifested in the energetic elements of a being's holo-somatic identity. Gregory Palamas gives a theological justification for this metaphysical principle as part of the incarnational architecture, in writing that "illumination or divine and deifying grace is not the essence but the energy of God."[69] Stated differently, the formal and concrete character of "holo-somatic" personality allots a being its transcendental properties, and in so doing, beauty transfigures one's lived experiences. The term holo-somatic already hints at these themes by theologically adapting the idea of holography. The prefix "holo-" or "hole" and "gram" or "to write" grounds and is synthesized with the concept of the body (soma). The body is understood to be a specialized manifestation of information as essential truth. This move is in concert with current scientific principles and past philosophical formulations. The Ancient Greek understanding of form as essence synthesizes the two themes of *eidos*: (image, idea) and (shape). As Balthasar states, "as form, the beautiful can be materially grasped and even subjected to numerical calculation as a relationship of numbers, harmony, and the laws of Being."[70] The idea of holo-somatic identity revises earlier substance-accident explanations for a being's spatiotemporal modification. Holo-somatic identity brings into a single consolidation what has been distinguished in part, in past theological and philosophical systems as *prima substantia* or *hypostasis*, the essential (*esse quid*) content, and *substantia* or *form* (*esse in se*).[71] Similarly, the "*acheiropoietic* act" aligns a

[69] Gregory Palamas, "Physical and Theological Chapters," 68–9, in Vladimir Lossky, *Vision of God*, Translated by Asheleigh Morehouse and preface by John Meyendorff (Crestwood, NY: St. Vladimir's Seminary Press, 2013), 33.

[70] Hans Urs von Balthasar, *The Glory of the Lord, A Theological Aesthetics*, vol. 1: *Seeing the Form*, Translated by Erasmo Leiva-Merikakis, Edited by Joseph Fessio, S.J. and John Riches, (San Francisco, CA: Ignatius Press, 2009), 118. Hereafter cited as *GL* 1.

[71] See *ST*. I, 29, 2 and Johann Auer and Joseph Ratzinger, *Dogmatic Theology: A General Doctrine of the Sacraments and the Mystery of the Eucharist*, Vol. 6, Translated by Erasmo Leiva-Merikakis, Translated and Edited by Hugh M. Riley (Washington, DC: The Catholic University of America Press, 1995), 224. Hereafter cited as *Mystery of the Eucharist.*

subatomic particle or holographic being's hidden essence to create an existing appearance or image in the created order [holographic identity]. We recognize something of this interpretative approach in Bonaventure's observation that the one "describes being in that it is whole, by reason of inner indivision; 'true,' in that it is intelligible, by reason of indivision between itself and its proper species; and 'good,' in that it is communicable, by reason of indivision between itself and its proper operation."[72]

The concepts of form, essence, gestalt, idea, and holograph all engage the substantial and spiritual aspects within the created order, whereas scientific determinism interprets only the phenomenon's physical properties. The two perspectives must be simultaneously maintained given our sensual and cognitive limitations. The artist and believer hold this tenuous balance together by symbolically making wonder part of their conscious and intuitive grasp of reality. Symbolic representations, Pavel Florensky explains, are revelatory and work toward a greater project. He writes, "the mystery of the world cannot be veiled by the symbol, but, on the contrary, manifest itself in its authentic substance, i.e., as mystery."[73]

Symbols are specialized semantic holograms that establish a relationship or at least set out the working conditions for a promising conjecture between a subject and an object.[74] Scientific research is a powerful tool that allows us to quantify something of the mystery of being by partially naming it through symbol and image. Before showing how holo-cryptic metaphysics works with regard to this common end, a few introductory comments on the areas of postmodern philosophy, Trinitarian identity, and Transubstantiation are necessary. Next, I will detail more specifically how to synthesize QMT into my own philosophical and theological model of the doctrine of transubstantiation.

At first glance, Alain Badiou and François Laruelle may seem unlikely dialog partners for this project. However, these postmodern thinkers offer profound insights into the nature of reality, the infinite,

[72] Bonaventure, *Breviloquium*. 1, 6, 2, in *The Complete Works of Bonaventure: Cardinal Seraphic Doctor and Saint*, vol. 2, Translated by Jose De Vinck (Patterson, NJ: St. Anthony Guild Press, 1963), 53–54. Hereafter cited as *Brevil.*

[73] Pavel Florensky, *Beyond Vision: Essays on the Perception of Art*, Edited by Nicoletta Misler, Translated by Wendy Salmond (London, UK: Reaktion Books LTD., 2002), 34. Hereafter cited as *Reverse Perspective.* See also Raphael Bousso, "The holographic principle," *Reviews of Modern Physics* 74 (2002): 825–874.

[74] See *ST.* I, 35, 1–2.

Nothingness, the void, and the concept of a hypothetical Transcendent ONE or rather its "necessary" denial as non(ONE). These thinkers help us create a bridge between mathematics, QMT, and theology and thus buttress our own exploration of the doctrine of transubstantiation. Where Badiou gives us several new foci to engage the metaphysics underlying a theological vision that attempts to discern the contemporary landscape, Laruelle's so-called anti-philosophy reimagines atheistic claims against the Transcendent.

These thinkers' rejection of a traditional notion of the Divine can engage fruitfully with my proposition that with a revised interpretation of Ur-Kenosis, QMT, and postmodernity, we can achieve a viable narrative of the doctrine of transubstantiation for a post-secular age. Although I reject the atheistic stand of many postmodern thinkers, we all still profit from seeing how their hypothetical, non-theistic, or non(ONE) descriptions of the "divine" initiate new avenues to think about how to incorporate advanced mathematical and scientific theories into our own Christian worldview and my own self-confessedly traditional, orthodox Catholicism. It would be imprudent to cease wrestling with theological problems or assume that we have already arrived at the final answers to difficult theological questions. Balthasar's perspicacity into Ur-Kenosis is only bolstered by later work of believers such as Ulrich and nonbelievers such as Badiou and Laruelle. The scholar Hollis Phelps rightly argues that although an atheist, Badiou never fully divorces the theological consideration from his own musings, as this ancient discipline *a priori* stages his rejection of its pronouncements: "Badiou's use of theological language, in this sense, implicates him formally in the discourse of theology."[75] Badiou's expansive oeuvre finds impetus in a philosophy of mathematics that attends to his systematic vision of the ontological. He is not here crudely equating number and being, as this primitive correspondence depends on its symbolic representation and logical formalization. This approach is instructive for my own interpretation of how a QMO can both exist and be mathematically stipulated.

Alain Badiou alleges that his understanding of subjective truth's intense manifestation is theoretically congruent with Saint Paul's own theory of individual Christian life and universal salvation. It seems clear that Saint Paul believed that one participates in Jesus Christ's mission and avails

[75] Hollis Phelps, *Alain Badiou: Between Theology and Anti-theology* (Durham, UK: Acumen Press, 2013), 12.

oneself of divine grace (Galatians. 2:19). Christ's redemptive act removes the disorders of sin and evil from Creation—finite beings see this outcome impartially in the paradoxical "subtractive" and "productive" non-subsistent relations that are our experiential grist.

Many Christians may find Badiou's proposed affinity to the work of Saint Paul problematic and perhaps even heretical given their own opinions on such matters. Certainly, Badiou's position that intimacy with the Divine is impossible given he denies God's existence is disconcerting to believers. However, the faithful can appropriate something of value from the atheist if his claims are read as a form of apophasis devoid of any personal relationship to the divine. The Eastern theology defense of negative theology follows from its measured judgment of the abilities of the finite mind. John Meyendorff explains that "[t]he whole of Byzantine theology—and particularly its 'experimental' character—would be completely misunderstood if one forgets its other pole of reference: apophatic, or negative theology… Rejecting the Eunomian view that the human mind can reach the very essence of God, they affirm the very essence of God and exclude any possibility of identifying Him with any human concept."[76] If Badiou's philosophy is to stake any theological ground, it can be found by seeing his mathematical philosophy as a reworking of the hermeneutic of apophasis in a new key, for "[t]he differences between religion and anti-philosophy is slight. Anti-philosophy is a rigorous and quasi-systematic extrapolation from an essentially religious *parti—pris*. Anti-philosophy is religion in philosophical guise, argued on philosophical terrain."[77]

The landscape where Badiou's most important "metaphysical" contributions are found in this regard is in his speculative study of the philosophy of mathematics.[78] Whether Badiou's insights into Paul's theological and mereology ordering of the universe are valid is an open-ended question. If an atheist can properly be said to have a "religious vision," it seems only fair to entertain the notion of whether a mathematical and scientific

[76] John Meyendorff, *Byzantine Theology: Historical Trends and Doctrinal Themes* (New York, NY: Fordham University Press, 1974), 12.

[77] Peter Hallward, *Badiou: A Subject to Truth* (Minneapolis, MN: University of Minnesota Press, 2003), 20. Hereafter cited as *A Subject to Truth*.

[78] Badiou rejects the term metaphysical as it treats being at the level of "empty generality," see Alain Badiou, *Briefings on Existence: A Short Treatise on Transitory Ontology*. Translated, Edited and Introduced by Norman Madarasz (Albany, NY: State University of New York Press, 2006), 45 and 57. Hereafter cited as *Briefings on Existence*.

worldview can also have enough explanatory range to gauge a theologoumenon.

Badiou has written extensively on the larger human experience with notable works on mathematics, philosophy, drama, cinema, music, and political theory. However, he sees in mathematical philosophy the best way to explicate a philosophy of being. Badiou claims that "mathematics provides philosophy with a weapon, a fearsome machine of thought, a catapult aimed at the bastions of ignorance, superstition and mental servitude."[79] A being's attributes are specified by philosophy, Badiou reasons, when it "exposes the category of truth to the unifying, metaphysical power of the One. It is no less true that philosophy exposes the power of subtraction in the study of mathematics. This is why every singular philosophy is less of an actualization of its metaphysical destiny than it is an endeavor, under the mathematical condition, to be subtracted from it."[80] Mathematics supplies Badiou with a language that describes the mutuality between a being's autologous self-organization and its non-subsistent correspondences. These relationships are described as a "count," as they evidence reality's horizon of infinite actual multiplicities but also depend upon subtracting from a theoretical "One," or in postmodern nomenclature, non(ONE). Individual beings can virtually and logically represent these processes in the language of math because "the shape of a syntax or a formal theory, in its linguistic connection is primordial."[81] Badiou's philosophy of number and ontology is grounded in three propositions: (1) privilege numbers over physical beings, (2) a series of inverses such as defining the Transcendent as a nonexistent ONE that is subordinated to all multiples, and (3) premising an actual infinity and rejecting potential infinity.[82] With these starting points Badiou sets out to illumine the meaning of being. In Badiou's own words: "[t]he thesis that I support does not in any way declare that being is mathematical, which is to say composed of mathematical objectives. It is not a thesis about the world as part of divine creation but about discourse. It arms that mathematics, throughout the

[79] Alain Badiou, *Theoretical Writings*, Translated and Edited by Ray Brassier and Alberto Toscano (New York, NY: Continuum International Publishing Group, 2004), 16. Hereafter cited as *Theoretical Writings.*

[80] *Briefings on Existence*, 42.

[81] *Briefings on Existence*, 112. See also Ian G. Barbour, *When Science Meets Religion: Enemies, Strangers, or Partners?* (New York, NY: HarperCollins, 2000), 2–3.

[82] See *A Subject to Truth*, 75.

entirety of its historical becoming, pronounces… what is expressible being *qua* being."[83]

An existing actual infinite and reality's identification as the multiplicity of beings is set against its absence or the void. This philosophical scenario illustrates how divergent many postmodern theories are to traditional metaphysical approaches. Lacking a strong investigation of the infinite burdens the subject's conscious experiences of reality: "in as much as the sphere of intelligibility instituted by the infinite manifestly poses no specific problems, neither in the axiomatic intuition nor in the proof protocols, the grounds for worry are always extrinsic, psychological or empiricist."[84] Where the Transcendent orientates the thoughts of a believer, for Badiou, human consciousness finds its significant source in how one conceptually represents the infinite and the void.

Thomas' theory of ontological exchange between an esse and essence within a finite being makes the infinite indispensable to his philosophical vision. Added to this natural and relative infinity, the human person Jesus Christ, I argue, holo-somatically realizes an actual infinity as part of the eternal hypostatic exchange of the Son. Any metaphysical formulation of the infinite is problematic, because full comprehension of it is impossible. It is debatably whether or not if one's conceptions of the infinite are not in reality ambiguous notions of the "very large."[85]

The meaning of the infinite cannot be simply equated with the idea of succession. Further, it is easy to confuse, conflate, or weigh disproportionately beings or numbers when compared. All conceptual representations must be balanced with our lived experiences, and yet, logical imprecision and impartial interpretations of being seem impossible to avoid.[86]

Because mathematical language uses a rigorous definition of a "mathematical being" it is able to address many of the shortcomings of imprecise speech in its use of measure and description. If language and its formal use is not part of a complete vision, a specious assessment of meaning threatens. Such a language risks subjecting its user to "disobjectivation,

[83] Alain Badiou, *Being and Event*, 8. See also *ST. I–II*, 2, 5.

[84] *Briefings on Existence*, 95.

[85] On the very large's critical relationship to the One see *Being and Event*, 145–46 and John Mullarkey, *Post-Continental Philosophy: An Outline* (New York, NY: Continuum International Publishing Group, 2006), 86–87. Hereafter cited as *Post-Continental Philosophy*.

[86] See *Theoretical Writings*, 101.

disorientation, interruption and isolation."[87] Theistic systems fail for many postmodern thinkers for this very reason. Christianity, these critics argue, introduces this dynamic by classifying an onto-theological rationale. According to her critics, the onto-theological principle introduces a bi-figuration of the real (*actual essendi*) and its formal description as concretely actual (form-matter) as evidence of a non-subsistent image of the Transcendent. This belief, Badiou argues, is erroneous and perhaps delusional: for no such direct access into being or the Transcendent is possible.

While Badiou understands being to be a formally intelligible antecedent to its symbolic or linguistic representation, it is not immediately clear how this starting point necessarily results in a thick metaphysics. The question remains open, as these symbolic markers exist as an interchange between being's representation as noetic (logical) and material (ontological). The concept of representation is certainly part of the metaphysical structuring of consciousness and subjective freedom, and it therefore determines how one experiences being and reality. Atheists such as Badiou reject this conclusion, arguing that the subject-object relationship "transcends" itself only to the degree that subjective suppositions create representations, or better stated, re-representations, of existing objects in the void and infinity's virtual display.

Being is the source, meeting point, and test for all overarching, universal, and metaphysical "hyperphagia." The universality of Being (e.g., Badiou's Actual Infinite) and its representation as a particular being—the infinite ordering or multiplicity of the plenum—disclose being's essential numeric nature. Of mathematical identity and becoming or transformation, Florensky writes that "we cannot think [of] a process without dividing it into a succession of steady states, into a succession of moments of unchangeability. In addition, we also cannot think [of] a continuum without dividing it into a discontinuous combination of point elements."[88] Or again as Paul Valéry muses, "modern man no longer works at that which cannot be abbreviated." Aesthetic, mathematical, and theological perspectives give different meanings to what a "point" is—the subtraction of the

[87] *Briefing on Existence*, 109. See also Alain Badiou, *The Subject of Change: Lessons from the European Graduate School*, Edited by Duane Rousselle (New York, NY/Dreseden, DE. Atropos Press, 2013), 84–85. Hereafter cited as *The Subject of Change*.

[88] Pavel Florensky, *The Pillar and Ground of the Truth: An Essay in Orthodox Theodicy in Twelve Letters*, Translated by Boris Jakim and Introduced by Richard F. Gustafson (Princeton, NJ: Princeton University Press, 1997), 345. Hereafter cited as *The Pillar and Ground of the Truth*.

given from the whole. In all causes, the idea of a point influences how one understands space and time and thereby one's theory of creation or a non-theistic reality.[89] How most theologians or theological systems understand a mathematical point is ambiguous, and more importantly so is how it functions in hyper-dimensional spaces. Thomas, for example, understands that the event of Creation's particular "point" of origins in Nothingness, space, and time must be open to include at least an implicit reference to the supernatural to what exists outside this mysterious point's domain. Christian theologians argue that the potential infinite division of space and time presupposes a supra-transcendent Being whose actual existence acts as an incomprehensible limit that defines and quantifies all—nothingness, geometric and metaphysical "points," and reality. A being's spatiotemporal ordering in light of the multiplicity or the plenum is properly seen as part of an individual's interior life. It is from an internal "stationary metaphysical point" that subjective recognition of the finite and divisive is tallied.

Without divesting their rooted position within the physical universe, the rational self objectifies herself from the immediate by retreating into interior mental spaces. A person's cognitive architecture is taken as a metaphoric immaterial mathematical point that is open to infinitely successive manipulation by the addition of a new context. Human consciousness simultaneously opens a being up to physical reality because of its imaging of divine exemplarity and the irreducibility of relational differing. Even Nothingness and death are a means to experience the Transcendent as Creation's reordering by Jesus Christ informs Creation with the meaning of its paradisal state. Conscious movements are commensurate with force of ontological differing: "The infinite and the finite come into accord in the ontological vacillation, in an '*essentiell* dimensions,' because the essence, which is posited as really distinct from being, as the dimension of its subservient transnihilation [crossing out of the essence's finitude and subsistence] in the substantializing of being."[90]

In my system, Creation and the human person are seen as analogous holographic refractions (*similitudinem aliorum ab ipso*) of the unqualified hypostatic relations and the simplicity of the divine's essence.[91] An analysis

[89] In later chapters, I show how the concept of the point is critical to how space and time are understood scientifically and in so doing clarify the universe's beginnings and operation.

[90] *Homo Abyssus*, 210.

[91] See *ST*. I, 14, 2.

of interactive differing (act—potency) reveals existing beings to be an impression of divine nature in the created order and the person of Jesus Christ. Ulrich uses the term "transnihilation" to refer to the isolation or possible eradication of immanent being's relation to the divine essence and therein, the finite being's positive dependence on the Transcendent. By extending the notion of "transnihilation," the practical application of Ur-Kenosis, I argue, allows me to argue for Nothingness and silence's theological import via divine exemplarity. This occurs when physical reality is part of one's conscious understanding of the *exemplum*: "[t]he exemplar can be observed by the senses (*oculis conspicitur*) and refers to that which one must imitate (*exemplar est quod simile faciamus*). The *exemplum*, on the other hand, demands a more complex evaluation (which is not merely sensible: *animo aestimatur*); its meaning is above all moral and intellectual."[92] I reinterpret Badiou and Laruelle's notion of the non(ONE) as a way to introduce a categorical but subversive application of the bold truth of transnihilation without fully accessing creation's holographic manifestation of the good, true, one, and beautiful.

We employ Ulrich's transnihilation as a neologism for "kenotic relational ordering" because the Spirit completes the exemplarity of the divine hypostatic identity throughout created order and ultimately in the human person as the image of the divine.[93] Thus, a philosophical and metaphysical event can either positively project internal desires against non-subsistent states as part of reality's preexisting motions and emerging harmonies or negatively can attempt to isolate and define itself in terms of inertia or meaningless infinite repetition. The apophatic encounter with the Transcendent is part of Nature and the imaginative thrust of all rational and "transfigurative" moments.[94] The mathematics of infinity, artistic gestures, and moral life all appeal to the idea that transformation can be quantified in successive iterations.

Postmodern thinkers such as Badiou distance themselves from all ontotheological propositions—even when thinkers for whom they may have a great appreciation make such moves. For example, Badiou charges Georg Cantor (1845–1918) with confusing mathematical and theological

[92] Giorgio Agamben, *The Signature of All Things: On Method*, Translated by Luca Di Santo and Kevin Attell (New York, NY: Zone Books, 2009), 18.

[93] See *Homo Abyssus*, 82.

[94] See David Morgan, "Secret Wisdom and Self-Effacement: The Spiritual in Art in the Modern Age," in Richard Francis, *Negotiating Rapture: The Power of Art to Transform Lives* (Chicago, IL: Museum of Contemporary Art of Chicago, 1996), 41.

perspectives—for being so bold as to see in the idea of an actual infinity a mathematical representation for personal Divine Being. Perhaps, even more problematic for postmodern philosophers is the fact that Cantor went so far as to find inspiration for his mathematical work in theology.[95]

Disavowing any relationship between an Actual Infinite Being and the Divine, Badiou is still left with the prickly question of their apparent reduction to a formal "static" equation. Badiou equates the actual "embodiment" of the infinite to the virtual representation of the One's own self-reflexive non-identity as non(One).[96] This equation can falsely suggest that being or Absolute Being is reducible to an axiomatic statement. Regardless of its usefulness, any virtualization of identity remains partial, and even creedal statements are only germane insofar as they articulate the deeper mysteries of being. Concurrently, finite beings exist as part of a horizontal but infinitely multi-branching relational environment that defies final definition. Consequently, the subtractive or differing relation must be understood as a positive force. Badiou argues that by abstracting or "subtracting" the concept of the universal from its theoretic delineation as an infinite series, some mastery results in how a finite being situates itself as bonded or a limited historical point within the universe's unending multiplicity and chaotic chance. Earlier philosophers Parmenides and Gregory of Nyssa adopted a similar logic that defined the infinite in being's relationship to nonbeing as a means to better understand the universe as ordered, but free. The difference between Nyssa's views and prominent postmodern thinkers becomes evident when actual infinity is taken as a formal analogy for an Eternal Personal God. The ideas of infinite space and nonbeing become the conceptual anchors by which Nyssa understands how the creature is ordered to the Transcendent.[97] Applied to the present discussion, the binary being-nonbeing principle translates in holo-cryptic metaphysics into that state of a being's non-subsistence—a

[95] See Joseph Warren Dauben, *Georg Cantor: His Mathematics and Philosophy of the Infinite* (Princeton, NJ: Princeton University Press, 1990), 232.

[96] I argue below that this virtualization becomes positive for the Christian when read in light of Christ's harrowing of Hell. With this redemptive action, the virtual rendering of formal atheism becomes a positive holographic witness to divine providence.

[97] See Hans Urs von Balthasar, *Presence and Thought: An Essay on the Religious Philosophy of Gregory of Nyssa*, Translated by Mark Sebanc (San Francisco, CA: Ignatius Press, 1995), 29. Hereafter cited as *Presence*.

void or Nothingness—whose principle is Ur-Kenotically based and reality's fundamental thesis.[98]

It is necessary to see the empty and its formal descriptions as a necessary and instructive step in various metaphysical applications of the infinite. Badiou's theory of the void, for example, attempts to strike a balance between Aristotle's previous definitions of matter (*hyle*) and its Platonic characteristic of unlimited or infinite motion. The spatiotemporal, Badiou tells us, opposes the void. Lacking matter, the void is understood to also be without time. Much like my own interpretation of nothingness, the void functions in Badiou's mathematical philosophy as a progenitor of networks of differing alliances. With the introduction of matter in the void, the spatiotemporal comes into its own. Given its source in *creatio ex nihilo* the void platforms the positivity of all beings—including time.

The paradox of the positivity of the negative is a mystery implicit in the Judeo-Christian Creation myths and the overt subject of many postmodern nihilistic formulations. Philosophers open to dialog with theology have at their best steered a middle course between the two extremes—Being and Nothingness. The theologian undertakes a further step. The space between being and nothingness is a point or "transnihilative mode" that holographically represents the Transcendent relationship to divine self-denial. Thinkers such as Aristotle take the relationship between being and nothingness as representative of the whole vantage of a contingent being. Ulrich critiques this position because each individual being's identity is fundamentally positive—it is, externally related to the Transcendent as its living image. This truth, he states, actualizes in history and consequently, "Being as being negates negation in a complete way. The other is no longer left 'outside.'"[99]

If such an interpretative strategy is to move beyond oblique gesture or bald critique of postmodern thought, we must show that a being's essential or informational identity is logically conducive to its mathematical justification in the concepts of nothingness, void, infinity, and non-subsisting or differing (kenotic) relationships.[100] If natural theology is to be taken seriously at all today, it must stake out a place with respect to mathematics and science's conclusion. In the ideal, a being's identity theologically brings about its openness to the other (Nothingness, void, etc.) by means

[98] See *Homo Abyssus*, 30 and 275.
[99] Ibid., 33.
[100] See Ibid., 430.

of "subtractive," "self-gifting," or "kenotic" gesticulations. Created being is not in any way imagined as a gnostic Disneyland that sees nothingness as a conceptual attraction for faithless nihilism. To grasp Nothingness, one must understand its role in relationship to absolute and relative perfection.[101] Again, we must remember the fact that the number zero is even and antecedently the mathematical proof that the number closest to Nothingness still promotes an algorithmic relation.

Denying the existence of an actual Transcendent Being (*Ipsum Esse Subsistens*), the scholar John Mullarkey comments that Badiou's philosophy avers that material being virtually and infinitely quantifies. Badiou argues that finitude requires one inverted qualitative measure with a historical rendering of the pure quantity of mathematical infinity. Mullarkey eloquently states that "The worlds of qualitative belonging, of dwelling within the world, of bodily incarnation, must be made spartan: Badiou'sworks subtracts every quality from such worlds in search of their pure immeasure, the paradox of quantitative infinities."[102] Aristotle's genuine intuition into infinity follows from his defense of finite being, and this systematization gives him the ability to see the infinite series as the sequential relating of individual finite being. Removed from the pure measure of an infinite and eternal Being, finite nature remains horizontal as the infinite is understood as a finite series." [103]

In the created order, an actual infinity is only comprehended in direct relationship to one's understanding of potential infinities. Thomas argues that the identities of these two infinities are quantifiably and qualitatively diverse. Consequently, the only way to critique actual infinity is to establish its identity in relation to potential infinity. Thomas appraises a finite being's partition of the infinite when an analogous proportion is recognized between these two concepts. He reasons that, "six and eight are proportionate because, just as six is the double of three, so eight is the double of four, for probability is a similarity of proportions. Now, … in every proportion because of some definite excess of one over the other, it is impossible for any infinite to be proportionate to a finite way by way of proportion."[104] Being's relationship to Nothingness is understood in light

[101] See *De pot.*, 1,2, resp. 2.

[102] *Post-Continental Philosophy*, 83.

[103] See Aristotle, *Physics*, Book III and IV, Translated by Edward Hussey (Oxford, UK: Oxford University Press, 1983), chapter 6, 206a–206b.

[104] *De ver.*, 2, 3, resp. 4.

of a being's correspondence to non-subsistence. These analogies shares similar properties with number, for, as Thomas argues, "the nature of light is spoken as being without number, weight and measure, not absolutely, but in comparison with corporeal things, because the power of light extends to all corporeal things." [105]

Badiou's theory of the metaphysics of the infinite is part of an ongoing philosophical discussion whose modern mathematical incarnation looks to the genius of Cantor's comprehensive proof of the infinite. Cantor gives infinity an ontological stability by quantifying it by placing or bounded it within a mathematical set.[106] A mathematical set gives numbers a rudimentary logical structure—a holographic image of the whole and the universal that conceptually houses infinite sequences. A set also makes possible the internal ordering (e.g., the ordinal integer) of numbers in their correspondence to its domain. The number of elements in a set determines its cardinality (domain). Thus, an unlimited number of various infinite sets and finite sets can be created. Logical relations such as union and intersection can define these sets.

The application of logical operators to mathematical sets implies that the set's brackets function as a boundary for the numbers within. Individual numbers are defined within the set by assuming their own unique "bracketing." Mathematical bracketing is defined and this interpretative layer implies numbers can exist in a virtual hierarchy or possess a "transcendental" identity.[107] Cantor explains, "[b]y a set we are to understand any collection into a whole M [*Menge*] of definitive and separate objects M [*Menge*] of our intuition of our thought."[108]

Cantor's theory of infinity advances that a singular or unique identity can be given to each infinite series or set. Just as a being's ontological existence cannot be reduced to its formal description, a number's categorical grouping within a set, for example, does not override its prior identity. Every infinite series, for example, can be put into relationship with the

[105] *ST.* I, 5, 6.

[106] On the stability of the empty set, see Alain Badiou, *Number and Numbers* (Malden, MA: Polity Press, 2008), 44.

[107] See R. L. Wilder, *Foundations of Mathematics* (New York, NY: John Wiley Sons, 1952), Chap. 5. Cantor mathematically determined that there was more than one type of infinity. In a similar way, Badiou develops the idea of the generic beyond its common mathematical definition.

[108] Georg Cantor, *Contribution the Founding of the Theory of Transfinite Numbers* (New York, NY: Dover 1955), 85.

number of zero or such analogates as nonexistence, non-subsistence, potency, and nothingness. Various conceptions of negation allow a better understanding of infinity and the theoretical construction of mathematical sets, just as modes of nothingness such as potency disclose the nature of being. The idea of zero implicitly categorizes and frames all numbers and mathematical sets because it presents a theoretic baseline to differentiate mathematical equations and objects.

It is only after one has grasped the identity of individual numbers that one can then group them according to their respective common properties. Every set can be related to any number of sets. This follows because the numbers of each set theoretically share at least one "morphism" or property that allows its bridging or mapping to other sets of other individual objects (numbers).

Furthermore, assuming shared identity, one can, *a priori*, proportionately and continually relate the numbers within each set. Every number of an infinite set is located in a theoretic non-dimensional space like a mathematical point. Without material constraints the mathematical set can be defined by abstract functions or operations. Thus, an operation can be applied to individual members of the set or the set as a whole any number of times.

No matter the numbers of iterations a set undergoes, it can theoretically be retroactively or reverse engineered to return it to its most primitive state. However, if numbers in a state are philosophically or metaphorically qualified the nature of their correspondence to other numbers becomes difficult to ascertain. Cantor's eloquent mathematical depiction of infinite series in sets gives us an example of this operation. Taking an infinite set of positive integers and zero, $A = 0, 1, 2, 3...n$, Cantor associates it to the infinite set $B = 1, 2, 3...n$. By hypothetically nullifying the potential or qualitative attributions of a number, Cantor arrives at a method to map the elements of sets A and B in a one-to-one relationship. The qualitative identity of each number is analogous to its "transcendental" capacities and categorization. Thus, set *B* is a proper subset of set A because each member can be mapped onto another member in its "opposite" set. The two sets are now quantifiable by making the hidden essential identity of each number in a one-to-one correspondence. The identity of the numbers of sets *A* and B remain unchanged but result in a new metastructure or topological arrangement that suggests a way to define the infinite span of the two sets.[109]

[109] See *Being and Event*, 156–159. Barring further information, it is not possible to prove if this one-to-one infinite correspondence results in a particular geometric structure.

It is only when two or more numbers or beings are placed in a relationship that a pattern is perceived among them. A pattern reflects elements of the interior potency of a being or number that may otherwise remain hidden. Cantor showed that not all infinite sets' elements could be mapped with other infinite sets (bijection). Cantor proves, for example, that the infinite set of real numbers (R) contains more elements than the set of natural numbers (N): R > N. The infinite set of natural integers and that of irrational numbers can be distinguished using the concept of the equinumerous as a meta-leveling structure (1-to-1) to quantify fundamental particles and offer a new hermeneutical take on the natural quantitative material elements of the bread and wine and their qualitative supernatural dimensions of the Eucharist.[110]

Mathematical sets of the ensembles of numbers are unique insofar as no mathematical set defines all other sets or groupings.[111] The open-ended nature of the relationship between mathematical sets and the coherence of infinite sequences and its various subclasses gives us an analogy for how a being or a group of beings' identities freely define in a variety of non-subsistent relationships. Given these insights, John Mullarkey's comments on Badiou's philosophy are instructive; he writes that "quantity, size or 'cardinality' has its own set of paradoxes that are generative: they can create genuine, unpredictable novelties or events."[112]

The emergent nature of finite beings finds its parallel in the generative capacities of concepts and their systematic and semantic description. Set theory makes a distinction between "equal" and "equivalence," such that "equal sets have the *same members*, while equivalent sets have the *same number of members*. Equal sets are, therefore, necessarily equivalent but the converse is, in general, not true. Furthermore, nothing is said in the definition of equivalence about the exact nature of the one-to-one correspondence between the sets-only that one exists."[113] The principles of inclusion and exclusion are used to impart elements in a set analogous to rational intention. In both cases these mathematical operations are

[110] This proof was a turning point in the history of mathematics and instrumental for my understanding of how various symmetries are found in vector space R, manifold space R, number field R, metric space R, and Hilbert space.

[111] See Robert Wall, *Introduction to Mathematical Linguistics* (Englewood Cliffs, NJ: Prentice-Hall, Inc., 1972), 175–76. Hereafter cited as *Mathematical Linguistics*.

[112] *Post-Continental Philosophy*, 85.

[113] See *Mathematical Linguistics*, 175.

compared to our ability to enter into and out of relationships given some determined metaphysical postulates or boundaries.[114]

A number can be theoretically placed anywhere within a set or in any other actual or hypothetical sets. These subsets can be likened to a non-subsistent grouping, as they invite displacement by future sets and subsets. Badiou calls attention to these factors by aligning a being's prospective membership in a set and subset as a mathematical prefiguring of a being's multiplicity and truth's endless progression under some "state" or defining order. Badiou argues that the mathematical set has a single essential property, that of multiplicity. There is no external or internal essential measure of the mathematical set other than that of an indifferent "differing." Logically, this means that there is "no other undefined primitive term or possible value for the variables besides that of sets. Thus, every element of a set is itself a set. This is the realization of the idea that every multiple is a multiple of multiples, without reference to unities of any kind."[115] A particular being or number's augmentation by means of multiplication is a mathematical proof for a being's exponential truth. Badiou argues this emergent characteristic holds in our political orchestrations. The infinite set becomes for Badiou a mathematical analog for the continual proliferation of a being's interior measure by means of adjudicating the conceptual extremes of the void and the One—a measuring reminiscent of potency's relationship to being. The mathematical bracketing of the mathematical set is likened to a being's external appearances and its members give another way to perceive how a being's interior properties mirror in the correlation of numbers.

Badiou sees in mathematical sets a parallel to a being's affiliations in the world. Every element of a group can be placed in more than one group because of the ontological and epistemological integrity of the respective numbers and beings and logical coherence of their interior relations. The difference between a number's placement in a set and the confines experienced by a living being must be stressed. The unbounded nature of consciousness suggests that no physical presence can ultimately overshadow one's irreducible freedom and hence, "one cannot refer to a supposed *inclusion* of the event in order to conclude in its *belonging*."[116] Implicit in this postmodern philosophical conceit is a refutation of the claim that a

[114] See *Being and Event*, 29, 367–71 and *Logics of Worlds*, 208.
[115] *Theoretical Writings*, 46.
[116] *Being and Event*, 202.

finite number's identity is fully explained by their existence on a given number line.[117] Every transfinite and finite number has an inimitable identity, for "there are always submultiples which, despite being included in a situation as compositions of multiplies, cannot be counted in the situation as terms, and which therefore do not exist."[118] Derrida expresses a similar perspective on signification's dependence on differing internal and external forces that engage meaninglessness when he writes that to "risk meaning nothing is to start to play and first to enter into the play of difference."[119] Given differing relations, there is no limit to the number of subsets that can be added to create an internal "hierarchy" within an existing set. In comparison to the endless addendum to a set's membership, one can subtract all members from a set and thus create an empty set. Every mathematical set is implicitly in relationship to the infinite, the complete, and nothingness.

The quantitative subtraction of elements from a mathematical set or grouping finds its theological key in the Ur-Kenosis or self-emptying of God in Jesus Christ's Incarnation. To wit, the generic set like the concept of the kenotic infinitely subtracts from the whole and the multiple. The undefined set, metaphysical differing, meaninglessness, or Nothingness is infinitely subordinate to what is conceived of in the notion of Transcendent's eternal self-emptying.

Generic mathematical categorizations and their theological counterparts are problematic, as a clear demarcation of infinite sequences, potency, and surplus meaning is difficult and perhaps impossible given our conscious and linguistic abilities. Badiou speculates that, "the generic is that subtraction from the predicative constructions of language that the universe allows through its own infinity. The generic is ultimately the superabundance of being such as it is withdrawn from the grasp of language, once an excess of determinations engenders an effect of indeterminacy."[120]

A generic set is inconsistent as it is receptive to an unlimited number of members or elements. Cantor proved that from all sets and subsets, at least one element could be derived to create a new set where at least one

[117] See David Hilbert and Paul Bernays, *Grundlagen der Mathematik* (Berlin, DE: Springer-Verlag, 1934), 15–17.

[118] *Being and Event*, 97.

[119] Jacques Derrida, *Positions*, Translated by Alan Bass (Chicago, IL: University of Chicago Press/Athlone Press, 1987), 9.

[120] *Theoretical Writings*, 107.

element will be found not in the original set.[121] Thus, "no multiple is capable of forming-a-one out of everything it includes."[122] Every generic set conceptually corresponds to a "supra-un-definable transcendent" empty set or hidden representation of nonbeing in Badiou's philosophy. Similarly, some theologians see nonbeing as a placeholder for being's identity under the influence of differing relationships (*omnis definitio est negatio*).[123] Being and nonbeing exist in a permanent relationship that means one never arrives at a final definition or "count" of a finite being.[124]

Non-subsistence is an extensive property as it can be associated with individual beings or numbers. This reinforces, for Badiou, the idea that being is intrinsically infinite or multiple.[125] The infinite is symbolically contained in the mathematical set as "[t]he postulates that every class can be well-ordered links the transfinite arithmetic of cardinal numbers, which also form an unlimited hierarchy and are defined in terms of the one-one correspondence between classes as ordered by various relations. Some of the notions defined in this way theory are of great importance in topology and other branches of pure mathematics."[126]

Speculating and developing upon these insights it is possible to conceive of how a number, being, set, subset. or group possesses depth or dimensional complexity. The subset is "clearly infinite and remains beyond the reach of completion. Nevertheless, it is possible to state that if it is completed, it will ineluctably be a generic subset."[127] However, by taking the definition of a proper subset and working backward, Cantor is able to show the subset relates to its most comprehensive expression as a set.

Cantor proposes a mathematical set that contains all possible members. The theoretical absolute complete set, like the notion of the Transcendent, is logical consistent, but Badiou argues, is a conceptual fiction. Badiou's argument against a set of all sets parallels his views on the existence of an

[121] See Alain Badiou, *Mathematics of the Transcendental*, Edited, Translated and Introduced by A.J. Bartlett and Alex Ling (London, UK: Bloomsbury Press, 2014), 17–20.

[122] *Being and Event*, 85.

[123] See Sergei Bulgakov, *The Bride of the Lamb*, Translated by Boris Jakim (Grand Rapids, MI: Wm. B. Eerdmans Publishing Company, 2002), 52. Hereafter cited as *The Bride of the Lamb*.

[124] See *Being and Event*, 42.

[125] See Ibid., 230.

[126] Stephan Körner, *The Philosophy of Mathematics: An Introduction* (New York, NY: Harper Torchbooks, 1960), 65. Hereafter cited as *The Philosophy of Mathematics*.

[127] *Theoretical Writings*, 113.

actual Transcendent Being (ONE). Where the faithful holds that the Transcendent creates from Nothingness, Badiou counters that this belief is another view should be revised given his categorization of the terms of zero, nothingness, and the void. He writes: "if a relation from One to Zero exists, the transcendental discriminates nothing, and its operation is void." The Transcendent is reduced to its own dialectical annihilation by non(ONE) just as the virtual set of non-existing totality exists in a divisive or subtractive relationship to predications of infinite multiplicity, nothingness, and the void. The subtractive operates excludes an actual Transcendent or totality as Badiou explains that "[w]e see that, for every situation, the transcendental only operates if it excludes for the field of relations everything which claims to count the zero as one. We will say that a situation is consistent, with regard to its appearing, only if its object **Zero**—which always exists—remains subtracted from the relational operations of the count which have their origin in the One."[128]

The number zero is an even number whose intrinsic identity can be understood as divisionally relational. In comparison the number one images the total or sum as it confirms by division the number it is associated with—for example, $1 \div 7 = 7$. The number one's reflective nature by means of self-definition ($1 \div 1 = 1$) means that the number one can be related to any number except zero in a non-subsistent manner. Given its properties the number one is extrinsically relational. Developing these propositions in a systematic matter lends to the formulation of philosophical and mathematical theories. Adapting his reading of Platonism to theories of axiomatic systems, Badiou sees confirmation of his own philosophical outlook. Axiomatic systems, Badiou observes, that truth is heterogeneous in the real, and, second, that reason, properly construed, can give a profitable account of this multiplicity.[129] Like the proposed philosophical reflective nature of the number one, mathematical axioms are fundamental assertions taken to be intuitively self-evident.[130] By way of the Zermelo-Fraenkel Axiomatic Theory (henceforth ZFC), Badiou intuits specific rules that allow an individual to freely distinguish subsets from the set that are members.

[128] Alain Badiou, *Mathematics of the Transcendental*, Translated by A.J. Bartlett and Alex Ling (London, UK: Bloomsbury, 2014), 262–63.

[129] See Alain Badiou, *Conditions* (2008), 10–11 and *Key Concepts*, 30.

[130] See Sean Bowden, "The Set-Theoretical Nature of Badiou's Ontology and Lautman's Dialectical Problematic Ideas" in *Badiou and Philosophy*, Edited by Sean Bowden and Simon Duy (Edinburgh, UK: Edinburgh University Press, 2012), 51.

ZFC's Axiom of Infinity argues for the existence of at least one infinite set. The infinite set is defined as one with 0 as a member and a succession of positive natural numbers (e.g., 0 + 1, then 1 + 1, then 2 + 1). This set will gives rise to other infinite sets if and only if they are equivalent to the original source set. The generation of countless new infinite sets and subsets and their relationship to their source in a preexisting series raises the question of how a being's identity depends on internal and external correspondences on reflexivity.[131] As all exist in a state of change, referential correspondences remain exterior and interior principles of a being or number's identity. By being intentional about the influences that help determine one's identity, a being can see how the infinite serves as a conceptual and holographic representation of the Transcendent.[132]

For Badiou, the puzzle of infinite recursion and finite being's experience of unending multiplicity is analogously explicated in the work of Paul Cohen (1934–2007).[133] Cohen shows how a generic set is logically consistent with ZFC. His mathematical model helps us to explain how an infinite set's subsets are reflexive to the members of original set from which they were derived. Cohen's understanding of an infinite set's "incalculable" sequential progression gives Badiou insight into a being's potential and relational capacities. The mathematical theorem that states that any number over one is either prime or the product of multiple primes inspires Badiou's conceptualization of a being's relationship to the universal or its metaphoric equivalent of a non-existing Transcendent. The production of a composite number from two or more primes simulates how multiple beings can share a single point of reference without sacrificing their individual identities.[134] Badiou's interpretation of infinite sets and finite numbers and its hypothetical relationship to the Transcendent are questionable. Bulgakov warns that one must scrutinize the communication of emergent behavior: "[t]he errors of evolutionism concerning genera and species

[131] On the linguistic representation of infinite reflexivity, see *Being and Event*, 150–60. *Logics of Worlds*

[132] On the generic set serving as a representation of the infinite see *Theoretical Writings*, 114.

[133] It is clear that Badiou expands Cohen's mathematical theory into areas clearly not envisioned by Cohen. It is left to future scholars to determine whether this metaphysical reading of Cohen's work is justified or indeed at what level any cross-disciplinary approach can develop that does not risk the autonomy of each. See Paul Cohen, *Set Theory and the Continuum Hypothesis* (New York, NY: Benjamin, 1966). See also *Theory of Subject*, 144, 170 and 273.

[134] *The Logics of Worlds*, 13.

consists in conceiving them as the sum or integral of an infinite series of differentials, of infinitesimal changes, so that the very existence of the whole or species is rejected in favor of these differentials, none of which contains the idea of the species as such. This is because evolutionary thought tends to reject prototypes [or] idea-forces."[135]

The scientific theory of evolution does not account for a being's categorical identity as a species or genera, Bulgakov suggests, as a being is more than the sum of its parts. Similarly, Peter Osborne critiques Badiou's metaphysical approach as it forwards an ontology that "is severed from all phenomenological relations to objects… only because Badiou decided to sever it, in advance. Then, he has the awkward task of restoring the connection between his set-theoretical mathematical entities, philosophically received ontological concepts (like nature and history) and the world."[136] The void's assertion over a being's identity in Badiou's philosophy is another example of the question Bulgakov and Osborne are bringing to this complex philosophical and theological issue.

The Christian who assumes Christ and the Incarnation are the source and final hermeneutical arbiter of individual beings denies all critiques of postmodern notions of autonomous identity and the void. Athanasius sums up this theme in stating "God was made man so that man might be made God."[137] A being's image, like a finite numeric succession's virtual reference to the infinite, obliquely presupposes its potential final completion in a preexisting relationship to perfect unity. For the person of faith this trajectory can only lead to the Transcendent. The transfinite person Jesus Christ bridges the chasm between the Transcendent and finite reality. Christ is the free point between an infinitely dense horizon of relative infinities and the Actual Infinity of the Spirit. Imaging Jesus Christ's infinite desire for the Father reflects in the human soul, which will always be "demanding a supply, always altering into the grandeur nature, and yet will never touch perfection."[138]

[135] *The Bride of the Lamb*, 182.

[136] Peter Osborne, "Neo-Classic. Alain Badiou's Being and Event," in *Radical Philosophy* vol. 142 (2007): 24.

[137] Athanasius of Alexandria, "On the Incarnation of the Word," in *St. Athanasius, select works and Letters*, Edited by Philip Schaft and Henry Wace, Translated by Archibald T. Robertson, vol. 4, *A Select Library of Nicene and Post-Nicene Fathers of the Christian Church*, Second Series (New York, NY: Christian Literature Company, 1892), 54.3, page 65.

[138] Nyssa, *Against Eunomius*, *NPNF2*, vol. 5, Book 1, chapter 22. See also *ST.* I, 45, 5.

Neither the physicality of the body nor the infinite desires of the human soul conclusively proves that a single diremption exists within a being. The body and soul are the geography where subterranean and supernatural routes are mapped toward the Transcendent. The physical universe's holographic manifestations of the Transcendent and the Spirit's omnipresence in the natural order entices our intellectual inquiry into Being, for: "the anima is 'born" [*natum*] for this—namely, to come into correspondence with all beings—the correspondence does not occur in each and every newly posited foregrasp of reason towards being; instead, in its a priori completeness, as an *actual spirit*, reason is always already the abiding power [*die waltende Macht*] of this convenientia." [139] After the incarnation, the hard edges and conceptual boundaries that exist between the Transcendent and immanent beings transform how the believer experiences all limiting circumstances.

No less is the case for postmodernity that the incalculable multiplicity of the natural order and the concept of the infinite series that denies the Transcendent undergird an analogous pseudo-sacred secular belief system. Without a perceived margin at the periphery or a unified theory, ceaseless repetition can serve to give a non-theistic version of the sacramental.

The celebration of these endless cycles and sequences becomes fertile soil for many forms of re-envisioned paganism notable in contemporary society. The difficulties with believing in an unseen Transcendent is replaced in the postmodern version of a faith system with the power and pleasure derived from the momentarily harnessing repetition and immediacy.

The inability to rehabilitate the part into the totality signals an ordering of the infinite by postmodern philosophies. This move will be accounted for in light of Christian renderings of multiplicity and the infinite. As Balthasar so astutely notes, in the "polyphonic melody of becoming, the soul perceives, as it were, an echo of the divine infinity to which she aspires."[140] [41] To be clear, it is no less a problematic philosophical concession to confine the infinite to a status of absolutely other. The Transcendent's image in the created order gives the faithful person a way to define how individuals participate in the whole of reality. The overarch-

[139] See *Homo Abyssus*, 278.

[140] Hans Urs von Balthasar, *Presence and Thought: An Essay on the Religious Philosophy of Gregory of Nyssa*, Translated by Mark Sebanc, (San Francisco, CA: Ignatius Press, 1995), 41. Hereafter cited as *Presence*.

ing category of the totality has its own pseudo-existence and hence is binary to anything it regulates and orders. The mereological drama rejects any attempt to eradicate individuality by denying difference or refuting the necessity of various modes of unity that umbrella concepts such as the collective or set provide—beings are not cloned nor is reality an edge-less monotony. Ontological difference is inseparable from the unified state that exists in a potential state of deconstruction. Comparatively, the set brackets the infinite series and in so doing ensures that its members do not relate along the lines of mathematical pantomime but inheres in a hidden or bared semantics of the continuum and its representation in a set.

The theological deciphering of existence's deeper meaning does not require us to fence in the infinite but rather to make sense of its paradoxical presence. I assume that Cantor is correct in stating that an anti-infinite ideology is conducive to a polytheistic mindset or perhaps a precursor to a new era of atheism, he reasons: "[t]he fears of infinity is a form of myopia that destroys the possibility of seeing the actual infinite, even though in its highest form [it] has created and sustains us." By making the number one a pivotal principle of identity, it is possible to hierarchically and recursively order all numbers from a single point represented by zero without risk of polytheism.[141] Thomas finds in monotheistic identity a similar argument for multiplicity's call for moral perfection: "[God] brought things into being in order that His goodness might be communicated to creatures, and be represented by them; and because His goodness could not be adequately represented by one creature alone, He produced many and diverse creatures… goodness, which in God is simple and uniform, in creatures is manifold and divided."[142] I develop these themes below, but applied to the developing argument suffice it to note for now two points: (1) the generic infinite set gives us a way to see how a particular historical event—say the Last Supper—finds its further actualization in the metaphoric "subsets" of all subsequent liturgies, making the truth in any one subset or Eucharist universally valid. Against the risks of abbreviating the liturgy to the lesser dignity of crude drama—historical cloning or "relativizing" its original institution—the mathematics of infinite subsets gives us a way to understand every liturgy as causally related and exceptional. (2) Every number can be thought of as a solitary point on a continuum, and every infinite set

[141] See Louis Claude de Saint-Martin, *Des Nombres* (Nice, France: Editions *des* Cahiers Astrologiques, 1946), 2–3. 136.

[142] *ST*. I, 47, 1.

can be thought of as *sui generis* whole. Just as every number and set can be re-represented, each liturgical sacrifice is both a discrete historical moment and eschatological holographic display. More broadly, the concept of an empty set gives us another way to envision the Nothingness that preamble Creation where all seeds of reality ("*Logos spermatikos*") are conceptually and then spatiotemporally born.

François Laruelle's anti-philosophy critiques traditional theological views of a Transcendent Creator by making the Divine an atheistic simulacra of its own nonexistence (non(One)). Ultimately all such sets or configurations formulate an analogous quantification of the ONE or its negative reflection of the non(One).[143] The One's self-reflexive relationship, or perhaps more accurately stated, hypothetical self-reflective state, identifies with its own self-alienation or "othering."[144] Laruelle appeals to the idea that the Transcendent's identity follows from understanding its logical opposite as (non)One.[145] Laruelle's interpretation dispossesses Christian Trinitarian dogma of its traditional framing of the Divine as Transcendent tri-communal personal exchange. Laruelle's position on the economic Trinity—the Eternal Logos' mission in Jesus Christ warrants a nuanced critique. Laruelle seems to espouse that the divine death on the Cross is answered by his theory of the non(ONE).

Jesus Christ's willingness to accept the Cross and death gives Laruelle a theoretical opening for his attribution of the Messiah as a self-proclaimed Antichrist. Theologians would most likely take exception to such a formulation. Except for the atheistic overtones in many postmodern theories and the fact that the biblical witness states, "God did not create death" (Wisdom 1:13), Christianity also proclaims the death of the finite-eternal person of Jesus Christ. There is more than malicious intent or superficial

[143] See Francois Laruelle, *Principles of Non-Philosophy,* Translated by Nicola Rubczak and Anthony Paul Smith (London, UK: Bloomsbury Press, 2013), 20–26. Hereafter cited as *Non-Philosophy.*

[144] The ONE's positing as (non)ONE undermines a traditional explanation of the monotheistic God and seems to suggest an unorthodox characterization of the Divine, as Laruelle writes, "I lay claim to the abstract—the Real or One—rather than an abstraction, The One is an abstract—without-an-operation of abstraction," François Laruelle, *Philosophies of Difference: A Critical Introduction to Non-Philosophy,* Translated by, Rocco Gangle (New York, NY: Continuum, 1986), 188. Hereafter cited as *Philosophies of Difference.*

[145] Philosophy takes on the transcendence and unity characteristic of the Divine for Laruelle, see *Non-Philosophy,* 26–28. See also François Laruelle, "*A Summary of Non-Philosophy,*" Translated by, Ray Brassier, in Pli: *The Warwick Journal of Philosophy,* 8 (1999): 141. On unity being the transcendental foundation see also *TL* 1, 6.

polemic, I think, behind Laruelle's philosophical retelling of Christ's Passion. Laruelle's view of "the Antichrist" can be interpreted as an apophatic explication of a person experiencing the unraveling of his human material embodiment and imminent death. He envisions Christ driven by a sense of hopelessness with a reworked "future-tense"-based eschatological vision.

The person and mission of Jesus Christ, the faithful believes, transform our own final experience of difference with death. In Christ our future death is read as a promised resurrected moment that is sacramental celebrated.[146] The humanity and divinity—the *communicatio idiomatum* of Jesus' person—is offered on the Cross and the grace therein won to the world at Pentecost.[147] The articulation of this mystery is part of the symbolic and signatory nature of the Eucharistic body. The Spirit shares the "sign of the Cross" in the upper room as a means of a post-lapsarian Creation's conversion. The sign, Catherine Pickstock argues, is never left behind as "the theological body turns everything into sign, in such a way that the distinction itself between a thing and sign can no longer be sustained."[148] There is much that needs addressing here, but this would not be useful for our current purposes other than to note that I will return later in greater detail to their significance for a holo-cryptic metaphysic.

Laruelle's "heterodoxy" of immediacy and immanence hopes to summon the "New Messiah" who works to bring about revision and revolution, rather than, the more traditional aims at Creation's renewal, the moral purification of all peoples—a conversion that completes its eschatological promise of integration.[149] Again, a generous interpretation of Laruelle's theory allows us to see it as an important editorial of traditional Trinitarian and Creation theologies from the postmodern perspective. The created order's genealogical and lateral progression is given an analogous vertical dimension when its limitations, effectual distances, and exceptional differences are seen as incorporated into the truth of Trinitarian life, Balthasar offers that: "[t]he distance between the Persons, within the dynamic process of the divine essence, is infinite, to such an extent that

[146] See John Meyendorff, *Byzantine Theology: Historical Trends and Doctrinal Themes* (New York, NY: Fordham University Press, 1974), 163–64.

[147] See *The Bride of the Lamb*, 404, *TD* 3, 222 and *ST.* III, 16, 4–5.

[148] Catherine Pickstock, *After Writing: On the Liturgical Consummation of Philosophy* (Oxford, UK: Blackwell Publishers, 1998), 261. Hereafter cited as *After Writing*.

[149] On the relationship between apophasis and philosophical deconstruction see *Philosophy and Non-Philosophy*, 201–210.

everything that unfolds on the plane of finitude can take place only within this all-embracing dynamic process."[150] With the divine-human union of the person of Jesus Christ, a unique transcendent-immanent tension is introduced into Creation. He holds this union in his person and this reality stretches of material embodiment to its limits without "metaphysically flattening out" or separating his eternal identity from his person.

Laruelle's non-theistic binary ordering of the Divine as non(One) does not spawn the transcendental or ecclesial dimensions of the finite order as we find in the historical person of Jesus Christ. However, it does offer a view of his eschatological nature read from the narrow philosophical template how a person can give birth to a militant "democratic" movement. The absolute difference realized in the concept of the One-non(One) correspondence offers a postmodern explanation that acts as an engine, measure, and means to explain material reality's motion from radical philosophical and political interpretations.

With this philosophical method, so-called Speculative Realist philosophers, such as Graham Harman (1968–) and Quentin Meillassoux (1967–), have also rediscovered that a new way to return to metaphysics is possible in a post-secular age. Favoring the ontological over the epistemological, speculative realists philosophers critique the existence of hierarchical systems.[151] Revisiting the presuppositions behind hierarchical, vertical, or transcendental "systems," the speculative realist puts aside past claims that define metaphysical truth aligned to classical theories of order. The traditional position that held human freedom exists on a different order than non-sentient beings is challenged. The move toward a unilateral approach to metaphysical status contests a Christian approach to transcendental consciousness.

Metaphysical and logical ordering of being is not defined fundamentally to an immoveable Transcendent source, but rather identity itself is only stabilized to the degree it continually tangles with the influences of relational differing. The conscious being only momentarily de-centers from the material continuum but this movement remains grounded and systematized by physical reality's "omnipresence." The speculative realist

[150] *TD* 5, 245. See also *TD* 4, 333, *GL* 7, 213–24, and 249. On the relational nature of the physical universe from a scientific perspective, see Lee Smolin, *Three Roads to Quantum Gravity* (New York, NY: Basic Books, 2001), 50–52.

[151] See Graham Harman, *Guerrilla Metaphysics: Phenomenology and the Carpentry of Things* (Chicago, IL: Open Court Press, 2005), 17–22 and 42.

is not without support from previous philosophical and theological investigations into this complex question. On the relative identity of a being, Whitehead comments that: "[t]he actual entities atomize the extensive continuum. This [space-time] continuum is in itself merely potentiality for division."[152]

Repeating the same act offers little new information. The emergence of the new requires creative change. New physical and conceptual motions inspire new disclosures and personal transcendental orientation. Infinite repetition without change makes it impossible for hierarchical, complex, or evolutionary systems that are necessary for nonmaterial informational states to be apprehended. It is not clear how one is even able to recognize any truth without some implicit or explicit objectification or meta-leveling of our physical identities.

Laruelle's metaphysical transliteration of the Transcendent's ONE as non(One) affords one way to promote the means to move beyond physical reality if only by proposing a hypothetical Transcendent. The essential in-differing of Laruelle's ONE as non(One) sets out a way for immanent existence to possess some differing and vertical relationships. The ONE's self-reflection as annihilating gesture relays and informs finite reality, Laruelle opines, as "unilateral duality." The ONE's self-antithetical conscious measure is hardwired into physical reality in an analogous fashion to divine exemplarity. The infinite interplay of ONE as non(One) dramatically unfolds and images in the differentiation, destabilization, and lateralization that are characteristic of finite existence.

Laruelle's position on the ONE-non(One) is not pure Gnosticism but does seem to owe something to this philosophy's sense of physical reality's hidden nonmaterial identity. Further, the virtual imaging of the ONE as non(One), I argue, leaves open the possibility of a Pure Transcendent Spirit and some principles of a metaphysics that takes serious traditional elements of Christology. These influences are sidestepped and only obliquely referenced. He writes that: "both the empirically given and the 'horizontal' ideality,... because the non(ONE) totally escapes objectification—it is a non-objectivizing or non-positional transcendence—and it implies the unilateralization, without recourse or return, of Difference, Being, etc. It is a radical depositioning of Being by the ONE."[153] The

[152] Alfred North Whitehead, *Process and Reality: Gifford Lectures* 1927–1928 (New York, NY: The Free Press, 1978), 67.

[153] *Philosophy of Difference*, 202.

conceptualizations of the Transcendent and finite reality reveal in impermanence and ceaseless relational difference. However, there is also a theological ecology in this system as the ONE, albeit hypothetical and virtual, cannot be completely divorced from reality or consciousness. Granted Laruelle's non-philosophy implicitly rejects a traditional understanding of divine providence for Transcendent lateralization does not afford an omniscient divine viewpoint. At best, the Transcendent can offer the colonizing gaze of divine blindness—a state of affairs that inspires spiritual theater from inauthentic players. This is not to suggest that the traditional belief in a divinely directed history must be made at the expense of the Divine's existence or human freedom's actualization.[154] Indeed, secondary causality and non-subsistent relationships are commensurate with the Transcendent's relationship to the non(ONE)—if one means by this correspondence what is entailed with Jesus Christ's crucifixion.

Laruelle's apparent philosophical guerrilla warfare against orthodox belief can be unsettling, in part, because it critiques those theological presuppositions dormant for much of modernity, namely, those Christological theories that depend heavily upon Aristotelian or Newtonian causality. Causality, Aristotle muses, looks to answer the question of "why" and ideally does so with reference to a being's perceived material, efficient, formal, and final arrangements. This schema demands discussion, as each of these four categories risks falling into an exercise of subjective exposition. The mathematical precision of Newton's quantification of the laws of motion reduces the latitude created by a subjective viewpoint. Newton's First Law of Motion orders an otherwise chaotic universe when it postulates that "[e]very body continues in its state of rest, or uniform motion in a straight line, unless it is compelled to change that state by forces impressed upon it."[155] Jesus Christ does not enter into Creation to deprive the human person of their moral agency nor renounce causality in favor of tumultuous destabilization. Rather, the scholarly work of Badiou and Laruelle requires of theologians some account for how the postulation of a divine-reality causal connection does not unwittingly advance its own heterodoxy. Finitude and material evolution exist as part of the same metaphysical spectrum as emergence and entropy. Paralysis, atrophy, and

[154] See *ST.* I–II, 114, 1, *De ver.*, 2, 3, 16.

[155] Isaac Newton, *Principia: The System of the World*, vol. 1, Translated by Andrew Motte (Berkeley, CA: University of California Press, 1934), 13.

impotency are not the necessary "by-products" of entropy given Nature's commitment to meaning and order.

Badiou and Laruelle offer philosophical theories that benefit alterity and the experience of the "other" that prohibit traditional Christian transcendental hierarchies without critique, for these are seen to eviscerate difference and potency. Consequently, these postmodern philosophers do not have a deep commitment to a traditional understanding of ontological difference as found in the work of most Christian philosophers. Although different presentations of the role the essential plays in a being's internal architecture, appearance, and external actions are offered, all are served by contemporary theories of the potential, difference, and reduction.[156]

Believing it is possible to have access to a being's inner meaning is seen by many postmodern thinkers to be a holdover of the medieval and modern thinkers' naïve faith in human objectivity and consciousness. Quentin Meillassoux argues that the central thesis of such thinkers is based on a strong correlation model. All such frameworks imply one can exceed the natural, physical, and immanent conditions of finite human existence. Meillassoux clarifies that "where the weak model of correlationalism de-absolutized the principle of sufficient reason by disqualifying every proof of unconditional necessity, the strong model pushes this disqualification of the principle of non-contradiction by reinscribing every representation within the bounds of a correlationalist circle."[157] Postmodern advocates rightly point out that as finite beings we share in the same immanent ordering and non-subsistent relationships that hold universally. Further, if this is the state of the human person and reality as whole cannot be exceeded, it is reasonable to assume that a pure experience of the Transcendent is incomprehensible and unattainable. Given this is the state of affairs faith must be posited in order to have any concord with the Transcendent. Balthasar summarizes this stating "we do not know unity is in truth, we are acquainted with unity only in

[156] See François Laruelle, Philosophies of Difference, *Philosophy of Difference: A Critical Introduction to Non-Philosophy*, Translated by Rocco Gangle (New York, New York: Continuum, 1986), 210. Hereafter cited as *Philosophy of Difference*.

[157] Quentin Meillassoux, *After Finitude: An Essay on the Necessity of Contingency*, Translated by Ray Brassier and with a preface by Alain Badiou (London, UK: Continuum Press, 2011), 42. Hereafter cited as *After Finitude*. See also Peter Hallward, 'Ethics without Others: A Reply to Critchley on Badiou Ethics,' in *Radical Philosophy* vol. 102 (July/August, 2000): 28.

the irreducible duality of universal unity and particular unity, and we can never make the two aspects coincide."[158] Transcendental reach or grasp of the Transcendent is possible for the person of faith. She sees her existence and discovery of truth as a way to image the resurrection of Jesus Christ.

Relational difference metaphysically grounds finite autonomy but also witnesses to one's holo-somatic identity. The stabilization of identity at the immaterial level makes possible one's self-awareness and hence a sense of freedom within a law-driven universe. Because truth's adjudication follows from our judgments concerning specific events and beings, human knowledge is inseparable from material alteration.[159] Postmodern philosophies agree that difference illumines being's varying ontological and accidental permutations and hence fosters the inexhaustible truth. However, they debate whether or not such influence will lead to us a meaning beyond the immediate experience: "[t]he tragedy of Difference is that it is condemned either to fall into Idealism and to arm the final primacy of reversibility and of Being; or to combat Idealism by resorting to the quasi-philistine platitude of a Greco-Occidental 'fact', or rather a finitude that it presupposes, the foundation and necessity of which it is unable to perceive."[160] I hold that Incarnation gives us a theological basis for claiming that relational differing speaks to one's ability to receive the other. This capacity to be receptive or transparently open is not limited. The receptive person can apprehend something about the Transcendent or abstract formal idealizations. In being open to the immaterial and spiritual a way suggests for one to more freely experience the truth of material reality. The rational supplementation of meaning in one's intellectual quest for exhaustive meaning is ongoing.[161] Theologians claim that the fragmentary nature of knowledge, the predominance of sin, cognitive indeterminacy, and deterministic immanence all provide evidence of the interplay between transcendent-immanent correlations. Human truth may be partial or metaphorically broken, in need of realignment and harmonized with one's subjective desire, however this process reveals to limited con-

[158] *TL* 1, 157.

[159] This informational informing is a principle of my holo-cryptic metaphysics.

[160] *Philosophy of Difference*, 187. See also Graham Harman, *Prince of Networks: Bruno Latour and Metaphysics* (Melbourne, AU: re.press, 2009), 72 and 122.

[161] See François Laruelle, *Philosophy and Non-Philosophy*, Translated by Taylor Adkins (Minneapolis, MN: Univocal, 2013), 71 and 173–75. Hereafter cited as *Philosophy and Non-Philosophy*.

scious some recognition of truth and therein, God's own revelation in the world.[162]

While neither Badiou nor Laruelle offer a satisfactory account of the logic that grounds the metaphysics that traditionally explains the Eucharist, they still give us insights into the nature of its physical foundations within existing philosophical, mathematical, and scientific paradigms. As unlikely as it seems, these thinkers may provide a theoretical basis for what the non-theistic "atheistic anti-liturgical" perspectives and practices may look like. The holo-cryptic interpretation of transubstantiation that follows insists that current scientific models for physical reality, which these thinkers critique, are helpful guideposts for understanding matter's spiritual infusion at transubstantiation. The configuring of the Eucharist to the Transcendent gives us a way to understand how information is represented and functions holo-somatically as a working example of the "pseudo-subsisting *ipsum esse*."[163]

Below, we develop how non-theistic transcriptions of the "non(One)" analytically parallel matter's destruction and "resurrection" as a holographic prolongation. This is not to say that scientific theory's chief goal is to illumine theological dogma or that the material order concertizes as some distortion of the essential. Science does not dictate theology's investigations any more than this work aims to contort science's objective and quantitative reading of reality. The following does not treat the doctrine of transubstantiation as a scientific theory, *per absurdum*. Simply put, my aim is to show that QMT allows for a better understanding of the miracle of consecration (*conversio mirabilis et singularis*) than either Aristotelian philosophy or Newtonian physics can postulate. Analogously, Darwin's theory of evolution outlines the general scientific processes responsible for life's ability to change, albeit from the non-spiritual point of view of material and biological interactions.

Darwin's theory of evolution offers a rudimentary analogy for how fields function in space-time.[164] Post-Newtonian physics shows us that the measurement of matter is more problematic than previously assumed. Measurement does not immediately define the identity of a being by com-

[162] See Hans Urs von Balthasar, *Theology of Karl Barth*, Translated by Edward T. Oakes, S.J. (San Francisco, CA: Ignatius Press, 1992), 204.

[163] See *Homo Abyssus*, 30.

[164] See Mark C. Taylor, *After God* (Chicago, IL and London, UK: The University of Chicago Press, 2009), 335. Hereafter cited as *After God*.

pletely reducing its identity to a numeric quantification in an analytically closed reality. Neither am I supporting those contrary "super-naturalistic" positions that see the universe as completely open and the divine as infinitely divided and spread horizontally throughout the universe as some pantheistic system would have us believe. The emergence of mathematical and physical symmetries (e.g., *AdS/CFT*) from nothingness or the void's interaction with infinite multiplicity that systematically comes together in QMT does not fully explain the miracle of transubstantiation. It does, however, offer a more plausible starting point for holo-cryptic metaphysics' understanding of the nature of material reality that divine action transforms at the Eucharist.

It is possible to describe the nexus between Transcendent and immanent reality by theological and quantum modeling. Similarly, when serious philosophers reject the existence of an actual divine being and do so in a logically consistent manner, they may share a methodological process with theologians who attempt to clarify the locus of ultimate meaning or divine action by means of conscious negative or apophatic measures. Regardless of the relevant findings of postmodern philosophy, scientific research, or theology, all ask deep questions and advance our knowledge of the nature of physical reality and consciousness. To fully understand a relationship it is necessary to distinguish between the ontological and epistemic factors and see the meaning at play, for "in spite of the fact that the views of philosophy sway from one system to another, we cannot dispense with it unless we are to convert knowledge into a meaningless chaos."[165] A range of possible solutions can be offered in order to overcome confusion or intransitive indeterminacy found in an investigation. Discovery and ordering of information are only possible when subjective consciousness is receptive, that is unencumbered.[166]

In 1918, the German mathematician Emmy Noether published a groundbreaking paper explaining how symmetry operates in formal systems. Acting as a constraint, Noether argues, symmetry mathematically defines how the conventional properties of an object remain under certain transformations. It furnishes a way to comprehend the intricate and

[165] Hermann, Weyl, *Space, Time, and Matter*, Translated by H. L. Brose (London, UK: Methuen, 1922), 10.

[166] See Robert Spaemann, *Persons: The Difference between 'Someone' and 'Something'* (Oxford, UK: Oxford University Press, 2017), 49.

probable nature of physical reality as information.[167] The essential or holosomatic identity of a being is ineradicable and fundamentally relational. That is, a being by definition always exists energetically in some virtual external or internal motions that define its identity: "[o]ne can try to leave behind the division between Being and essence, between the infinite and the finite; one will then say that all Being is infinite and immutable (Parmenides) or that all is movement, rhythm between contraries, becoming (Heraclitus)."[168] Frank Wilczek summarizes these ideas, stating: "symmetry, in general, is Change Without Change."[169] Furthermore, a being's identity is ordered to change as it does so in a way consistent to its interior properties and its environment—since the full nature of this correspondence is known to be symbolically mapped.[170]

Where mathematics is the foundational language of contemporary science and physics, I accord a similar role to the Trinitarian Ur-Kenotic identity and therein the divine's work in physical reality and sacramental disclosure in the Eucharist. Ancient literature shows that the word kenosis predominately means absence, being empty-handed, or the void. These three definitions of the kenotic are further used to refer to the existential states of vanity, being bereft, and feeling all is fruitless—those destitute or negative states associated with silence and absence.[171] In the Bible, the prophet's hunger and thirst, like the void, freights an image of all that wages against God's plan—but in the end one transmuted by divine grace to their providential designs. The word kenosis also refers to the empty cistern or pit into which Joseph is lowered when abandoned by his brothers (Genesis. 37:24). Here, again, the concept of kenosis in some way

[167] See *Emblems of Mind*, 164 and 166–67.

[168] Hans Urs von Balthasar, *My Work in Retrospect*, Translated by Brian McNeil, C.R.V. (San Francisco, CA: Ignatius Press, 1993), 112. Hereafter cited as *My Work*.

[169] Frank Wilczek, *A Beautiful Question: Finding Nature's Deep Designs* (New York, NY: Penguin Books, 2015), 309. The distinction of various communications of symmetry does not reject their coordination in a particular being—time- and spatial-dependent variables work in conjunction at the most fundamental levels given the uniformity of electrons, protons, and so on. See "the core theory" in ibid., 293–94 and 297–305. Hereafter cited as *A Beautiful Question*.

[170] See G.J. Chaitin, *Algorithmic Information Theory* (Cambridge, UK: Cambridge University Press, 1987).

[171] See *Theological Dictionary of New Testament*, Edited by, Gerhard Friedrich and Geoffrey W. Bromiley (Grand Rapids, MI: Williams Eerdmans Publishing Company, 1976), 659–662 and *Doctrine in the New Testament*, Translated by John Bowden (Eugene, OR: Wipf and Stock Publishers, 1981).

negates created space by paradoxically positing a positive predication of empty space.[172] Ur-Kenosis is closely associated with the universe's creation and its subsequent recreation as symmetrically fallen and redeemed.

The human observer plays a role in defining how the kenotic is grasped theologically as an interior principle and mathematically in the function of subtraction. For my part, I seek to extend the polycentric notions of the kenotic as an analytical register in the physical spaces within an atom or, on a cosmological scale, in a black hole. Where scientific theories mimic the Ur-Kenotic by attempting to order reality by various negentropic tendencies, theologians and mystics consider how "emptiness" acts as a prolegomenon for exuberance in the created order. The Spirit's presence is non-evasive yet gives structure to the silent and vacant Nothingness: "in the beginning God created the heaven and the earth. In addition, the earth was without form, and void [*tohu vabohu*]; and darkness was upon the face of the deep. In addition, the Spirit of God moved upon the face of the waters" (Genesis.1:2). The Spirit's presence is not equiprimordial to Heidegger's concept of Nothingness (*das Nicht*), which ties humanity to a fundamental state of fear, anxiety, and discordant and egotistical self-care.[173] Rather, this Nothingness makes possible the actualization of God's plan as Creator, where "Being itself unveils its final countenance, which for us receives the name of Trinitarian love."[174]

The Divine's providential plan announces itself in the dimensional and "space-less" mystery of Nothingness, where Creation's circular Providential pattern is first spatiotemporally initiated as part of being's infinite fecundity: "The ontological vacillation always implied the radicalized positing of the unmediated reality, which refused the fissure that passes through being's transnihilation, that is, its kenosis. In the same way, God too was robbed of all necessity in this speculative aberration, insofar as the bonitas divina was driven back into the essentia divina by being logicized, and it fragmented in the space of the divine intellect into the infinite variety of possibilia."[175] Like a symbol, individual beings have the capacity to exist in relationship to the Ur-Kenotic Transcendent, as these finite examples are communally and infinitely defined by their signification. Zero degrees are

[172] Some relation to the informing of empty space is discerned in String Theory's concept of *AdS/CFT* dualities. I return to this question in Chap. 4.

[173] See Martin Heidegger, *Being and Time*, Translated by Joan Stambaugh (Albany, NY: State University of New York, 1996), 312–320.

[174] *GL* 1, 158.

[175] *Homo Abyssus*, 215.

part of the greater whole of the compass, and nothingness is paradoxically a sign of the Transcendent's generative work in the created order. Through Nothingness the Divine brings into existence the created order—our means to positively and negatively participate in the triune truth of the Divine Persons. Being's inexhaustibility and signification's unending deferral and creation of meaning are positive modes of finite being's revelation, whereas, through sacrifice, contemplation, or other practiced metanoia, one apophatically references the Ur-Kenotic love of the Divine Persons.

The free and rational commitment to the Transcendent is revealed in the human heart and externally in moral action. Existentially, these two modes communicate the identity of divine-human symmetry. There is much that a deconstructive philosophy shares with prior theological articulations of meaning's unraveling and reconstitution. "The sign of this excess must be absolutely excessive as concerns all possible presence—absence," Derrida states, but further adds, "must still signify [that] the trace is produced as its own erasure."[176] Death is the "erasure" of one's spiritual imprint in a body. The weight of a being or idea's development works against its own inevitable schizophrenic undermining—a position reconcilable with the Church's warning of the self-dissolution that occurs when one freely chooses evil by consciously sinning. What is absent in Derrida's account is the further caveat that the individual is never excluded from redeeming grace given the Transcendent's Providential plan put into effect with "kenotic" consent. The Incarnation is an axial point that makes possible a deeper sense of human freedom and the personal nature of the reception of the Eucharist.[177] The truly free person holo-somatically images the immortal divine in one's soul or informational essence.

The Pure Divine Spirit acts as divine creative agency from Creation's Nothingness to the present moment. Trinity and Creation are not reversible—the created order does not perfectly and necessarily manifest the Divine, as pantheists would have us believe. The created order is not a chimera of indefinite teleology or embodied divinity.[178] Empirical reality

[176] Jacques Derrida, *Margins of Philosophy*, Translated by Alan Bass (Chicago, IL and London, UK: University of Chicago Press, 1972), 65.

[177] See Catherine Pickstock, *After Writing: On the Liturgical Consummation of Philosophy*, (Oxford, UK: Blackwell Publishing, LTD., 1998), 264.

[178] For a postmodern argument to the contrary, see Michel Foucault, "Nietzsche, Genealogy, History," in *Aesthetics, Method, and Epistemology*, Translated by Robert Hurley (New York, NY: The New Press, 1998), 370–376.

gains metaphysical traction with the Spirit's hypostatic Trinitarian identity and the historical arrival of Jesus Christ.[179] The Divine's exact relationship to the Creation is beyond our comprehension, in part because it is grounded in Trinitarian Ur-Kenosis and in part because its full manifestation remains restricted until the Eschaton. However, the human person analogously participates in the divine act of creation through acts of charity and creative gestures. Because a continuum exists between causes and effects (*causa aequat effectum*), we can retroactively trace events back from changes in creation from their most immediate disclosure in our lives back to their source in the Divine Act. Thomas notes that the continuum between cause and effect is fundamentally formal, not material, because the Transcendent creates as a mental act. As Thomas writes, "a creature is said to be a non-being only in comparison with God...a creature is known by God only in so far as it compared with Him."[180] Divine revelation makes known the soul is the body's measure just as the physical body helps inform the soul's state. In this way, the difference between the actions of the essential and existential aspects of the human person can be further posited of a finite being's non-subsisting relationships in the created order given all these correspondences analogously reflect the absolute freedom of divine hypostases.

When aligned to divine purposes, the prophetic kenotic word reflects the human person's experiences of emptiness, darkness, and silence as a suitable medium for meaning's impregnation in Nature's cycle of growth and decay. John's Gospel begins with this vision: "In the beginning was the Word, and the Word was with God, and the Word was God. The same was in the beginning with God. All things were made by him; and without him was not anything was made that was made. In him was life; and the life was the light of the world. In addition, the light shines in darkness; and the darkness comprehended it not" (John 1:1-5). In the New Testament, kenotic emptiness implies that reality has Christological and eschatological dimensions: "God is light, and in him is no darkness at all" (1 John 1:5).

[179] See Sergius Bulgakov, *The Comforter*, Translated by Boris Jakim (Grand Rapids, MI: William B. Eerdmans, 2004), Chaps. 4 and 5.

[180] *De ver.*, 2, 3, resp. 18. See also Thomas Aquinas, *Truth. Vol. 1*: Questions I-IX, Translated by Robert W. Mulligan, S.J. (Chicago, IL: Henry Regnery Company, 1952), 76.

1.3 Transubstantiation

Having outlined the general circumstances of the debate between post-modern philosophy, mathematics, science, and theology that I will develop, I turn now to detail how the central holo-cryptic metaphysical concepts and processes demanded by my own speculative reading of reality at its most counterintuitive—as the medium participating in the liturgical event of transubstantiation. Hildebert de Lavardin, Archbishop of Tours (d. 1133), is credited with the earliest recorded use of the term "transubstantiation" in a sermon given in 1079 C.E.[181] He introduces the term transubstantiation to refer to the "change" that occurs at the consecration, namely, the conversion of the substance of bread and wine into the body and blood of Jesus Christ.[182] While the early Church universally assented to what happened at the Divine Liturgy, considerable discrepancy exists in regard to how theologians continue to attempt to describe this mysterious event or the meaning of the word substance. A rich theological vocabulary developed in eastern Orthodoxy when speaking of this sacramental heart of the Divine Liturgy. Saint John Chrysostom, for example, speaks of the moment of Eucharistic change in terms of "transformation" (*μεταρρύθμησις*),[183] where Saint Cyril of Alexandria uses the term "transposing" ((*μετίτησις*)[184] and Saint John Damascene writes about the *μεταβολή* or "metousiosis" (*μετουσίωσις*) for the process of the bread and wine's change into the body and blood of Christ at the divine liturgy,[185] whereas the always insightful Gregory of Nyssa employs the term "transelementation" (*μεταστοιχειωσας*) to begin to explain this mystery. Nyssa focuses his discussions on the miracle of the Eucharist around the change (*μεταπιοειν*, *μεθισταναι*) or transformation (*ἐπισκηνωσις*) that occurs at the sanctification of the bread and wine and its faithful reception.[186] The

[181] The theologians Stephen of Autumn (d. 1139), Gaufred (1188), and Peter of Blois (d. circa 1200) adopt the term transubstantiation.

[182] See *The Sources of Catholic Dogma*, Edited by Henry Denzinger and Karl Rahner and Translated by Roy. J. Deferrari (St. Louis, MO: Herder Book Company, 1954), no. 848 Hereafter cited as *DS*.

[183] John Chrysostom, "Homily on 2 Corinthians 3:18." *NPNF1*, vol. 12, in *Catholic Logos Edition*, 12, 7, 5.

[184] See Cyril of Alexandria, *The Gospel of Saint Luke with a Commentary by Saint Cyril of Alexander: Sermons 81–156*, Books 8–14, Translated by R. Payne Smith, M.A. (copyright Elaine Phang, 2017), 251–255.

[185] John Damascene, *An Exact Exposition of the Orthodox Faith*, *NPNF* 2, 4, 13.

[186] Nyssa, *Against Eunomius*, *NPNF2*, 2, 3.

Eternal Logos' kenotic incarnation in the flesh (John. 1:14) logically justifies the Lord's indwelling (ἐπικηνωσις) in the sanctified species.[187]

With the Council of Trent's adaptation of the term transubstantiation, theological supposition on the issue of the Eucharistic change is given one of its primary interpretative loci in western Christianity. Most importantly, the notion of transubstantiation is linked in the West to Aristotelian hylomorphism, the old medieval form-matter distinction at the heart of Thomistic metaphysics.[188] This development traces its roots to a Platonic-Augustinian synthesis with view to the question of individuation, reason, and thereby an ecclesial interpretation of the covenant: "within the Thomistic discursive logic, which concerns us here only as the act-potency counterpoise to the dialectic of the Augustinian hylomorphism, the act-potency analysis governs metaphysics simply because, as it governed Aristotle's analytic or logical discursive hermeneutic of Thomas. In this Thomistic analysis, the act—potency correlations are actual in a free affirmation, of covenantal."[189]

The logic of hylomorphism and the covenantal relationship of Sacred Scripture historically anticipate the Eternal Word's transfiguring of human language by divine fiat at Creation. The Logos' power to transfer divine meaning through all physical mediums and divine-human holo-somatic symmetries is historically realized by the incarnate Lord. These developments do not alter the fact that the Transcendent *Ein-Sof* remains inconceivable and unknown outside of its negative appropriations, subtractive gestures, and conceptual differing.

Negative appropriations of the Eucharistic mystery are fundamental to the Church's self-identity. In the Council of Trent's thirteenth session, the participants stated that the ecclesial penalty of anathema is to be imposed on any Roman Catholic who denies the doctrine of transubstantiation: "whoever denies that wonderful and unique conversion of the whole substance of the bread into the Body and the whole substance of wine into the Blood, with only the species of bread and wine abiding, which conversion of the Catholic Church most aptly calls 'transubstantiation' ***anathema*** sit [outside the Church]."[190] However, the actual definition of this term by

[187] See *Mystery of the Eucharist*, 213.

[188] Hugo of Cher seems to be one of the first theologians to draw the comparison between hylomorphism and the Eucharist, approximately 1230.

[189] Donald J. Keefe, S.J., *Covenantal Theology: The Eucharistic Order of History* (Novato, CA: Presidio Press, 1996), 167. Hereafter cited as *Covenantal Theology*.

[190] See *DS* 884 and *DS* 1652.

the Council thankfully remains open to further theological development. Consequently, my interpretation of the mystery will develop approaches favored by Orthodox theologians—chiefly Gregory of Nyssa's (335–395) philosophical insights concerning the concept of Nothingness, Sergei Bulgakov's (1871–1944) theology of the Spirit, and Pavel Alexandrovich Florensky's (1882–1937) theories concerning infinity's partition of created reality and artistic presentation in the icon. These thinkers will be discussed given the work of such western theologians and thinkers, such as Thomas Aquinas (1225–1274), Erich Przywara (1889–1972), Ferdinand Ulrich (1931–2020), and Hans Urs von Balthasar (1905–1988).

By developing the thoughts of important thinkers, a new theological understanding of transubstantiation is possible, one that can be put into conversation with serious scientific and mathematical reproductions of reality. This exchange means interpreting the philosophical concept of the essential traditionally approved by theologians as analogous and commensurate with a virtual or holographic rendering of information as found in the theories of black hole entropy and quantum metaphysics.[191] This "amounts to treating quantum mechanics as a theory about the representation and manipulation of information transfer in our world."[192] This re-envisioning of being's relationship to truth moves beyond simplistic estimations of the real provoked by hard deterministic and materialistic philosophies—however quixotic such a project may sound—by arguing that a traditional theological understanding of the soul-body union is analogous to its definition as a holographic representation of information. I hold to the hylomorphic tenet that material being is "free" insofar as it is informed by essential interior forces that holographically or holosomatically image the Transcendent [holographic identity].

Although this book coheres with the doctrinal teachings of a Roman Catholic theology of the Eucharist, I argue that the mathematical and scientific quantification of physical reality as holographical informational structures should be given some prerogative if we are to justify theology's claim that it is based on an accurate picture of reality. Our Theo-philosophical focus stresses that the study of the question of transubstantiation and the mathematical-scientific approach to physical reality

[191] See Stephen Hawking, "Creation by Black Holes," *Commun. Math. Phys.* 43 (1975) 199.

[192] Jeffrey Bub, Quantum mechanics is about quantum information, *Foundations of Physics*, 35:2 (2004): 557.

advanced by QMT shares "tools" of human reason and experience. The mathematical language of QMT and GTR chronicle an observer's description of physical being. The theoretical observer of the postclassical scientific world and the believer participating in the liturgy must similarly account for how they understand their role as free subjects. The scientist in the lab must be shown to be "objective" in her work, whereas the individual believer is asked to appropriate objective truth as a subjective expression of her personality. One expects that at the experiential level, the highly abstract theoretical tenets of the doctrine of transubstantiation are subjectively revealed in the diversity of perspectives and gifts of the people of God: "[t]here are diversities of gifts, but the same Spirit. In addition, there are differences of administration, but the same Lord. In addition, there are diversities of operations, but it is the same God which worketh all in all" (1 Corinthians. 12:4-2). To be clear, I am not attempting to give an exhaustive account of quantum metaphysics or the doctrine of the Eucharist. We enter into this study hoping it inspires future conversations and scholarship. By incorporating insights of modern and postmodern philosophical, theological, and scientific advances, I am attempting to counter both reductionists and revisionists of matter and thereby provide insights into transubstantiation as a metaphysical event with larger ontological, noetic, and cosmological import. Indeed, properly situated the doctrine of transubstantiation reveals the fundamental nature of the created universe and free and conscious personhood in it. While contemporary thinkers may reject out of hand past Aristotelian form-matter conceptual frameworks, they are nonetheless hesitant to jettison the mathematical explication of physical reality that leans on a similar logical framing. There is much that the theory of QMT can recommend about the nature of reality that has profound implications for the doctrine of transubstantiation.

Theology's encounter with science is not only possible but also necessary. We look to the theories of black hole entropy and QMT as modes of informational systems that can be put into a notional relationship with the doctrine of transubstantiation. The choice of one's particular interpretative strategy influences the nature of the information collected. The GTR and QMT outline a way to understand a being's informational identity "holographically," that is, as an essential and super-essential, supramundane entity—a material being that is, at the metaphysical core, an energetic signature of free and probabilistic nature. Traditional Theophilosophical approaches have stressed the essence's interior qualities as embodied in a material substrate. However, with the holographic principle, we have a way to speak of the disclosure of a being's essence in the external world beyond a being's limited accidental properties. It is important that one does not imagine this externalization of an individual's essential structure to be equivalent to projecting a three-dimensional image onto a screen. This verisimilitude is only valid if one holds to what I think are inadequate classical interpretations of matter, energy, and space-time. Whereas past scientific accounts tend to conceive of space as a receptacle for material bodies, Post-Newtonian theory advances allow us a more encompassing and holistic view. Matter and energy in this view become one with the spatiotemporal—not as a pantheistic resolution but rather as individuated and free beings holo-somatically instantiated: "the most general notion of holography has to do with the complete reduction of a bulk description of a physical system in particular space-time to a description of the same purely in terms of boundary data and information."[193] The idea of holographic boundary will be developed below, for the present to suffice comment that it forces us to synthesize what we have previously kept conceptually separate, namely, mathematical and theological margins. Past theological explanations of the soul's physical embodiment influence my holo-somatic explorations of boundary conditions. The question of boundaries is an old question that needs to be articulated in light of our vision of a "holo-cryptic" universe. Frontiers and boundaries can be physical or conceptual, without breadth and infinitely expandable along any number of trajectories. Past executions of this concept are restricted by the introduction of matter and its perceived asymmetrical correspondence to energy. However, once matter and energy are understood to be symmetrical, past understandings of embodiment and boundaries are challenged.

Being is fundamentally relational given being's nonmaterial prototype and its "fragmented" or "probable" manifestations within the world. Reality is perceived as thoroughly polar in nature. This perspective risks becoming antinomic when matter is divorced from its essential properties. Matter's entrance into the relative "nothingness" that is a black hole gives

[193] Norman Sieroka and Eckehard W. Mielke, "Holography as a principle in quantum gravity?—Some historical and systematic observations," *Studies in History and Philosophy of Science Part B Studies in History and Philosophy of Modern Physics* 46 (May 2014): 171, https://doi.org/10.1016/j.shpsb.2013.12.003. The authors have critically evaluated this postulate in regard to a variety of theories of quantum gravity—topological quantum field theory (*TQFT*), String Theory, and Supergravity.

us the best scientific evidence for believing that material beings have what may be called a holographic essence. In this way, matter's translation into a physical informational hologram by gravity's force in a black hole parallels the mind's own abilities to conceive the paradoxical or even one's own nonexistence. Where black holes deconstruct matter leaving only its essential properties, the mind creates intelligible forms or phantasms based on one's interactions with physical reality in the world. Expanding this analogy to its cosmic and theological manifestations, the event of transubstantiation bounds the Divine's identity within bread and wine, whereas the black hole contains matter's identity within its horizon. Both are examples of how information holo-somatically communicates spatiotemporally and dimensionally.

Having already provided an introduction to and overview of some of the past theological, mathematical, and scientific representations of the semantic plenitude of being, I turn now to look at how Post-Newtonian and postmodern philosophical developments ask us to re-evaluate our existing metaphysics. Thus, I argue that the QMO's entropic configuration supports my own view that previous hylomorphic interpretations of transubstantiation are better understood from matter and energy's conversion into holographic constructs.

1.4 Nothingness and Ur-Kenosis

Liturgical space is a privileged interpretation of spatiotemporal boundaries. Theologians argue that religious architecture and dedicated areas can be understood as sacred, that is, the analogous holographic or spiritual meeting point of the supernatural and natural orders. Ritual dramatically enacts how the Transcendent encounters the immanent through the use of natural elements (water, oil, bread, and wine). By the Father's directives, Jesus Christ's redemptive action and the Spirit's eternal mediation of both infuses the entire created order with grace. This "sacramentalization" of the ontological and noetic or conceptual components of the spatiotemporal as history is born in the grace won by Christ's Cross and His Resurrection. The Incarnation's inseparable link to the Atonement (1 Corinthians 15:3, Romans 5:8-10, Colossians 1:19-22) transforms Creation, or in my reading of postmodern terms, Jesus Christ's redemptive actions "reboot" Creation. Creation and its recreation or restoration as Creation 2.0 are part of a single spatiotemporal continuum because both find their center in the person of Jesus Christ as a localized experience of universal renewal and restoration: "therefore, if anyone is in Christ, the new creation has come: The old has gone, the new is here!" (2 Corinthians 5:17). The ontogenesis of this redemptive movement is part of a single movement because both are analogous disclosures of an eternal Ur-Kenotic decision—part of the self-sacrificial or loving identity of the Divine. The Transcendent-immanent drama of Creation dictates that our experience is fundamentally relational (a *commercium* or exchange). As materially embodied souls, our existence emulates and participates in the work of divine causality. This logically follows given that the Son's Incarnation "ensouls" itself in the one material body that restores Creation as part of eternal Trinitarian thought processes: "in the new supernatural rhythm in which God becomes incarnate right down to the lowest depths and out to the farthest bounds, the physical is 'divinized', permeated with God's Pneuma, transfigured and 'transferred' into the kingdom of the Son, and hence of God."[194] The human person is "theonified" (συναποϑεωτηϑη) by being called into communion with the Divine.[195] This perfection is first realized in the person of Mary: "only the Mother of God, through whom the Word was made flesh, will be able to receive the plenitude of grace and to attain an unlimited glory, by realizing in her person the holiness of which the Church is capable… Alongside the incarnate divine hypostasis there is a deified human hypostasis."[196] The human person first proportionally models this same process as part of their interior topology; for "the soul is bound up with the body whole with whole and not part with part: and it is not contained by the body but contains it as fire does iron, being in it energies with its own proper energies."[197] This mysterious meeting of human identity and the Transcendent's exemplar causal force finds its personal actualization in the fact that the human being, Jesus Christ, is also a theophany—a perfect holo-somatic actualization of the Divine.

Created beings are identified with the Transcendent by essentially imaging the uncreated actuality of Christ's nature. All created beings,

[194] *TD* 2, 412.

[195] Nyssa, *The Great Catechism*, *NPNF2*, vol. 5, 1, 5, Translated and Introduced by Philip Schaff, D.D., LL.D., and Henry Wace, D.D. (Grand Rapids, MI: Wm. B. Eerdmans Publishing Company, 1954).

[196] "Pangaea," in *In the Image and Likeness of God*, Edited by John H. Erickson and Thomas E. Bird (Crestwood, NY: St. Vladimir's Seminary Press, 1974), 196.

[197] John Damascene, *An Exact Exposition on the Orthodox Faith*, *NPNF2*, 1, 13.

rational and non-sentient alike, image the Creator.[198] This interior space (*spatium*) is actualized within reality in a movement that is singular to each free being, as individuation is inseparable from the Ur-Kenotic Transcendent hypostatic identity. The multiplicity of being's imitation of an actual infinity lacks this transcendental dimension in postmodern thinkers such as Badiou, and in his writings, he distances himself from a traditional Christian understanding of personhood. Matter's extensive properties for the Christian are a means to share in the life of Jesus Christ through personal embodiment. In this way, an individual's material identity is interior and subsumed by divine love. Every being shares in this enigmatic process of embodiment that is simultaneously a call to relationship and equally a "personalization" of identity: "The prospection of the finite substance's coming toward itself from its rootedness in the always already posited reality thus occurs with a view to the entity's further development as a 'whole.' This wholeness, however, does not mean that the essence is able to posit itself over against the concrete enactments 'in each moment' in a univocal manner or to sustain itself as a permanent structure that is univocally closed in on itself."[199]

The universal operation of the principle of non-subsistence or relational differing means that individual existence is *a priori* a concretion of a larger communal praxis. The implication of individual identity is ascribed to the attempt to explicate the meaning of beings. The multiplicity of beings is foundational to conscious subjectivity. This extensive communication is a free spatiotemporal movement as being is temporally finite by *Chronos*' measure, but virtually part of those infinite supernatural spaces (Αἰών) that ascribe to Creation 2.0's final *metanoia* an eschatological fulfillment. The universal virtualization of the eschaton finds its particular spatiotemporal manifestation at transubstantiation. The Spirit holo-somatically preserves Jesus Christ's identity in the consecrated species.[200] [77] In short, there is no path to Jesus Christ except through the Spirit's relational differing of Creation 2.0 as there is no way to the Father except through the Son (1 Corinthians 12: 3-5). History's linear unfolding is possible because it takes as its interior teleological principle, one that holo-somatically premises the Trinitarian relations.

[198] See *TL* 2, 256.
[199] See *Homo Abyssus*, 155.
[200] See *ST.* III, 77, 1–8.

Finite being experiences the infinite in the transitory moment. To take subjective responsibility for the meaning we desire to see in the world is to advance one's spiritual maturity as a historical fact. The Eucharist ultimately provides the sustenance for this journey, and this is experienced in all physical objects for "it is not until the other enters into the space of the subject that, like Sleeping Beauty, it awakens from its slumber—at once to the world and to itself."[201] Our actions influence others and define us as part of the supernatural realm's infinite orchestrations of the material and historical. An infinite exchange takes place between these two spheres of influence, an exchange made possible by the Absolute Groundless identity or Pure Nothingness of the Spirit who projects in dyadic representations (potency, nonbeing, etc.) that open space for created freedom. The relationship between being and *creatio ex nihilo's* predication of nonbeing has been variously interpreted throughout history, but it is beyond my purview here to assess this philosophical and historical development. Rather, my focus is more limited to explicating how traditional scriptural and theological interpretations of Creation are expounded and situated in our Post-Newtonian world of QMT. I discuss classical views only insofar as they relay my own thesis that being's non-subsistent nature requires its denomination as a holographic construct within a universe whose reconciliation of the quantum realm and GTR demands that we revise how being, infinity, and Nothingness can be understood as analogous principles of Jesus Christ's divine-human nature, sacramentally revealed in the event of transubstantiation.

From certain Ancient and Medieval theological interpretations, Creation can easily be justified by some of the canons of classical scientific theories. The theological premise that Creation and its recreation by the *Creator Spiritus* are seamless must be commensurate with scientific theories that factually describe physical reality. Any formal explication of reality from a theological, scientific, or philosophical perspective is congruent as long as each makes truthful statements about the same event or theme. Thus, I hold that the identification of divine identity finds its parallel in scientific circles and vice versa.[202] As Plato and his disciples argue, the irreducible change and motion discerned in the universe requires a constant Transcendent point or points of reference. The phenomenological continuity between the traditional symbolic chronicles of the *sacramentum*

[201] *TL* 1, 67.
[202] See *Covenantal Theology*, 154.

tantum and res tantum of the Eucharist foreshadows how information functions in QMT.

Jesus Christ welcomes death as fully human and fully divine. Christ experiences the phenomenon of nonbeing as part of his identity as uncreated divine essence and equally as part of his existential and material nature as a human being. For Christ, being and nonbeing communicate their true supernatural dimensions in his person, as the material elements of his embodied nature are "spiritually" infused.[203] Consequently, Creation is both a gateway to the Transcendent via Christ and simultaneously a gifting of finite freedom that does not override our own existential state. The being as *completum et simplex* participates in Ur-Kenosis as its fundamental existence as finite or non-subsistent. As a finite being, one's encounter with "Nothingness" or the threat of nonbeing prefaces one's death and therefore one's potential resurrection. "The corpse becomes quite simply the preeminent emblematic property,"[204] writes Walter Benjamin. The resurrected body of Jesus Christ becomes a point of infinite singularity within the created order. The Christian is called to be in solidarity with this event because it is the mediation point of all Creation and the means by which the liturgy frames a participant's communal praxis as part of a single continuum that is salvation history. The extension of one's relational capacities engages one's interior non-subsistent trace of nonbeing or potential infinity as a mode of holo-somatic emulating of Christ's resurrected state and creation's own kenotic eschatological trajectory. The person's mirroring of the integral state of the human-divine natures of Jesus Christ is key to their holo-somatic participation in salvation history.

Proceeding from the haptic experiences to imaginative embellishments, human consciousness circles the edge of the terrifying abyss of the meaningiful and meaninglessness. Our conscious experiences grasp and make sense of our sensual encounters and this at once risks infinite multiplicity and endless emptiness. The door that opens consciousness to the infinite never loses sight of its grounding in the void nor its innate contrapuntal desire to transmute all that is silent, vacant, or monotone in reality. The faithful are able to hold this paradox together when they, like the

[203] The dual transcendence of the essential and existential differentiates from earlier philosophical approaches, see Etienne Gilson, *Being and Some Philosophy*, 2nd ed. (Toronto, ON: Pontifical Institute of Medieval Studies, 1952), 161.

[204] Walter Benjamin, *Origin of German Tragic Drama*, Translated by J. Osborne (London, UK: New Left Books, 1977), 218.

mystic or abstract artist, plumb the depths of the created order. For both practitioners, some latent sense of the whole or total is necessary. This idea of aggregation for the abstract artist realizes when they derive from the notion of the whole some boundary condition that reflects in surface tension and pattern: "in the spirit of the second Mosaic law, [he] ordered artists to banish iconography and to use the painted surface as end in itself. Artists were expected to paint but they were not expected to paint images."[205]

The work of abstract artists, such as Ad Reinhardt (1913–1967), Robert Rauschenberg (1925–2008), Franz Kline (1910–1962), Robert Motherwell (1915–1991), (1879–1935), and Pierre Soulages (1919–), takes up the challenge of painting reality's anti-meaning as art. These artists wrestle meaning out of the void by de-emphasizing the traditional elements of the pictorial arts—color, perspective, light, and an allegiance to realistic depiction. The abstract painting or contemplative image that visualizes the whole and part in terms of its source and beauty is consciously appropriated. These artists and the prayerful person attempt to fully experience physical reality and creatively present its pure form. The sensual, imagination, and conceptual capacities work to discover the self-reflexive unity of created object, and thereby offer some radiant expression of its truth, goodness, and physical splendor.

The pure and absolute elements of the physical object and Nature, Reinhardt argues, reveals in the deconstructive or apophatic mode the abrupt brush stroke, the vibrant color chosen over more somber possibilities, the re-envisioning of geometric spaces, and so on. The dignity of the viewer of abstract art or the spiritual individual possesses a singular and local perspective that implies something about reality as a whole. Thus, abstract art finds its place within "a long tradition of negative theology in which the essence of religion, and in my case the essence of art, is protected or the attempt is made to protect it from being pinned down or vulgarized or exploited."[206] The black hole, the canvas of the abstract work of art and the crust of the Eucharist bread all attempt to sensually

[205] Hans Belting, "Beyond Iconoclasm: Nam June Paik, The Zen Gaze and The Escape From Representation," in *Iconoclash*, Edited by Bruno Latour and Peter Weibel (Cambridge, MA: MIT Press, 2002), 391.

[206] Quoted in Robyn Denny and Phylis Ann Kallick, "Ad Reinhardt," *Studio* 174 (December 1967), 267. See Donald Kuspit, "Concerning the Spiritual in Contemporary Art," in *The Spiritual in Art*: Abstract Painting: 1880–1985, Edited by Maurice Tuchman, et al. (New York, NY: Abbeville Press, 1986), 319.

and consciously celebrate the light and the shroud that cloaks the abyss that separates the material, the given, and the Transcendent. The Eucharist stands for the eternal fullness of the created order, as it is the real presence of the Eternal Logos in the divine-human person Jesus Christ and the supernatural light that holo-somatically drives this revelation. I agree with the assessment that Reinhardt's so-called last paintings artistically present a similar inclination as there are "[n]o lines or imaginings, no shapes or composing or representings, no visions or sensations or impulses, no symbols or signs or impastos, no decoratings or colorings or picturings, no pleasures or pains, no accidents or ready-mades, no things, no ideas, no relations, no attributes, no qualities—nothing that is not of the essence."[207]

The absolute and pure is referenced in all sacramental reality and all superlative representational art. Reality is "kenotically" structured; that is, it emerges through differing relations that fundamentally reveal matter's conspansive symmetry with energy.[208] Unending motion means that unity exists in the physical world in creative and intelligible compositions of likeness and dissimilarities, *res* and its interpretative differing.[209] There are numerous Biblical passages that see a fundamental correspondence between nature and grace, a dynamic that means that finite being has the ability to convey grace.[210] The soul-embodied person testifies to the sacramentality of the body. There are ecclesial and sacramental dimensions to the human person in her call to spiritual perfection and to encourage others to live peaceful, just, charitable, and loving lives in the world, as Paul testifies (Ephesians 4:3-6).

As rational creatures searching for certainty, we enter into a lifelong journey to discover the mysteries of our universe. Imaging and transforming the physical nature and energies of being's core identity within our consciousness is a moral action—a celebration of the truth. Thus, the person creatively situates herself between complacent self-absorption and transitory appearances.[211] This is both a risk and a discomfort, as it tempts us to see as an inevitable conclusion that although conscious beings we

[207] Ad Reinhardt, "Art-as-Art (1962)," in *Selected Writings of Ad Reinhardt*, Edited by Barbara Rose (New York, NY: Viking Press, 1975), 56.

[208] Christopher M. Langan in his *Cognitive Theoretic Model of the Universe* (CTMU) coins the term conspansive.

[209] See Etienne Gilson, *The Spirit of Medieval Philosophy*, Translated by A.H.C. Downes (New York, NY: Charles Scribner's Sons, 1936), 100.

[210] On the Incarnation and grace see *GL* 4, 404.

[211] See *TD* 2, 212.

will never exceed our identity as earthbound creatures. As Christians, we are encouraged to remember Job's insight into our source and return to "dust" should inspire further commitment to moral change and the will to love God, our fellow creatures, and respect for the proper non-rational domain of Creation. The spatial and temporal components of this tension must be integrated and thereby personalized. The open mind and heart are not abstract ideals but signs that the "transubstantiation" of the individual being has begun.[212]

As Absolutely Groundless, the Spirit ushers in Christ's redemptive moment without discarding the created order's freedom or the centrality of Christ's existential place in Creation.[213] Consequently, the free exchange between the Transcendent and created order is understood to be holosomatic in nature. The holographic principle of theoretical physics analogously interprets the idea of divine exemplarity in light of how QMT defines the material order. Creation introduces the Absolute Goodness of the Divine Trinitarian Persons within the material order. Classical scientific and materialistic portrayals are misleading, as they distort the Divine's bearing in the spatiotemporal order. The Transcendent is not finally explained, measured, or proven as part of the material order. Rather, sacramental reality opens the human person up to an infinitely greater truth of divine life.

[212] See *Homo Abyssus*, 189–90.
[213] See *TL* 2, 78.

CHAPTER 2

Ur-Kenosis and Nothingness

2.1 Pure Being

The tri-personal union of Father, Son, and Spirit is equally the undivided One (*simplicitus Dei*). The concept of Ur-Kenosis is seen as an analogy for the eternal hypostatic relational "differing" that defines the union of the Three Divine Persons. The elimination of hypostatic relationships within the Trinity would make heterodox the Divine Persons' "interwoven" monotheistic identity in shared divine essence.[1] Past theologians have described the Trinity's point of source as the Father, the I AM, who acts as a metaphoric font that eternally begets as Son—the *fons et origo totius divinitatis* (Romans 8:32 and John 3:16). This timeless confirmation of the Son or Eternal Logos is the foundational Ur-Kenotic or self-emptying sharing of divine identity.[2] The Father's giving of Godself as the Eternal Logos is the "Ur-primordial," free, and absolutely complete event that is infinite and "reckless in its magnanimity."[3] The Ur-Kenotic sharing of

[1] See *ST.* I, 40, 2–4. On the dangers of viewing divine relations anthropomorphically see Yves Congar, O. P. *I Believe in the Holy Spirit*, Translated by David Smith (New York, NY: Crossroads, 1997), 155–64 and Hans Urs von Balthasar, *Theo-Drama*, vol. 2: *The Dramatis Personae: Man in God*, Translated by Graham Harrison (San Francisco: CA: Ignatius Press, 1990), 257. Hereafter cited as *TD* 2.

[2] See Thomas Aquinas, *Truth. Vol. 3: Questions I-IX*, Translated by Robert W. Schmidt, S. J. (Chicago, IL: Henry Regnery Company, 1954), 312. See also *ST.* I, 40, 1–4.

[3] On the specific role of the Son's generation in this Trinitarian act, see *TD* 4, 323.

M. P. Fusco, *The Physics and Metaphysics of Transubstantiation*,
https://doi.org/10.1007/978-3-031-34640-8_2

Father is not self-canceling but an act of absolute love or simple unity that realizes as an absolute renunciation or sharing of Godself to the Eternal Logos. The procreant Ur-Kenotic event confirms the divine substance or essence between the Divine Persons without introduction of division into Trinitarian simplicity.[4] Upon receiving all from the Father, the Eternal Logos or Son shares all received with the Spirit, who returns all to the Father—an endless "cycle" whose monotheistic and simple identity among the three Divine Persons is partially conceived by finite consciousness as differing relations among persons.[5]

The Divine Abba initiates the only begotten Son, arming as Fern points out, the Amen Abba as the self-reflective moment that is aware of its Absolute Goodness—the Great *A Se*. This move, Fern continues, is Agape actualized—the *Shekinah*: the Glory that is the ever-greater Spirit who is the paradoxical ever-greater whole. The Father is the Son's eternal transcendent "point" of origin and by extension that of the Spirit.[6] Of the infinite "differences" within the Trinitarian communion, Pavel Florensky quotes Athanasius writing that the "procession (ηεκπορυσις) of the Spirit is distinct from the 'birth (*gennēsis*)' of the Son. Of the three personal properties of the Divine hypostases, *agennēsia*, and *gennēsis*, and (*ekporeusis*), the first two are spiritually wholly comprehensible, while the last one turns out to be only the sign of some sort of spiritual experience that is yet to be understood."[7] [1]

I am not staking a claim here in the Filioque controversy but agree with Bulgakov when he sees the Spirit's role as the divine principle of unity-in-difference. He states that "[t]he Divine Sophia is God's *exhaustive* self-revelation, the fullness of divinity, and therefore has absolute content. There can be no positive principle of being that does not enter into the fullness of sophianic life and revelation."[8] [2] God's desire and will (*per*

[4] See *TL* 2, 135.

[5] See *ST*. I, 47, 3.

[6] See Richard L. Fern, *Nature, God and Humanity: Envisioning an Ethics of Nature* (Cambridge, UK: Cambridge University Press, 2002), 155–156 and Thomas Aquinas *Commentary on the Gospel of Saint John*, part 2, chp. 14, lect. 3, no. 1906, Translated by James A. Weisheipl and Fabian R. Larcher (Petersham, MA: St. Bede's Publications, 1999), 221. On Christ's passivity and the Father's generation, see *TD* 3, 183–84, John D. Zizoulas, *Being as Communion: Studies in Personhood and the Church* (Crestwood, NY: St. Vladimir's Seminary Press, 1997), 16–19 and 41–46. Hereafter cited as *Being as Communion*.

[7] *The Pillar and Ground of the Truth*, 86. Designations for the Divine are understood apophatically.

[8] *The Bride of the Lamb*, 39.

modum voluntatis) are identical to God's tri-personally shared essence by what we understand to be a conscious construct (*per modum intellectus*). Human desire's stake in consciousness finds its roots in the mystery of Trinitarian relations. The Father and Eternal Logos communicate in the Absolute Spirit: "possession of its own end and infinity, are thus far from being mutually exclusive that, on the contrary, they convey exactly the distinctive character of the Spirit: total reflection."[9] [3] Lacking the spatiotemporal configurations of created differences, the reflective capacities of divine causality (hypostases) in the Person of the Spirit are perfectly simple. Consequently, self-giving among the Divine Persons sanction their unique identities. Because the Persons share their personhood in Trinitarian life, it defines the nature of all lesser finite sacrifices.

The mutual investments of the Divine Persons in each other lead to an immutability that is not absurd formlessness or reckless infinity. Rather, the unity-in-difference among the Divine Persons (holo-somatically) models all finite, life engendering, personal relationships of love. The supra-sacrifice of divine sharing (*Vor-opfer*) gives a fallen Creation its providential path without recourse to preexisting ensconced ideas that limit individual freedom or a mereological essentialism that may deny universal purposes.[10] This divine framing gives us insight into how history projects into a trans-spatiotemporal future that gifts formal or essential forces with a discernible ontological weight or force. Consequently, one can speak of reality's metaphysical description as being ontologically verified by its innate mathematical structure. Similarly, the essence of unleavened bread orders its physical properties, and both are open to the reorientation initiated by the words of consecration given its imaging of Jesus Christ own Last Supper. The Transcendent directs the natural order to the New Covenant by physical and supernatural influences. Finite freedom is possible in reality's natural and historical unfurling into a sacramental reality because Creation never overrides its preexistence in Nothingness. The deterministic elements of the created order overshadow but never eradicate Nothingness' continued ability to clear spaces in the physical order under the Spirit's direction.

[9] *Presence*, 35.

[10] Mereological essentialism argues that an object cannot undergo any change without changing identity. See Roderick M. Chisholm "Mereological Essentialism," in *Person and Object: A Metaphysical Study* Person and Object: A Metaphysical Study (London, UK: George Allen and Unwin Press, 1976).

The interior freedom that Mary exemplifies in her fiat to the Lord immediately corresponds to the Nothingness that the created order wombs at the Spirit's direction. The Eternal Logos incarnates in the person of Jesus Christ, the Second Adam in Mary's womb. This verifies the original covenant in her person as well as setting out humanity's free relation to the new covenant. She is the Second Eve, the sinless woman who serves to echo the Father redemption of Creation from eternity. Through angelic invitation, the Spirit opens up all of reality to this trans-spatiotemporal drama. In this divine encounter, Being's imaging, the non-substantial, personal, and free nature of human nature is shown in its perfected state.[11] With Mary's encounter with the Transcendent, a way opens for her to define her place in the created order and humanity unreservedly. Just as the Divine never abuses human freedom, created freedom never overtakes the place of divine providence for what we take as innovation, eternity already has "taken better measures to avoid such an inconvenience, and therefore, indeed, he has actually done it."[12] [4] The concept of divine emanation discloses the Transcendent's absolute state, that my idea of the holo-somatic expresses. Mary's response opened the way for all of reality to recall its birth in an act of divine freedom and promised Christological in-forming.[13] The mereological ordering of the universe is first and foremost a personal event because Mary's "yes" announces the bridge between Transcendent and immanent. This is made possible in Abraham's yes and confirmed in the ecclesial community. Mary's love gives witness to this paradox of unity-in-difference in her pain and weeping when shadowed by her son's Cross. Balthasar writes, "[h]ere, finally falling silent, the Word is empowered to make his whole body into God's seed; thus, the Word finally and definitively becomes flesh in the Virgin-Mother, Mary-Ecclesia. In addition, the latter's psycho-spiritual answer is more fruitful than all attempts on the part of the sinful world to fructify itself—attempts that are doomed to sterility."[14] [5] Mary's yes to the Father reverberates Nothingness's and the Cross' ordering to the Divine's creative plan.

[11] See *Homo Abyssus*, 124–26.

[12] G.W. Leibniz and S. Clarke, *The Leibniz-Clarke Correspondence: Together with Extracts from Newton's Principia and Optics*, Edited by H.G. Alexander (Manchester, UK: Manchester University Press, 1956), 29.

[13] See Athanasius, "On Luke X. 22 (Matt. XI. 27)", *NPNF2*, vol. 5, 3. See also Norman Russell, *The Doctrine of Deification in the Greek Patristic Tradition* (New York, NY: Oxford University Press, 2006), 183–84 and *GL*, 1, 563 on the sacrament of reconcilation.

[14] *TD* 4, 361.

The Church's apostolic character is inseparable from the blessed and virginal Mary, who is Mother of God. She is the mediation point between finitude and ecclesial unity that the Second Vatican Council's document on the Constitution of the Church says: "[t]his is why the Church, even in its apostolic activity, rightly looks up to her who gave birth to Christ, since Christ was conceived by the Holy Spirit and born of the Virgin in order to be born and grow through the Church also in the hearts of the faithful."[15] Mary's love of the Father and her son echoes to this day in the Church's charitable and sacramental work and thus continues to contribute to those prayers and actions that work for Creation's essential and material "conversion" to its Edenic state.[16] The city of New Jerusalem will be the seat of the Risen Lord, the metaphoric tabernacle, to which Mary as the New Eve (*Theotokos*) consented in her radical obedience to the Father's will.[17] [34] These truths are hidden and remain outside the scientist's direct preview and may not immediately concern the artist or philosopher's gaze.

An object's essential recrudescence or informational structure is inextricable from its recurrent concealment in every new appearance. This ongoing disclosure can be symbolically represented in the concept of non-being, words and numbers. Following Plato's lead Badiou comes to a similar conclusion.[18] Indeed, mathematical theory, Badiou shows, even illuminates being in conjunction with infinity, inconsistency, multiplicity, and the predications of nothingness and nonbeing. The former atheist defines the meaning of these manipulations and projections of nonbeing by conceiving them as a geometric shape that harmonizes numbers that are made possible by means of the act of counting. Badiou incorporates the boundless properties of nonbeing as a hidden part of the mathematical sequence much like zero-dimensional points exist latent on an infinite number line and are the foundation to geometric shapes and topologies. Like infinity, multiplicity's extent and variety defy complete explanation, and this

[15] *Lumen Gentium: Dogmatic Constitution on the Church*, nos. 65. vatican.va/archive/hist._councils/ii_vatican_concil/documents/vatii_const_19641121_lumen-gentium_en.html. Hereafter cited as *LG*.

[16] See *Homo Abyssus*, 192 for a discussion of matter's ordering to a divine plan.

[17] See John Damascene, "*An Exact Exposition on the Orthodox Faith*," in *NPNF2*, vol. 9, Book 1, 1 section 3, Translated and Introduced by Philip Schaff, D.D., LL.D., and Henry Wace, D.D., (Grand Rapids, MI; Wm. B. Eerdmans Publishing Company, 1983).

[18] See Plato: *The Dialogues of Plato*, Translated by B. Jowett, M.A. Introduction by Professor Raphael Demos, vols. 1–2, (New York, New York: Random House, 1937), *Sophist*, 254e.

incomprehensibility fathoms something of a spiritual vision. However beautiful and compelling an artistic image, icon or picture is, an uneven and incomplete perception of the divine results.[19] As an atheist, Badiou denies the validity of sacramental and theo-anthropological interpretations of reality. I would still wager that his own Maoist and mathematical philosophy leans on a similar narrative to explain reason's engagement of physical reality. He does offer that beings and numbers have a "non-relational" ("non-rapport") correspondence with an actual Transcendent Being. Consciousness can create some mental image of the ONE's presence by its presumed dialectical relationship with the non(ONE). The horizontal and lateral organization of finite beings as an aniconic representation of potential infinities can also be forwarded as evidence of a similar viewpoint in many postmodern philosophical systems and artworks.[20] [7]

The infinite assortment made possible by the ongoing budding of beings whose representation will always remain partly unknown when ultimate meaning is understood to be buried in the simplicity that is the Transcendent.[21] This hypothesis finds its theological resonance in the fact that the Spirit lovingly unifies and "observes" the unending gifting or sacrifice of the Divine Persons. The Spirit's Person and Absolute Freedom mediates essential and substantial "in-division" within the Trinity without disturbing Trinitarian supra-unity—that is the Spirit eternally asserts the Father and the Eternal Logos.[22] In this eternal unfolding, the Transcendent offers in Godself something that formal systems such as mathematics can only infer and notionally circumscribe—the infinite interior possibilities of Divine love actualized in the existential order in finite acts of love and self-giving. Unlike mathematical subtraction or the act of dividing physical matter, the Father eternally realizes the Son as an everlasting spiritual moment of the Divine Person's self-reflective identity (*ein Loslassen des*

[19] See Ibid., 254a.

[20] See Alex Ling, *Badiou and Cinema* (Ediburgh, UK: Edinburgh University Press Ltd., 2011) and Thomas Crow, *No Idols: The Missing Theology of Art* (Sydney, AU: Power Publications, 2017), 11.

[21] See *TL* 1, 252. On the correspondence of the infinite, discontinuity, quantum and the symbolic, see *The Pillar and Ground of the Truth*, 351.

[22] See W. Norris Clarke, *Person and Being* (Milwaukee, WI: Marquette University Press, 1993), 82, Joseph Cardinal Ratzinger, "Concerning the Person in Theology," *Communio* 17 (1990): 438–454 and Maximus the Confessor, *On the Lord's Prayer*, in *Philokalia: The Complete Text,* Compiled and Translated by G. E. H. Palmer, P. Sherrard, and K. Ware. In Chap. 4, we will see in this Trinitarian interplay a theological metaphor for the hyper-dimensional spaces that quantum mechanics uses.

Gottseins). Ur-Kenosis is an eternal process that purely reflects the tri-noetic identity and relational capacities of the Absolute Freedom of the Divine Persons.[23] Because the Father's actions are free but in unison with the Son and the Spirit, we can infer that the essential substratum (the divine essence) of the Divine Persons is a unanimous and harmonious self-reflective ratio (Ephesians 3:10). The Father personally initiates the Son (*per modum intellectus*) and the Spirit (*per modum voluntatis*) as a pure and real "conscious" eternal act. The Son and Spirit's response to this eternal gifting defines the truth of thanksgiving: "whatever the nature of the '*generatio per modum intellectus*' [generation by way of understanding] may be, it is the result of an unfathomable gratuity behind which no thought can probe, a gratuity that, as such, is the primal ground of the mystery of God."[24] [8] The gratitude that reveals in the Trinitarian life finds its finite parallel in the created order's own communication in the worship of God's chosen People.

The personhood of the Father is only notionally prior to the Son and Spirit, as all three Divine Persons mutually arm the Transcendent.[25] The Father's self-reflectivity is a singular act for the Trinity but one shared without division individually by the Son and Spirit. Consequently, the indissolubility of Divine thought and will does not mean that the Father commands the Son or the Spirit. Rather, the Father's role as the provenience of Divine causality reinforces and reappears in the Son and the Spirit's conscious agency as an eternal act of Godself. Supported by its own metaphysical structuring, divine thought is analogously ontological as divine consciousness and will as "the one freedom of the divine essence that is possessed by each of the hypostases in its own specific way; this means that the unity of the divine will is also the result of an integration of the intention of the hypostases."[26] [9]

[23] See Bonaventure, *Sent.* I, d.27, Book 1, q.3. 21 in Bonaventure, *Commentaria in Librum Primum Sententiarum*, Translated by The Franciscan Archive, fransscian-archive.org/bonvrnture/1.Sent.html. Hereafter cited as Sent. I.

[24] *TL* 2, 140. See also *Sent.* I, d.18, q.2.

[25] On the difference between divine generation and becoming, see Hans Urs von Balthasar, *Theo-Drama, Theological Dramatic Theory*, vol. 5: *The Last Act*, Translated by Graham Harrison (San Francisco, CA: Ignatius Press, 1998), 512. Hereafter cited as *TD* 5. and See also Hans Urs von Balthasar, *The Glory of the Lord*, vol 7: *Theology: The New Covenant*, Translated by Brian McNeil C.R.V. and Edited by John Riches (San Francisco, CA: Ignatius Press, 1989), 397. Hereafter cited as *GL* 7.

[26] *TD* 5, 485.

The desires of the Son and Spirit have an intrinsic substantial bond to the Father as a common point of origin—the three wills exist in a superpositional state.[27] The Spirit interminably assents to and clarifies the Father's will and the Son's love in the created order—Divine Personhood unifies finite conscious action, wisdom and love in Jesus Christ.[28] Ur-Kenosis gives us a way to conceptualize how the Divine Persons exist by the metonymy of the Father—the Eternal Word never ceases to be thought and spoken by the Father, actualized by the Son, and echoed by the Spirit (John 1:1-14).[29] The eternal propagation of the Word ("epanadiplosis") is not properly an actualization of divine potential but rather a captured moment of Ur-Kenotic equilibrium in God's perpetual, flawless self-determination—nothing new is added to the Trinitarian will. The initial state and the final state of divine imitability are bounded by eternity (Acts 14: 17). The Second Person's denomination as Eternal Word (*Verbum mentis*) secures Trinitarian identity because the Son's actions are perfectly symmetrical or holo-somatic to the Trinitarian identity. The conceptual and syncategorematic conjunction between divine "Suprathought" and Supra-word (*Überwort*) can never fully be understood by finite beings. Given our limitations it is impossible to fully understand how the Divine Persons relate without any trace of subordination (Matthew 11:27). Holo-cryptic metaphysics does offer one way to comprehend how the synthesis of thought and word, image, and splendor of the Trinity are virtually imaged and synthesized without metaphysical parataxis in the Eternal Logos who is Jesus Christ.[30] When we want to stress the Eternal Word's consubstantiality with the Father, we employ the term "Son," accenting the coeternal nature that is the created order's "radiance" (Hebrews 1:3). The Spirit safeguards freedom and the "intellectual" bond between the Father and Son and this gives a certain license to the individual finite consciousness set on understanding the Divine. The intent

[27] For a theological discursive see *TL* 2, 134 and for a scientific account of superposition see Chap. 4.

[28] See *ST.* III, 3, 3. Each Divine Person is absolutely free but in an uninterrupted or entangled relationship with the Father. The mystery of Trinitarian reciprocity upholds the Son and Spirit's portrayal as two different holographic projections (*Umlegung*) or apportionment.

[29] See also *TL* 1, 167 and *ST.* I, 42, 4.

[30] See Hans Urs von Balthasar, *You Have the Words of Eternal Life*, Translated by Dennis Martin (San Francisco, CA: Ignatius Press, 1991), 11–13.

and thought of each Divine Person do not signal any division or dissipation in the Trinity but project indivisible concord.[31]

In knowing, God wills what he knows and this desire is already real as an eternal realization of divine existence as love (Colossians 1:17). Love is a simulacrum for the Ur-Kenotic relationship that holo-somatically registers in Creation as divine thought or exemplarity, as Maximus notes: "[t]he freedom of God's ideas opens the way for us to recognize that the existence of things is absolutely underivable from their ideas in God."[32] [10] The holo-somatic nature of Christ's consciousness is a finite manifestation of Trinitarian self-consciousness. He exemplifies the relation between the Divine Persons qualitatively and personally, not quantitatively or in a reductive fashion—even his historical life is inseparable from its eschatological conclusion. For finite persons, "the revelation of God, in which man's fulfillment of the true knowledge of God takes place, is the disposition of God in which He acts toward us as the same triune God that He is in Himself, and in such a way that, although we are [persons] and not God, we receive a share in the truth of His knowledge of Himself."[33] [11] Our idea of personhood implies some comprehension of the divine essence and tri-hypostatic relations. It is impossible for a finite creature to fully grasp "I Am Who AM's" relationships (*circumincession*).

The Divine Persons are transparent insofar as their "intra" and "extra" relations are part of the same eternal orchestration. The Trinitarian relations are free and real, yet wedded to a common ipseity by means of kenotic arrangements and rearrangements. Divine relations equally and simultaneously distinguish and unite each of the Divine Persons without impediment in infinite malleable association. The unique identity of each of these three Divine Persons, the Christian theologian argues, should never be understood as conflated, subsumed, or abstracted as found in collective or hive activity. The hypostatic relations do not introduce an unbridgeable distance as all is holo-somatically shared.[34] Early theological discussions on the "non-compostable" divine substance and relational

[31] See *ST.* I, 30, 3 and *TL* 2, 131.

[32] Maximus the Confessor, *Ambigua* (PG 91, 1081A).

[33] Karl Barth, *Church Dogmatics* 2/1, Edited by G.W. Bromiley and T. F. Torrance, T. H. L. Parker, W. B. Johnston, Harold Knight, Translated by J. M. L. Haire (New York, NY: Charles Scribner & Sons, 1955), 51. See also Augustine, *Letters*, *NPNF1*, Letter CXLVIII, 4 and Augustine, *On the Holy Trinity*, *NPNF1*, vol. 3, Book 5, Chap. 2.

[34] See *Homo Abyssus*, 158. The One is taken to be a postmodern equivalent to the idea conveyed in the concept of *Actus Purus*. See *TL* 2, 137, *ST.* I, 4, 2.

properties thread a narrow philosophical path that remarkably arrives at an analogous description of the finite person existing (*suppositum*) in non-subsistent relations.[35] [12] The meaning of person and relation here draws upon their past philosophical renditions in an Aristotelian-inspired scientific worldview. This is especially the case in those Trinitarian models that monolithically elucidate the "substantial" identity of the Divine based on materialistic and deterministic portrayal of reality. When divine identity is seen as analogously modeled on the QMO's structure in the natural order, we can stress the relational attributes of the Divine Persons without risk of introducing some rarefied "fourth."[36]

Divine immutability need not signify a lack of change. It can instead reflect the continual movement that eternally inspires Nature's own evolution into ever-greater meaning.[37] The tri-mutually beneficial nature of the relationship among the divine community frames our interpretation of the mereological. The natural order's identity and a being's evolutionary path holographically image divine identity.[38] Gilles Deleuze argues that Leibniz, like Thomas before, directly ties divine causality to the notion of infinite play and similitude of the parts in correspondence to the whole. [13] Leibniz's ordering of these concepts under the principle of similitude is an important advancement for early QMT theorists and holo-cryptic metaphysics. The intractable unity of the parts (monads) in Leibniz's philosophical system is categorized by a principle of unity (e.g., unity in difference). Otherwise, postmodern thinkers such as Deleuze would take exception to Leibniz's monadology, one may presume, if the intractable unity that a monad offers works against his certitude that difference always opens up a fissure within the conceptual and ontological orders.

The principle of similitude finds an analogous application in a being's relationship to nonbeing and its subsequent categorical placement in a universe that holo-somatically images the economic Ur-Kenotic Transcendent. Ur-Kenotic communion does not abrogate the idiosyncratic nature of individual finite identity but relationally configures non-subsistence to its image of theoretic nonexistence, for by its scholastic

[35] See *ST.* I, 76, 1 and *TL* 2, 129.

[36] See *TL* 2, 87–95.

[37] See *ST.* I, 31, 2, *TD* 4, 323–34, and *TD* 5, 91–93.

[38] See *ST.* I, 30, 3. On the Trinitarian reciprocity among the Divine Persons, see *ST.* I, 31, 1, *TL* 2, 77–78, 292–93. On a differing exposition of this theme, see W. Norris Clarke, *Explorations in Metaphysics* (Notre Dame, IN: University of Notre Dame Press, 1994), 108 and David S. Cunningham, "Participation as a Trinitarian Virtue: Challenging the Current 'Relational' Consensus," *Toronto Journal of Theology* 14 (1998): 7–25.

definition "a relationship is a respect of one to another so that one is relatively opposed to the other."[39] [12] The Trinitarian exchange hints at how difference and diversity are inclusive factors of the created order's promotion of unity. Creation involves in a supernatural "differing of difference" for where the ONE is Ur-Kenotic unity, "we, being many, are one body in Christ" (Romans 12:5). Divine identity acts as a holographic supplement to reality because from the Divine's perspective, "[t]here is one body, and one Spirit" (Ephesians 4:4). Even the possibility of the ONE disrupts the complex identity of created beings. Without arriving at a traditional Transcendent view of the Divine as ONE, Badiou nonetheless eloquently articulates how difference's identity is counted as part of its own differing. This state witnesses to the inevitable failure of attempting to subtract all beings or numbers out of existence. Whenever one attempts to subtract all to arrive at Nothingness or zero, difference reveals its own symmetrical relationship to difference. Badiou states that "there is no graspable inconsistency which would be subtracted from the count and thus a-structured."[40] [6] The Transcendent ONE supersedes its finite grasp, procedural manipulation, or conceptual formalization. Even the possibility of this fact leads many postmodern promoters to surmise the existence of an antagonistic relationship to what is equally seen as impossibility—the ONE's imaging as non(ONE). However, there is no way to circumvent the hostility that exists between finite beings and the non(ONE) because, Badiou presupposes, their opposition is nomothetic. It is not trivial that relational differences and formal subtraction are instrumental to directing the positive and additive or receptive capacities to their "mathematical" grounding.

Ur-Kenotic identity inspires a positive multiplicity or plenum of the created order. In comparison, postmodern philosophers' hermeneutical approach can result in presenting difference in a "positive light." Badiou's theory of mathematical subtraction and differing counting elicits the exponential beauty of difference, just as a theological critique of evolution leads us to the concept of the "magis" and the beauty of the embodied and ontological. This "concurrently implies that the one and the multiple can quite happily be at the same time both different and the same."[41] [7]

[39] This question is dramatically revealed in Christ's Passion, see *ST.*, III, 48, 1–4.

[40] *Being and Event*, 52.

[41] Alex Ling, *Badiou and Cinema* (Edinburgh: Edinburgh Press, Ltd., 2011), 60. It is beyond my purposes here to explore how this logic pertains to all visual arts. Thus, the mathematical representation of beings and ideas is logically consistent with such a project.

Secular interpretations of history and philosophical metaphysics emulate relational differing. Set theory's grouping and situational ordering as multiple representations in subgroups and subsections can be seen as an abstraction of social, political, or ecclesial ordering as "any situation (any presented multiple) admits sub-situations, as much as any situation is itself a sub-situation of another situation (*ad infinitum*)."[42] [14] Theologians would doubtless agree with the gist of Badiou's historical organization of being's excessiveness, although at the minimum with the caveat that all created finite differences and similarities are called to an unspoiled conformity to the ONE in the incarnate person of Jesus Christ.

The Father's begetting of the Son and the Spirit coordinates divine simplicity as an *ad extra* economic or historical event. The Son's eternal situation within Trinitarian relationships finds temporal equivalence in Jesus' desire to do the Father's will.[43] Jesus' obedience to the Father is part of the Son's eternal choice to redeem Creation and return it whole with the Spirit to the Father (1 Peter 1:19). Jesus' obedience to the Father reflects his personal identity is not based on a will-to-power as would be advanced by the dictator. Following the Father Jesus' love opens a path for conversion from our own distorted sense of self-identity. A way presents itself for an individual to accept and live the truth of Jesus Christ's person and mission analogously as holo-somatically ordered by the gifts of the Spirit.[44] Jesus Christ's humility presents us with an example of how a finite being's dependence on God does not exclude human freedom but instead gives us a way to be authentic through the judgments and choices we make. The Son's eternal birth in the "generative causality" of the Father endures as the inner compass of Jesus' life. Consequently, Jesus' freedom is infinite as is ours if we find its source in the fathomless depths of Trinitarian communion.[45]

The Son incarnates in the spatiotemporal realm of history as Jesus Christ—a self-emptying made without loss of eternal and essential affiliation within the Trinitarian life. Christ can be seen as a material and universal "transducer" for Reality because "*Christus medium tenens in omnibus*" [Christ connecting the middle in each thing and among all things]. In the self-gifting of the Trinitarian Persons, Balthasar finds the model for all finite

[42] See *Key Concepts*, 50.

[43] See *TD* 3, note 68, pages 228–230.

[44] See *TL* 2, 32.

[45] The reciprocal relationship between Trinitarian Persons grounds their freedom, see Joseph Cardinal Ratzinger, *Behold the Pierced One: An Approach to a Spiritual Christology*, Translated by Graham Harrison (San Francisco, CA: Ignatius Press, 1986), 41.

"sacrificial," "donative," or "differing" correspondences. The Father is the font, the generative self-utterance of the Son who with the Spirit underlies Trinitarian Life and created structures. The unique identities of the Divine Persons paradoxically confirm the simplicity of their monotheistic identity. The Trinitarian Persons eternally exist in the divine essence—a reality that accounts for all differences and distances in the physical universe.

2.2 Divine Groundlessness

The Spirit's metaphoric parabolic movement between Son and Father delineates their Divine Personhood in eternity and in Jesus Christ's historical life in the physical and infinitely expanding hyper-dimensions of the created universe.[46] The Spirit actualizes harmony among the Divine Persons in the Trinity. The singular identity and missions of each of the Divine Persons is not an expression of "difference" but rather the communication of the supra-ordination of Trinitarian freedom. Trinitarian selflessness (*Selbstlosigkeit*) is the active mode of Ur-Kenosis. It is characterized by a perfect personal and eternal transparency and receptivity.[47] Love motivates and encompasses the life of the Divine Persons. However, divine love's different heuristic or revelatory positions are reveled in the Spirit eternal status relative to the Father and Son's interior and economic movements.[48]

At Creation, the Spirit's eternal love actualizes perfect Trinitarian "unity-in-difference" that is holo-somatically imaged in created being and ecclesial existence.[49] The Father's generation of the Son demonstrates how love unifies by creating "space" or a "lacuna" for created freedom.[50] The Father ratifies for the Son and Spirit a "dramatic space," a euphemistic "playing room" (*der Spielraum*) for their omnipotent and supra-abundant

[46] See *TD* 5, 89–91. On the essential predication of the Divine see also *ST.* I, 39, 2.

[47] See Hans Urs von Balthasar *Love Alone is Credible*, Translated by David C. Schindler (San Francisco, CA: Ignatius Press, 2004), 83. Hereafter cited as *LA*. This does not imply a logical contradiction (see *ST.* I, 17, 1), but rather the perfection that is love, see *ST.* I, 4, 1.

[48] See *TD* 3, 199 and 522–23. Modalism and Polytheism are rejected, in part, for their failure to keep this mystery of divine unity and incarnate God in proper tension. Only with faith can one begin to perceive truth's bond with the multiplicative, individual identity and inclusivity.

[49] The ecstatic character of the Holy Spirit's self-reflectivity is subsumed under the idea of spiration, see *TD* 4, 324–7, 467–70, *TL* 2, 151–54, Barth, *Church Dogmatics*, 2/1, 493, and 512 and Gregory of Nyssa, *Commentary on the Song of Songs*, Translated by Richard McCambly (Brookline, MA: Hellenic College Press, 1987), 208.

[50] See *TD* 5, 91–95.

conversation.[51] Consequently, when finite beings intentionally grant the other space (*Geschehenlassen*) a means for personal liberty, reciprocal self-reflection and mutual growth are made possible.[52] As Rainer Maria Rilke writes, "Once the realization is accepted that even between the closest human beings infinite distances continue to exist, a wonderful living side by side can grow up, if they succeed in loving the distance between them which makes it possible for each to see the other whole against the sky."

The spaces between the real relationships of the Divine Persons do not add anything to God but furnish a way for Divine intimacy to flourish. Consequently, the concepts of an absolute freedom that is complete love justify our consideration of the Trinity's relationship to Nothingness. There was only the Divine before Creation—God created everything—even Nothingness and all its metaphysical, physical, and philosophical predications. The non-subsistence that exists between the subject and object, for example, is a created "fluidity" that gives one finite and particular analogy that can be traced back to the Creator's introduction of Nothingness. Every being radiates the force of divine exemplarity, and Balthasar writes of this encounter highlighting its ability to forward potency's resolution, such that "in the creative mirror of the subject, the object sees the image of what it is and of what it can and is meant to be."[53] [15]

The eternal identity of specific Divine Persons is conceptually deployed in spatial equivalences. The idea of a positive absence, or Nothingness, is the principal tool or medium that ensures the natural order's metaphysical coordination as desirably by the Creator. The concept of dimensional branes and anti-branes in string theory presents an analogous theoretical formulation of this difficult hypothesis. Heidi Ann Russell explains that "One possible explanation for the big bang ordered by String Theory is that it was caused by the collision of our brane [our universe] with a parallel brane, and that these collisions happen in a cyclic pattern."[54] [16] I assume a scientific reworking of theological Nothingness separates these

[51] See *TD* 2, 216–17.

[52] See *TL* 2, 29–33. On the transcendent receptivity in his earthly mission, see *TD* 4, 109 and *TD* 5, 304–305.

[53] *TL* 1, 77–78.

[54] Heidi Ann Russell, *Quantum Shift: Theological and Pastoral Implications of Contemporary Developments in Science*, Forward by George V. Coyne, (Collegeville, MN: Liturgical Press, 2015), 86. See also Brian Greene, *The Hidden Reality: Parallel Universes and the Deep Laws of the Cosmos*, loc. 118-19 of 347. EPUB eBook Ed. (New York, NY: Alfred A. Knopf, 2011). Hereafter cited as The *Hidden Reality.*

two branes. This view is not immediately endorsed by those Judeo-Christian views that promote that the Transcendent creates without pre-existing matter for: "[b]y faith we understand that the world was created by the word of God, so that what is seen was made out of things, which do not appear" (Hebrews 11: 3). However, even prior to the Creation of material reality, the correspondence of the Divine Persons manifests a privileged spatial topology where "negative" spaces confirm the positive horizon of Divine freedom and pure subsisting as Creator.[55] While it is beyond the scope of my present purposes here, Karl Barth's interpretation of the *logos asarkos* ("Word without Flesh") offers one way to conceive of how the Transcendent scores divine beneficence by means of advantaging the economic Trinity's correlations to elements of negativity in the created order. The Trinity's presence in salvation history is perfectly realized in the person and work of Jesus Christ. The relationship between grace and predestination frames Barth's theory of concept of *logos asarkos.*[56] The distance or Nothingness that separates the Divine from Creation is not eradicated but invested with meaning by the economic Trinity. Divine "supra-freedom" and "supra-love" includes the Son's incarnate mission as a mysterious revelation of divine impassibility.[57] The infinite stretching and displacement of the Eternal Logos' hypostatic relations and essence is a historical occurrence and a mode of eternal self-reflective intelligibility. The Father's relationship to the Son through the Spirit exists in a perfect union with the person of Jesus Christ.[58] Jesus Christ is the physical embodiment of the hidden transcendent Eternal Word holographically and harmoniously present in an existing person in history (Colossians, 31). As Jesus Christ never ceases to look to the Father or recall his incarnate source in the Spirit, his human self-awareness is tri-relational mode of divine intelligibility.[59] The Spirit makes possible Jesus' "heno-tradic" experiences

[55] See Hans Urs von Balthasar, *Epilogue*, Translated by Edward T. Oakes, S.J. (San Francisco, CA: Ignatius Press, 2004), 89–90. Hereafter cited as *E*.

[56] See Karl Barth, *Church Dogmatics*, vol. IV: *The Doctrine of Reconciliation* Part 1, Translated by G.W. Bromiley, Edited by G.W. Bromiley and T.F. Torrance (London, UK/New york, New York: T & T Clark International, Continuum imprint, 2004), 52. Hereafter cited as CD IV/1.

[57] See *GL* 5, 631–32.

[58] See *ST*. I, 30, 2, and *TD* 5, 93–95.

[59] On the necessity of a triune Creator, see *TD* 5, 66–109, *ST*., I, 45, 2 and Augustine, *Tractates on John*, *NPNF7*, 20, 9.

of the Father.[60] The Spirit's presence extends from the Transcendent Trinity's innermost depths to the phenomenal margins of their holo-somatic revelation in the person of Jesus Christ.

The Son's eternally chooses to subtend the eternal kenotic relations temporally in the person Jesus Christ is an act of complete freedom.[61] The Son and Jesus Christ share in their person a single self-reflective movement. This event conveys trans-dimensionally for the Son and spatiotemporally for Jesus Christ. The Incarnation and resurrection of Jesus Christ opens up a new avenue to understand self-reflective consciousness. In the move from eternity into history and subsequent return to the Father, Jesus Christ's mission diagrams the fundamental arc of subjective apperception. Balthasar writes "that departure point, the created person, too, should no longer be described chiefly as subsisting in itself, but more profoundly (supposing that person to be actually created in God's image and likeness) as a 'returning (*reflexio completa*) from exteriority to oneself' and an 'emergence from oneself as an interiority that gives itself in self-expression."[62] [17]

Theologians such as Robert W. Jenson critique the *logos asarkos*, claiming that the preexisting Son is none other than Jesus Christ. Thus, the distinction between the Incarnate Lord and the Transcendent Son reveals in Jesus' eternal and historical redemptive mission.[63] Robert W. Jenson also acknowledges the need to conceptualize the existence of Jesus Christ prior to the Incarnation. Jenson defends his position with reference to scripture's use of repetitions in the narrative patterns in both the Old and New Testaments—the full picture of the story of Jesus' birth from Mary

[60] The term "heno-triadic" from the Greek word "*henos*" for one and "*tri*" for three from the Latin offers a shorthand way to talk about divine nature and action that remains Trinitarian and simple.

[61] See *Homo Abyssus*, 58–59. See also *GL* 7, 147–161.

[62] Hans Urs von Balthasar, *Mysterium Paschale*, Translated and Introduction by Aidan Nichols, O.P. (Grand Rapids, MI: Wm. B. Eerdmans Publishing Company, 1990), 28. Hereafter cited as *MP*.

[63] On Barth's perspective of this theory see Bruce McCormack, "Grace and Beauty: The Role of God's Grace: Election in Karl Barth's Theological Ontology," in *The Cambridge Companion to Karl Barth*, Edited by John Webster (Cambridge, UK: Cambridge University Press, 2000), 100. Note also "[w]e have to take care to avoid any 'Christological restriction' that would want to ground the reality of creation in some 'prior' reality of redemption or grace; there is no 'before' or 'after', and, even if the goal is the first thing intended by the agent, in *ordine executionis* we must first posit a natural (non-divine) subject as the possible recipient of grace," *TD* 3, 257.

demands one reference and understand Jewish scriptures.[64] Barth's explication seems to walk this same theological edge when he writes that "YHWH does not stand above the covenant, but in it, yet He is also not under it."[65] [18] The Son and Jesus' identity are two equal and self-reflective holo-somatic instances of Trinitarian life. The Incarnation is part of the eternal identity of God, for: "[w]hen the Word of God becomes flesh...He both showed forth the image truly, since He became Himself what was His Image; and He re-established the similitude after a sure manner, by assimilating [the human person] to the invisible Father through means of the visible Word."[66] [19]

Unlike postmodern approaches that argue that the *logos asarkos* or analogously, the non(One) can be known outside of revelation, I maintain that all such theories severs Creation's holo-somatic consonance to an existing God in errand of more palpable abstract notions.[67] [56] Indeed, the Son is known in Jesus' person and mission, the logic of the *Deus pro nobis* whose sanctifying acts attest to their Trinitarian source. The conceptualization of the faith, expressly concerning the *logos asarkos*, allows us a natural point of reference (as sinful creatures who distance themselves from the Divine) to understand God's revelation of love in the crucified Savior: "Now I rejoice in my sufferings for your sake, and in my flesh, I complete what is lacking in Christ's afflictions for the sake of his body, that is, the Church" (Colossians 1:24). This apparent paradoxical teaching finds logical justification in reason's capacity to expand the bounds of our intuitive presuppositions without escorting irresoluble contradictory statements, as well as the fact that identity reveals in ontological differing: "reason can [only] achieve the analogical ascent to God only insofar as it in turn nihilates the univocal ontological vacillation through the *non-subsistent relationships.*"[68] [20]

The boundless freedom of the Spirit and Divine Logos and Jesus Christ is eternal. The Spirit maintains the nonlocal Jesus and the Son as a simple and singular identity. Further, the Spirit ensures that this divine

[64] See Robert W. Jensen, *Systematic Theology*, Vol. 1: *The Triune God* (Oxford, UK: Oxford University Press, 1997), 140–41. On Israel's preparatory role to the New Covenant, see *GL* 1, 628.

[65] *Church Dogmatics*, IV/I, 25.

[66] Irenaeus, *Against Heresies, NPNF1*, vol. 5, 16, 2.

[67] This question is made coherent with establishing a personal relationship with Christ. See Karl Barth, *Church Dogmatics: The Doctrine of Reconciliation*, Translated by G.W. Bromiley, Edited by G.W. Bromiley and T.F. Torrance: IV/I, (London, UK/New York: New york: T&T Clark International: A Continuum Imprint, 1956), 51–54.

[68] *Homo Abyssus*, 203.

expediency manifests in the Eucharist. The Spirit's universal presence in Creation and in the Trinity realizes the economic Trinity without undermining divine immutability.[69] The Spirit bridges the eternal dialog between the Father and Son and reveals this eternal bond in the life of Jesus Christ (*opera trinitatis ad extra communia*) and sacramentally in the Eucharist.[70]

The Ur-Kenotic crossing of the unassailable ambit between the Divine and Creation is an internal principle of Jesus' knowledge of the Father (John. 18:26).[71] Ur-Kenotic hypostatic relationships echo in Christ's experiences of finitude and non-subsisting relations in the created order: Jesus is head of Creation (Ephesians 1:10), the "concrete universal," an archetype for all reality as: "[t]he Father creates the world after the primordial image of the Son; that being not enough, however, he creates the world in the Son as well, by locating its continuance in him."[72] [21] Making Jesus the center of one's life allows one to "holo-somatically" share or image his identity as Eternal Logos. Balthasar asserts that "in this two-one of the image (*Bildnatur*) of the Son, the paradoxes of man are solved: the finite image is exceeded, but not destroyed; the Incarnation is a function of the being of the Word (*Wortseins*); the creaturely image (*Bildnatur*) is an expression and conveyance of the eternal and infinite image (*Bildnatur*)."[73] [24]

The consubstantial communion among the Divine Persons is an interminable overflowing Trinitarian dialog that seamlessly echoes in Creation

[69] See *TD* 3, Hans Urs von Balthasar, *Theo-Drama: Theological Dramatic Theory,* vol. 3: *The Drama Personas: The Person in Christ*, Translated by Graham Harrison (San Francisco, CA: Ignatius Press, 1992), 523. Hereafter cited as TD 3. See also Hans Urs von Balthasar, *Theo-Logic Theological Logical Theory*, vol. 3: *The Spirit of Truth*, Translated by Graham Harrison (San Francisco, CA: Ignatius Press, 2005), 20–31. Hereafter cited as *TL* 3.

[70] Heresy intimidates when one's understanding of divine personhood comes from a deficient grasp of the person Jesus Christ. In 1148, the Council of Rheims set out that in the immanent Trinity, the relations are real against contrary readings forwarded by theological schools such as Sabellius' Christology. On a theological response to these issues see also *ST.* I, 73, 3, *TD* 2, 175, *TD* 4, 389–394 and St. Maximus the Confessor, "*On the Lord's Prayer*," in *The Philokalia*, vol. 2, complied by St. Nikodimos of the Holy Mountain and St. Makarius of Corinth, Eds. and Translated by G. E. H. Palmer, Philip Sherrard, and Kallistos Ware (Boston, MA: Faber & Faber, 1981), 287.

[71] See *TL* 2, 289–316. For an early theological explanation of Christ's place in Creation, see Irenaeus' Doctrine of Anakephalaiosis.

[72] Adrienne von Speyr, *Colossians*, Translated by Michael J. Miller (San Francisco, CA: Ignatius Press, 1998), 32. Hereafter cited as *Colossians.*

[73] Hans Urs von Balthasar, "Persönlichkeit und Form," *Gloria Dei* 7 (1952): 10. See also *Mystery of the Eucharist*, 322.

with Christ's Incarnation and all that is entailed in its liturgical celebration. Christ is the New Adam (Romans 5:14) who makes known the invisible God (John 1:18). With the Eternal Logos, the Spirit fulfills the Father's designs for Jesus Christ.[74] On the spiritual notion of causality (*causaliter*), Augustine states, "Scripture has said that the earth brought forth the crops and the trees, in the sense that it received the power of bringing them forth. In the earth from the beginning, in what I call the roots of time, God created what was to be in times to come."[75] [37]

Relational kenosis or "difference" enacts union in the Trinity and vouchsafes the Transcendent Son's entrance into the fragility of human existence as a spiritualization of the creative impulse and matter's burgeoning properties. If the person of Jesus Christ is the apex of the created order, the universe as a whole orchestrates the material toward some Eucharistic manifestation. The Eucharist is a theological analogate of evolution's physical potential. Supernatural forces are able to bring forth an integrated being from material sources because such a causal move is analogously a holo-somatic manifestation of the Son's Ur-sacrificial embodiment of all that follows from spatiotemporal and material constraints. The question of beings' creative and emergent properties thus becomes a question of their dual contextualization by those opposing and annihilating forces in the created order and concatenated with their supernatural perfection by Ur-Kenosis. Subjective freedom acts as the midpoint in this metaphysical balancing act.[76]

The Son respects the freedom of every creature and this demands each person must do the same. Commitment to a spirit of mutuality is critical to truth's disclosure. Subjective truth must be tested, as the True never reveals completely to any individual or milieu. Thus, "we know each thing more perfectly the more fully we see its difference from other things."[77] [22] [If truth's disclosure and discovery were otherwise, the creative and rational capacities of human identity and the embryonic nature of the material order would remain unobserved, impotent, and dormant. The rational mind must continually embrace reality as any one experience

[74] See *TD* 3, 525–529.

[75] Augustine, *The Literal Meaning of Genesis*, Translated by John Hammond Taylor (New York, NY: Newman Press, 1982), 153.

[76] See *Homo Abyssus*, 316–319.

[77] Thomas Aquinas *Summa Contra Gentiles*, Book I: God, chp. 14, Translated with Introduction and Notes, Anton C. Pegis, F.R.S.C. (Notre Dame, IN: Notre Dame Press, 1975). Hereafter cited as *SCG*.

remains partial: "It is not able to receive all beings into itself in a single enactment of the movement of finitization; it must enact and undergo the giving of being in every moment anew, and it is never able to close the ontological difference in 'absolute knowledge.'"[78] [20] Partial knowledge does not need to make some concession to nihilism because intellectual differentiation depends on an *a priori* belief in a meaningful whole. It is equally certain that any concept of totality implies the Transcendent, for: "[t]he intellect… has an operation that extends to universal being (*ens in universali*)."[79] [12] Knowledge of a particular finite object presumes a boundary exists that separates it from the infinite and unbounded. A similar principle is operative in a material being's relationship to supernatural grace. The Father's redemptive desires for Creation require a person's openness to the Spirit and Godself's gift of divine grace. Creation's call to perfection in divine grace coincides with preexisting proclivities inherent in created matter to accept form and the human person's original perfection of the human person (*status naturae purae*). Furthermore, one must avoid restricting Jesus Christ's freedom by presuming that the Spirit's ordering of grace is given prior to one's free consent. The Spirit's uniform manifestation in the universe means that grace actualizes in matter. The Spirit does not impose grace but offers it to engage personal freedom (*ordine executionis*) and thus presupposes the positive nature of non-subsistent relations.[80] The Spirit guarantees Christocentric naturalism—a Eucharistic celebration writ large—the Spirit invigorates all Creation toward its eschatological fulfillment.[81]

Fern argues that this specialized naturalism operates in correspondence with the threefold work and threefold offices of Jesus Christ. Briefly put, the Spirit's work continues that of the chthonic Christ as Logos, the incarnation of the charisma Christ who is the eternal sacrificial Lamb, and the charismatic Christ, the Lord who makes possible Creation's ultimate transubstantiation in grace. The essential informs the material, and thus the concrete appears to exist in terms of the metaphoric or pseudo-neutrality. In the Spirit, this movement in the created order is personalized in regard to eternal Trinitarian identity. However, in relationship to the Incarnation,

[78] *Homo Abyssus*, 369.

[79] *ST*. I, 79, 2.

[80] *See TD* 3, 257–259, GL 7, 177–178, Augustine, *On the Holy Trinity*, *NPNF1*, vol. 3, Book 4, Chap. 1, 2.

[81] See *TL* 2, 156.

the Spirit discloses how divine receptivity or objectivity is an irreducible element of divine identity.[82]

Unlike the Divine, finite beings have a dualistic relationship to Nothingness. Nothingness is a tool forged by divine imagination to convey divine truth and beauty in the infinite distances separating material reality from its transcendental source. Unlike the eternal supra-motion of the hypostatic unity of the Divine Persons, physical space and time are open to conscious observation, successive iterations, and allotted measure. The quantification of physical reality is made against the backdrop of a Creation that every future division and difference was potential at reality's point of origin. Representative of a physicist's position on this issue, Einstein comments that the "distinction between past, present, and future is only a stubbornly persistent illusion."[83] [26] The spatiotemporal order mimics the eternal and supra-dimensional divine life and thus Thomas reasons: "Since creation is without movement, a thing is being created and is already created at the same time."[84] [12] Creation is neither a physical nor a conceptual addendum to the Transcendent, for "God is all (otherwise he would not be God), and yet the world exists, and this real world cannot be added to God's reality as an increment."[85] [31]

The non-necessity of created being also advocates for spatial images of the indispensability, power, and freedom of the Divine: "His dominion is an eternal dominion; his kingdom endures from generation to generation. All the peoples of the earth are regarded as nothing. He does as he pleases with the powers of heaven and the peoples of the earth. No one can hold back his hand or say to him: 'What have you done?'" (Daniel 4:35).[86] The appellation of Trinitarian Nothingness can also help contextualize the absence of empirical beings prior to Creation. Holding the idea of nothingness as a trilateral thought, a divine directive for creation is issued.[87] The Father freely "begets" Creation as a shared thought of the

[82] See Richard L. Fern, *Nature, God and Humanity: Envisioning an Ethics of Nature* (Cambridge, UK: Cambridge University Press, 2002), 151–161 and Hans Urs von Balthasar, *Explorations in Theology, vol. 4: Spirit and Institution*, Translated by Edward T. Oakes, S.J. (San Francisco, CA: Ignatius Press, 1995), 237–238. Hereafter cited as *ExT.*, IV.

[83] Albert Einstein, *Correspondence Avec Michele Besso* 1903-1905, Edited by Peter Speziali (Paris, FR: Hermann, 1979), 51.

[84] *ST.* I, 45, 2, resp. 3.

[85] *See GL* 5, 225–228.

[86] *TL* 1, 106.

[87] See *ST.* I, 45, 6.

"Nothingness" that "germinates" unity among the Divine Persons. Holding to no measure, the Divine life is self-reflectively simple, and this mysteriously premises a holographic density to Nothingness.[88]

2.3 Holographic Dimensions of Nothingness

The Divine maintains a nonphysical bond or spiritual presence in reality. Realization of this "contact of power" is found in the ontological differing and consciousness of the person in Jesus Christ. Christ embraces the point where Creation is called into existence as a metaphysical principle of his finite nature: "The peculiar nature of the human and Divine life is separated, and without common ground, and their distinguishing properties stand entirely apart, so that those of the latter are not apprehended by the former, nor conversely, those of the former in the latter."[89] [25] The concept of divine ideas genealogically grounds the meeting of Transcendent and immanent as a "virtual imitation" exists between the divine exemplars writ into every being—and divine consciousness (*generatio per modum intellectus*).[90] Basil explains that "he who receives the Father virtually receives at the same time both the Son and the Spirit; for it is in no wise possible to entertain the idea of severance or division, in such a way as that the Son should be thought of apart from the Father, or the Spirit be disjoined from the Son."[91] The physical order's natural telos evolves by means of the principles afforded by divine exemplarity: "The form of the world is due to the wisdom of the supreme Artificer, matter came to the creator from without, and thus the world results from a double origin."[92] The names God gives being are eternal, essential realities; they are written into the natural order holo-somatically for the theologian and as information for the scientist. The power of divine naming, like divine agency, endorses teleological ends without obstructing the "autonomous" movements of material bodies—for: "Christians, as Christians, aspire to something greater

[88] See *The Bride of the Lamb*, 44–46.

[89] Nyssa, *Against Eunomius*, *NPNF2*, vol. 5, 1. Translated and Introduced by Philip Schaff, D.D., LL.D., and Henry Wace, D.D. (Grand Rapids, MI: Wm. B. Eerdmans Publishing Company, 1954). Hereafter cited as *Against Eunomius*.

[90] See Basil, *The Hexaemeron*, *NPNF2*, vol. 8, Homily 1, 5. Translated and Introduced by Philip Schaff, D.D., LLD., and Henry Wace, D.D. (Grand Rapids, MI: Wm. B. Eerdmans Publishing Company, 1983). Hereafter cited as *Hex*. See also Aquinas, *SCG*, Book 1, Chap. 29, 1–6.

[91] See Ibid., Basil, "*Letter* 384," *NPNF2*, vol. 8.

[92] Basil, *Hexaemeron*, "*Homily* 2, 2", *NPNF2*, vol. 8, 2, 2. See also *AE*, 168, 231–32.

than a 'natural' immortality. They aspire to an everlasting communion with God, or, to use the startling phrase of the early Fathers, to a theosis."[93] [28]

I am not proposing changes in Catholic doctrine based on scientific developments, be it with regard to transubstantiation or any other doctrine. In this sense, my reflections need to be carefully distinguished from those of pantheists, who reject traditional Catholic teachings regarding the nature and acts of God in favor of what they may take to be a more scientifically informed and congenial doctrine. Science's early views on this question of nonphysical forces are exemplified by Isaac Newton's doubts concerning nonlocal influences in his findings on the force of gravity. Newton writes, "[t]hat gravity should be innate, inherent and essential to Matter, so that one Body may act upon another at a Distance through a Vacuum without the mediation of anything else, by and through which their Action and Force may be conveyed from one to another, is to me so great an Absurdity, that I believe no Man who has in philosophical Matters a competent Faculty of thinking can ever fall into."[94] [29]

For many ancient philosophers, the ingrained formalization of physical reality is found in the concepts of Prime Matter or substance. Prime Matter only becomes intelligible when materially predicted or when consciousness imposes a structure upon that which is without inherent properties: "where no shape is perceived, no place, no size, no recokoning of time or anything else knowable is there: and so it is inevitable that our apprehensive faculty seeking as it does some object to grasp, must fall back from any side of this incomprehensible existence."[95]

Prime Matter and my own theological conceptualizations of the Nothingness do not presuppose the existence of prejacent matter or even a spatiotemporal receptacle.[96] A number's filiation to a geometric object highlights the foundational relationship between the spatiotemporal and the concepts of "pure space" and the eternal. Physical matter is under-

[93] Archpriest George Florovsky, "The Immortality of the Soul", fatheralexander.org/booklets/english/immortality_soul.htm. See also *ST.* I, 105, 2, 1.

[94] Bernard I. Cohen, *Isaac Newton's Papers and Letters on Natural Philosophy*, 2nd edition (Cambridge, MA and London, UK: Harvard University Press, 1978), 302–03.

[95] Nyssa, *Against Eunomius*, *NPNF2*, vol. 5, 1, 26. See also Wolfgang *Smith*, *The Quantum Enigma: Finding the Hidden Key* (San Rafael, CA: Angelico Press/Sophia Press, 2011), 93–94. Hereafter cited as *Smith*.

[96] See David L. Balas, "Eternity and Time in *Contra Eunomium*," in *Gregor von Nyssa und Die Philosophie*, Edited by H. Dorrie, M. Altenbeurger, and U. Schramm (Leiden, NL: E.J. Brill, 1976), 133–34.

stood to be bounded within the spatiotemporal, whereas abstract mathematical objects are derived from space-time.[97] An existing being holo-somatically images the Divine by distancing itself from Nothingness and determining Prime Matter, for "the givenness of being in the world is not explicable from the negative potentiality of 'Nothingness.'"[98] [30] The Judeo-Christian concept of creation ex nihilo states that the universe was without quiddity, for "the fact that the world is created out nothing means only that the world exists in God and only by God, for the world does not have within itself the ground of its own being. In itself, the world is groundless; it is established on top of an abyss, and this abyss is 'nothing.'"[99] [2] Prime Matter receives a particular identity when its "form" or informational structure is materially educed "not only in the negative sense (*omnis definitio est negatio*) but also in a positive sense-consists in a specific theme, a special color, sound, or word: an idea in creation as it is given from all eternity in Sophia."[100]

The concept of substance (*substratum* or subject) points to that which subsists as a singular being.[101] Matter's relation to substance is philosophically equivalent to the mathematical idea or epistemic state of a being prior to its axiomatic formalization by a proof.[102] Thomas comments that: "mathematical entities do not subsist as separate in the order of being: because if they subsisted, they would have goodness in them, namely, the goodness of their esse itself."[103] [12]

What Prime Matter lacks hints at its future purposes as it gives us a way to speak of being's form prior to its material and spatiotemporal individuation or how numbers quantify the ontic character of phase states. The measurement of a physical being implies that symmetry exists between the representational and phenomenal realities. The measurement of a

[97] See Wolfgang Smith, *Smith*, 86.

[98] Niels C. Nielsen, Jr. "Eric's Philosophy of the *Analogia Entis*," *The Review of Metaphysics*, vol. V, no. 4, June 1952, 603. In the next chapter, I will develop how by extension the concept of nothingness is used as a description of the lowest possible entropic informational state in the quantum system.

[99] See *The Bride of the Lamb*, 7 and on Aristotle, *Metaph.* XII, c.10 (1075a 16), see Aquinas, *Commentary on the Metaphysics of Aristotle*, Translated by John P. Rowan (Chicago, IL; Henry Regnery Company, 1961), 402–403 on 5, 1: 1070–1078.

[100] *The Bride of the Lamb*, 84.

[101] On Aristotle, XII, *Metaph.* I, 2419, see Ibid., 854.

[102] See *ST.* I, 86, 1. For a scientific viewpoint, see Cynthia Sue Larson, "Primacy of Quantum Logic in the Natural World," *Cosmos and History: The Journal of Natural and Social Philosophy* 11, no. 2 (2015): 329.

[103] *ST.* I, 5, 3, resp. 4.

particle's phase state, for example, represents its temporal and spatial location within a larger system. This spatiotemporal event can be abstracted to include information about the particle's geometric shape and evolution over time.[104] The phase state can also be graphed as a series of points in way that highlights the space in which a phase state inhabits and predict this space's own development. The idea of phase state is a cipher akin to the manner in which zero functions in numeric equations that involve addition and subtraction. In such mathematical statements, zero is provisional to all future quantification.

The universe as a whole can be construed as a continuum of zero points (e.g., without spatial-temporal actualization) receptive to any future numeric values and potential dimensional boundaries. Accordingly, these points correspond to the idea that zero point energy or vacuum energy yields infinite positive and negative values. Zero-dimensional points or movement, such as "spatial nothingness," presumes that associative identities are possible and non-contradictory even when material reality is absent.[105] Individuated matter's malleable properties follow from Nothingness' capacity to differentiate and to fold ontologically and infinitely, thereby drawing topological identities from Prime Matter (*Numerus stat ex parte materaie*), as well as conceptually rationalizing how "the unfolded [goes] into the always enfolded, which is shown in the unfolded to be non-unfolded."[106] [31]

Reality herself is only known when both the ontic and noetic folds are unified against their own potential absence. The free subtracted or "kenotic" movement away from nothingness fixes what numbers mysteriously realize by formally quantifying some potential territory or matter's particular placement and function in the universe. It is not "in the power of time to define for each one the measure of nature, but nature abides self-contained, preserving herself through succeeding generations: and time has a course of its own, whether surrounding, or flowing by, this nature, which remains firm and motionless within her own limits." [25]

[104] Dean Rickles, Penelope Hawe, and Alan Shiell, "A Simple Guide to Chaos and Complexity," *Journal of Epidemiology and Community Health* 61 (2007): 933.

[105] On how the principle of non-contradiction is known through sensations and perceptions of being, see *Metaph. 5,* lect. 6, 599 in *Commentary on Aristotle's Metaphysics,* Translated and Introduced by John P. Rowan and preface by Ralph McInerny (Notre Dame, IN: Dumb Ox Books, 1961), 221.

[106] *GL* 5, 210.

One can conceive (*excogitavit*) of numbers as consonant with divine exemplarity by seeing each phenomenal instantiation as the formal or essential individuation of matter. For "[f]rom the plenitude of its 'virtual quality' [holo-somatic nature] the infinite is measuring everything else as greater or lesser to the degree that it approaches the whole or recedes from it."[107] By divine directive this universal truth is actualized in individual beings for "created things are confined within the fitting measures, as within a boundary, with due regard to the adjustment of the whole by the pleasure of a wise Creator; and so, although human reason in its weakness cannot reach the whole way to the contents of creation."[108] In Aquinas' commentary on Aristotle's *Metaphysics*, he offers a commensurate observation: "But the infinite is not said to exist potentially in the sense that it may sometimes have a separate actual existence alone; but in the case of the infinite, actuality and potentiality are distinguished only in thought and in knowledge…the capacity of the infinite for being never comes to an end; for when it is actually divided it is still potentially further divided." Just as divine exemplars preclude their complete observation in the material object, mathematical axioms and symmetries can realize any number of potential geometries in the future. Again, retroactively, we are only able to make such a statement about regularities by presupposing the "existence" of some undistinguished and all encompassing "Primal Matter," "Substance," "Mathematical Zero," or "Nothingness."

It is the free mutuality between ontic and noetic realms or the Ur-Kenotic collation between ideal or formal explanation and physical reality that supports the ongoing interpretation of holo-cryptic metaphysics. Ur-Kenotic differing can be conceived but not finally defined in a particular context like Prime Matter or Nothingness, as Ulrich notes: "[t]his 'pure spirit' is the reason that has got caught up in the ontological vacillation and that negates the incarnational act of being's exinanitio [kenosis]. The intelligibility stops at the unfathomable darkness of matter and necessarily separates itself from it, according to the constellation within the ontological difference—ideality and reality having been separated from each other!"[109] [20]

[107] See Dun Scotus, *Opera Omnia,* Editors, C. Balie, et al. (Vatican City: Vatican Scotisitic Commission, 1950), 25 Vols., "Quodlibetal Questions", q. 5.57.

[108] Ibid., 5, 1, 26. See also John Duns Scotus, Questions on the Metaphysics of Aristotle, Translated by Girard J. Etzkorn and Allan B. Wolter, O.F.M., vol. 2, Book 69 (New York, New York: Franciscan Institute Publications, St. Bonaventure University, 1998), 506–515.

[109] *Homo Abyssus*, 194.

Thomas provides a detailed account of transubstantiation by adapting Aristotelian hylomorphic materialism, that idealistic and nominalist approach favored by many scientific and historical methodologies that seem to theoretically parallel Scotus' Eucharistic theology. John Duns Scotus' theory of theological adduction evidences how the immaterial and the physical correlate with his view on the glorified Lord in the Eucharist finding parallel in Jesus Christ's sacrifice on the Cross. Thomas' metaphysics of theological realism and Scotus' nominalist interpretation speak to how the mystery of material identity and historical change must be accounted for in the doctrine of transubstantiation.

The ritualistic sanctification of bread and wine is most definitively a historical act, but paradoxically one categorized as much by spatiotemporal context and supernatural presence.[110] [33] Where scientists quantify a material object by universal laws, I describe the interior reality of the Eucharistic species with that supra-universal force of the Spirit's "holographic" presence. Thomas' distinction between a principal and an efficient cause opens a way for us to conceptualize the Transcendent's universal efficacy within a localized bandwidth of instrumental actions. The manifestations of a principal cause at work within defined parameters can, on deeper inspection, be traced back to the mystery of Divine Simplicity—perfect self-reflection. Analogously, a principal cause immaterially influences efficacious movements in Nature. Thomas explains that "the sign of a hidden effect [exists] in virtue of the fact that it is not only a cause but in some sense an effect too, inasmuch as it receives its initial impetus from the principal agent. And this this is why the sacraments of the New Law are causes and signs at the same time. Hence, too it is that, as the usual formula puts it, they effect what they figuratively express [*eciunt quod figurant*]."[111] [12]

Transubstantiation evidences a comparable logic to the time-space symmetry evolution in the translations of a QMO in an electromagnetic or thermal system. The QMO reveals as a conceptual construct when properly explained by a theory just as a mutual equilibrium exists between the QMO's motion and its representation as a holographic projection on an extreme boundary when observed. Analogously, the Transcendent conceived as eternal *fons* is fundamental to every Eucharistic consecration as it is source of the Spirit's Absolute Freedom and the identity of Jesus Christ. The Ur-Kenotic life of the Trinitarian Persons preambles and informs each

[110] See *DS* 886.
[111] *ST.* III, 62, 1, resp. 1.

celebration of the Eucharist. The formal, efficient, and material individuation of a being, Aristotle contends, is a force (*metabolē kath'ousian*) that forces the nonexistence into extant [finished *ousia*].[112] "Simple immediacy is itself an expression of reflection and is related," Hegel argues, "to the difference from what is mediated."[113] [27] At the Eucharist, the simplicity of Godself's self-reflective nature informs material reality at the consecration as part of the Son and Spirit's missions.

The historical reenactment and impersonation of Christ's Passion at the liturgy calls forth the faithful's participation. The ecclesial body's spatio-temporal state is relative to eschatological time, and the holo-somatic presence of the heavenly kingdom is recognized at the moment of consecration (Luke 22:15).[114] [33] The Eucharist represents, memorializes, and embodies Christ in a transient (*actio transiens*) bloodless liturgical reenactment (*immolatitio modo*) that sacramentally participates in the heavenly liturgy.[115] The liturgical reenactment (*actio transiens*) is non-localized, as it repeats without disturbing the perfect unity that exists between Jesus Christ's manifestation in the Eucharist and in the permanent reality (*res permanens*) of the Son.[116] The human person never exceeds the spatiotemporal limits until created relational differing is fully interpreted as moral perfection in light of divine unity. For "eternal life [will] be the perfect fulfillment of the eternal enhancement (*Steigerung*) contained in being itself; it would be comparative of life becoming permanent condition."[117] [15] The congregation's ability to intuit eternity is possible because the partnering of the temporal with the spatial suggests this relationship could continue forever. The act of creation configures the Eucharistic space

[112] See Jean-Luc Marion, *God Without Being*, Translated by Thomas A. Carlson (Chicago, IL: The University of Chicago Press, 1991), 87. Hereafter cited as *God Without Being*.

[113] Georg Wilhelm Friedrich Hegel, *Science of Logic*, Translated by A.V. Miller (Atlantic Highlands, NJ: Humanities Press International, 1969), 69.

[114] See Jean Danielou, *The Bible and The Liturgy*, (Notre Dame, IN: University of Notre Dame Press, 2002), 168. Hereafter cited as *The Bible and Liturgy*. On the title "Lord" have Eucharistic overtones, see Oscar Cullmann, *The Christology of the New Testament*, rev. ed., Translated by Shirley C. Guthrie and Charles A. M. Hall (Philadelphia, PA: Westminster Press, 1963), 209.

[115] See *ST.* III, 82, 2, 3 and Karl Rahner, *The Celebration of the Eucharist*, Translated by Angelus Häussling (New York: Herder and Herder, 1968), 18.

[116] See Council of Trent, Session 22 "*Doctrina de ss. Missae sacrifice*," cap. 1, "*De institutione sacrosanct Missae sacrifii*", and *ST.* III, 82, 1–5.

[117] *TL* 1, 199.

around the restorative moment of the Cross as well as its eschatological fulfillment in the timeless attendance of the divine court.

When patterned rationally, adjuvant connections become the conceptual and thematic nodes of an integrated whole. Conceptual formulations of space and matter nominate a variation of the idea that Nothingness is analogous and theoretically symmetrical to the concepts of the divine exemplar and the grace unfolding in history. This correlation of the Transcendent and the divinely initiated spatiotemporal created order is discerned throughout salvation history and perfectly manifested at the moment of transubstantiation, for: "Eternity is a circulation in which beginning and end join in unity. By the same token everything that has a ground, every truth claim that needs grounding occurs within this order, but the order is sustained by the ultimate ground which is love."[118] Eternity's relationship to the material and spatiotemporal is foreshadowed in the concept of boundless time (*ʿôlam*).[119] Difference is a necessary component if one is to discern reality's sprouting properties.[120]

Quantum informational systems and the Divine Liturgy both suggest nonlinear correlations and virtual representations of the infinite and eternal. The Mass manifests these various spatiotemporal extremes, Thomas argues, by being a commemorative and representation of the Cross, an oblation that is a means to participate in Christ's Passion.[121] There are secular analogies for the liturgy in those systems that see revolutionary acts and positions as a stimulus to historical change. Those thinkers having an allegiance to the broad tenets of the postmodern school of Speculative Realism, for example, intimate that universally differing forces organize reality by democratic, horizontal, and egalitarian forces.[122] More specifically, this marriage of certainty and interpretation clears a way to group the concept of the infinite. The Ancient Greeks conjectured that the infinite is characteristically formless—a thesis that imbues their circular depiction of history. René Guénon causally links the indefinite properties of the infinite and matter's necessary pairing of the analytically "inexhaustible" and

[118] See *TL* 1, 272.

[119] See Thorleif Boman, *Hebrew Thought Compared with Greek*, Translated by Jules L. Moreau (Philadelphia, PA: The Westminster Press, 1960), 151.

[120] See Peter Gärdenfors, *Conceptual Spaces: The Geometry of Thought* (Cambridge, MA: M.I.T. Press, 2000).

[121] See *ST.* III, 48, 6.

[122] *Philosophy and Non-Philosophy*, 45–51.

reflexivity's "perpetual multiplicity."[123] [35] Matter's spatiotemporal edifice is encompassing enough to order a being's potential "actuality."[124]

Thomas sites being's inestimable and potent nature in the concrete by distinguishing between its information or essential ordering and its material identity (*actus essendi*). The theory of hylomorphism argues that the identity of being consists of the substantial union of prior undifferentiated Prime Matter (*materia prima*) and its primitive essential imprint (*forma substantialis*).[125] According to this reading, the concept of Prime Matter (*materia*) relates to an existing being's essential and essential properties because "in order to explain a bit more distinctly how temporal, contingent, or physical truths arise form eternal, essential or metaphysical truths, we must first acknowledge that since something rather than nothing exists, there is a certain urge for existence or (so to speak) a straining toward existence in possible things or in possibility or essence in itself; in a word, essence in and of itself strives for existence."[126] This "real" distinction infinitely interprets the relationship between a being's noetic and ontic facets.[127]

The essential and reflexive properties of being are seen as congruent with the "supra-ordering" of a mathematical postulate. Thus, a being's interior-exterior identity reflectivity communicates for Baiou: "The postulate is that no signifier finds its place in a mathematical text by random chance, and that even if it is true that its mathematical character derives from its role within the formal texture of a demonstration, this texture should also be considered, in its overdetermination, as the retroactive analysis of this very non-random character."[128] A being's formal or essential definition is limitless—information can be theoretically placed in any number of interminable material architectures—when identity stabilizes

[123] René Guénon, *The Metaphysical Principles of Infinitesimal Caluculus*, Translated by Michael Allen and Henry D. Fohr, Edited by Samuel D. Fohr (Hillsdale, NY: Sophia Perennis, 2004), 118. On the soul's capacity for the infinite, see Nyssa, *The Lord's Prayer, The Beatitudes* IV, *PG* XLIV, 1244A–1248C.

[124] See Ernan McMullin, "From Matter to Materialism. … In addition, (Almost) Back," in *Information and the Nature of Reality: From Physics to Metaphysics*, Edited by Paul Davies and Niels Henrik Gregersen (New York, NY: Cambridge University Press, 2010), 33. See also *Quantum Shift*, 48–49.

[125] See *ST.* I, 44, 2.

[126] G.W. Leibniz, "On the Ultimate Organization of Things," in *Philosophical Essays*, Translated by Roger Ariew and Daniel Garber (Indianapolis, IN: Hackett Classics, 1989), 150.

[127] See Etienne Gilson, *Elements of Christian Philosophy* (Garden City, NY: Doubleday & Company, Inc., 1960), 130–31.

[128] *Theory of the Subject*, 148.

around a particular number, one experiences what J. R. Oppenheimer described as "explaining magic by miracles." Although a material being's final evolution (*natura naturata*) is indiscernible in the moment, it is also the case that conceptual and perceptual limits partially define another being's spatiotemporal boundaries. Thus, the philosopher Gilles Deleuze wages a theoretical war against permanence, the "specter of the one" that denies the hierarchical nature of an object's topology.

Unless otherwise stated, when looking at space from a localized perspective, it is idealized as a flat Euclidean space.[129] However, when local space is considered in relation to its cosmic dimensions as outlined by GTR or with its extreme quantification as part of theories of quantum gravity, the implicit question of spatial curvature is made explicit. Indeed, reality itself is irreducibly bound to the time's communication in spatial folding and unfolding; as John Archibald Wheeler and Charles W. Misner writes, "[t]here is nothing in the world except empty curved space. Matter, charge, electromagnetism, and other fields are only manifestations of the bending of space. Physics is geometry." Collectively a being's internal qualities spearhead a body's shape, for "the given is not in space; the space is in the given [;]…extension, therefore, [it] is only the quality of certain perceptions."[130] Because space is given, a being's internal qualities are externally displayed and apprehended by subjective perceptions. The subject's experience of an object's qualitative and quantitative or numerical properties depends upon one's conscious ability to distinguish and conceptualize both the corporeality and physical dimensions of other beings.[131]

Deleuze's line of thought inverts both a traditional Christian interpretation of the spatiotemporal and also risks missing the mark that the background dependent metaphysics of QMT suggests. The concept of essence rigorously categorizes the "proto-physical" or sub-physical informational

[129] See Leonard Mlodinow, *Euclid's Window: The Story of Geometry from Parallel Lines to Hyper space* (New York: Touchstone Books, 2001), 17–23 and 53–54. Hereafter cited as *Euclid's Window*.

[130] Gilles Deleuze, "Empiricism and Subjectivity: An Essay on Hume's Theory of Nature" Translated by Constantine V. Boundas (New York, NY: Columbia University Press, 1991), 91. Although it is beyond the scope of this book, Deleuze's idea of the virtual seems compatible with some aspects of my understanding of the Divine Form (*Gestalt*) and holographic information. On Deleuze's understanding of the relationship between representation and difference, see Gilles Deleuze, *Difference and Repetition,* Translated by Paul Patton (New York, NY: Columbia University Press, 1994), 117.

[131] See I. Rice Pereira, *The Nature of Space* (Washington, DC: The Corcoran Gallery of Art, 1968), 49–50.

structure of phenomenal beings in Neo-Platonic and Christian systems. The preexisting or exemplary form that individuates a physical being already presupposes the being's existence—that is, the Divine's plan for Creation guarantees its fulfillment since "existence [esse] is attributed to the quiddity [essence], not only the existence but the very quiddity is said to be created, since it is nothing before having existence, except perhaps in the Creator's intellect, where it is not a creature but the creative essence itself. … At the same time, as He gives existence, God makes that which receives existence, so He has no need to cause preexisting things."[132]

The realist would argue that the observation of the QMO presupposes its existence and analogously the faithful believe that the eternal presence of the Spirit secures the actual metaphysical existence of the real presence of Jesus Christ in the Eucharist beyond its simplistic material equivocations—the dignity of actual existence takes for granted the eternal essence of the Transcendent. No individual being or subjective perception accounts for the full meaning of eternity. Indeed, any final definition of either term forces reason's essential and embryonic properties.[133] Bulgakov highlights how the essence of time is a lived event that acts as a lens for our understanding of the eternal: "in itself, time does not exist. It is the 'subjective form' of temporality, just as temporality, in turn, is a mode of becoming, of becoming eternity or becoming being."[134] [2] Emerging out of Nothingness, temporality is a dedicated relationship within the domain of potentiality. Ulrich explains this complex theme, proposing that one should not aim to "illuminate the ontological temporality of being," as is the case in the work of many existential philosophers. This approach, he argues, suggests that being is overtaken in time. Rather, if one begins with the hypothesis that being is gifted, it is possible to define the temporal on the basis of being.[135] The faithful commit their time to participate in rituals and thus endow spatiotemporal reality with personal meaning. The Eucharist is at core "dimensionless" insofar as the groundless Spirit's holographic aura throughout reality "pushes out" the previous interior ordering principles of the bread and wine and commutes its material existence by the nonlocal Eternal Word and the historical person Jesus Christ.

[132] The interior meaning of a subject, Scholastic theologians argue, is distinguished through sensual perceptions (*proper sensibles*) involving one or more of the senses (*common sensibles*). The importance of conscious perception is confirmed in the theory of quantum mechanics and I will return to this theme.

[133] See *Homo Abyssus*, 173.

[134] *The Bride of the Lamb*, 71.

[135] See *Homo Abyssus*, 162.

References

1. Pavel Florensky, *The Pillar and Ground of the Truth: An Essay in Orthodox Theodicy in Twelve Letters*, Translated by Boris Jakim. Introduced by Richard F. Gustafson (Princeton, NJ: Princeton University Press, 1997).
2. Sergei Bulgakov, *The Bride of the Lamb*, Translated by Boris Jakim (Grand Rapids, MI: Wm. B. Eerdmans Publishing Company, 2002).
3. Hans Urs von Balthasar, *Presence and Thought: An Essay on the Religious Philosophy of. Gregory of Nyssa*, Translated by Mark Sebanc (San Francisco, CA: A Communio Book, Ignatius Press, 1995).
4. G.W. Leibniz and S. Clarke, *The Leibniz-Clarke Correspondence: Together with Extracts from Newton's Principia and Optics.* Edited by H.G. Alexander (Manchester, UK: Manchester University Press, 1956), 29.
5. Hans Urs von Balthasar, *Theo-drama: The Action*: Vol. 4, Translated by Graham Harrison (San Francisco, CA: Ignatius Press, 1995).
6. Alain Badiou, *Being and Event*, Translated by Oliver Feltham (New York, NY: Continuum, 2010).
7. Alex Ling, *Badiou and Cinema* (Edinburgh, UK: Edinburgh University Press, Ltd., 2011), 60.
8. Hans Urs von Balthasar, *Theo-logical, Theological Logical Theory: Vol. 2 Truth of God*, Translated by Adrian J. Walker (San Francisco, CA: Ignatius Press, 2004).
9. Hans Urs von Balthasar, *Theo-drama: Last Act*: Vol. 5, Translated by Graham Harrison (San Francisco, CA: Ignatius, 1998).
10. Maximus the Confessor, *Ambigua* (PG 91, 1081A).
11. Karl Barth, *Church Dogmatics* 2/1, Edited by G.W. Bromiley and T. F. Torrance, T. H. L. Parker, W. B. Johnston, Harold Knight, Translated by. J. M. L. Haire (New York, NY: Charles Scribner & Sons, 1955), 51.
12. Thomas Aquinas, *Summa Theologiae* (Garden City, NY: Doubleday Company, 1964–1976).
13. Gilles Deleuze, *The Fold: Leibniz and the Baroque*, Translated by Tom Conley (Minneapolis, MN: University of Minnesota Press, 1993), 46.
14. Alain Badiou: *Key Concepts*, Edited by A.J. Barletta and Justin Clemens (Abingdon, NY: Routledge, 2014).
15. Hans Urs von Balthasar, *Theo-Logic: Theological Logic Theory vol. 1: Truth of the World*, Translated by Adrian Walker (San Francisco, CA: Ignatius Press, 2005).
16. Heidi Ann Russell, *Quantum Shift: Theological and Pastoral Implications of Contemporary Developments in Science*, Forward by George V. Coyne, (Collegeville, MN: Liturgical Press, 2015).
17. Hans Urs von Balthasar, *Mysterium Paschale: The Mystery of Easter*, Translated and introduced by Aidan Nichols, O.P. (Grand Rapids, MI: Wm. B. Eerdmans Publishing Company, 1990).

18. Karl Barth, *Church Dogmatics: The Doctrine of Reconciliation* IV/I Edited by Rev. G.W. Bromiley and Rev. Prof. T. F. Torrance, Translated by G. W. Bromiley (London, UK: T&T. Clark International, 2004).
19. Irenaeus, The Ante-Nicene Fathers: Translations of the Writings of the Fathers down to A.D. 325, Series 1, Vol. 1: The Apostolic Fathers with Justin Martyr and Irenaeus: "Against Heresies." Edited by Alexander Roberts, D.D. and James Donaldson, LL.D. Revised and chronologically arranged, with brief prefaces and occasional notes by A. Cleveland Coxe, D.D., *Catholic Logos Edition* (Buffalo, NY: The Christian Literature Company, 1885).
20. Ferdinand Ulrich, *Homo Abyssus: The Drama of the Question of Being*, Translated by David C. Schindler (Washington, D.C.: Humanum Academic Press, 2018).
21. Adrienne von Sepyr, *Colossians*, Translated by Michael J. Miller (San Francisco, CA: Ignatius Press, 1998).
22. Thomas Aquinas, *Summa Contra Gentiles*: 4 vols., Translated and Introduced by Anton C. Pegis, F.R.S.C. (Notre Dame, IN: University of Notre Dame Press, 1955).
23. Ferdinand Ulrich. *Homo Abyssus: The Drama of the Question of Being*, Translated by D.C. Schindler (Washington, DC: Humanum Academic Press, 2018).
24. Hans Urs von Balthasar, "Persönlichkeit und Form," *Gloria Dei* 7 (1952). See *Glory of the Lord* 7.
25. Nyssa, Gregory, "Against Eunomius", Dogmatic Treatises, etc., *Nicene and Post-Nicene Fathers of the Christian Church*, Vol. 5: Second Series, Translated with prolegomena and explanatory notes under the editorial supervision of Philip Scha, D.D., LLD., and Henry Wace, D.D. Logos Catholic Edition (New York, NY: The Christian Literature Company, 1893).
26. Albert Einstein, *Correspondence Avec Michele Besso 1903–1905*, Edited by Peter Speziali (Paris, Fr: Hermann, 1979).
27. Georg Wilhelm Friedrich Hegel, *Science of Logic*, Translated by A.V. Miller (Atlantic Highlands: Humanities Press International, 1969).
28. George Florovsky, *Collected Works* 14 Vols. (Belmont, MA: Nordland, 1972), 240.
29. Bernard I. Cohen, *Isaac Newton's Papers and Letters on Natural Philosophy*, 2nd edition (Cambridge, MA: Harvard University Press, 1978), 302–03.
30. Niels C. Nielsen, Jr., "Eric's Philosophy of the Analogia Entis," *The Review of Metaphysics*, 5, no. 4 (June 1952): 603.
31. Hans Urs von Balthasar, *The Glory of the Lord: A Theological Aesthetics*, Vol. 5: Realm of Metaphysics in the Modern Age. Edited by Joseph Fessio, S.J. and John Riches, Translated by Oliver Davis, Andrew Louth, Brian McNeil C.R.V., John Saward, Rowan Williams, and Rowan Williams (Edinburgh: T&T Clarke, 1991).

32. Dean Rickles, Penelope Hawe, and Alan Shiell, "A Simple Guide to Chaos and Complexity," *Journal of Epidemiology and Community Health* 61 (2007): 933.
33. Jean Daniélou, S. J., *The Bible and The Liturgy* (Notre Dame, IN: University of Notre Dame Press, 2008).
34. John Damascene, "An Exact Exposition on the Orthodox Faith", *The Nicene and Post-Nicene Fathers*, vol. 2.
35. René Guénon, *The Metaphysical Principles of Infinitesimal Calculus* (Chicago, IL: Kazi Publications, 2002).
36. Thomas Aquinas, *Commentary on the Gospel of Saint John*, Translated by James A. Weisheipl and Fabian R. Larcher (Petersham, MA: St. Bede's Publications, 1999).
37. Augustine, *The Literal Meaning of Genesis*, Translated by John Hammond Taylor (New York, NY: Newman Press, 1982), 153.

CHAPTER 3

Holographic Matter

3.1 Holo-somatic Identities and the Material Order

Prior to reflecting on the Spirit's role in infusing the bread and wine with Christ's physical and spiritual body at the moment of transubstantiation, I turn to non-theistic interpretations of information's material transmission. Claude Shannon and Warren Weaver's seminal 1949 book, *A Mathematical Theory of Communication*, quantifies information's correspondence and transmission in physical matter.[1] Information's semantic and syntactical meaning and its logical quantification can be distinguished. Information is defined in terms of its ability to articulate difference. Mark C. Taylor explains, according to Shannon and Weaver, "information, in the strict sense of the term, is inversely proportional to probability: the more probable something is, the less information it conveys; the less probable it is, the more information it conveys."[2] [1] This differential and differing is what characterizes information and is also critical to our framing of reality by Ur-Kenosis.

Information is reproduced and disseminated by five different material properties and processes. Information is related to a (1) source, (2)

[1] See Claude Shannon, "*A Mathematical Theory of Communication*," *Bell System Technical Journal* (July and October issues 1948).

[2] *After God*, 15.

M. P. Fusco, *The Physics and Metaphysics of Transubstantiation*, https://doi.org/10.1007/978-3-031-34640-8_3

encoder, (3) channel, (4) recorder, and (5) destination. Loosely translated holo-cryptic metaphysics sees each of these stages analogously to a personal or material being's hylomorphic status, that is, existing as an informational state in a physical and spatiotemporal medium. The soul's relationship to the body is taken to be complementary to information's encoding given its formal and logical properties. Because information exists within a defined system it is taken as a physical model for the material and emergent characteristics of the Eucharist. Further, the categorization of an informational system given its source gives us a physical analogate for the idea of a Creator.

The recording and reproduction of information depends primarily on the material properties of the system. Internal and external factors can distort the physical transmission of information. It is possible to design an informational system with feedback loops and other recursive functions to help ensure the optimal transmission. The recording and transmission of a particular signal (a vinyl record, a computer disk, light, etc.) depends upon fundamentally on its physical properties, whereas the holo-somatic nature of a rational being extends beyond its immediate underlying physical processes to include an exterior reference point. Consequently, it is imperative that holo-somatic metaphysics sets out principles that verify the validity or relative truth of a particular statement or historical event.

According to holo-cryptic metaphysics, the concepts of probability and chance act as placeholders for the more traditional notion of potency. Individual freedom and choice finds an analogy in the emergent behavior of artificial intelligent programs' use of probability and appropriation of open source information. The conscious being's self-reflective nature is ultimately meaningful, I argue, given their implicit correspondence to the Transcendent. Shannon's theory of information promotes a particular reading of final causality, as information transmission includes a destination. Again, in order for information to reach its endpoint, the respective material chosen to transmit the signal must be adequate for its purposes. In terms of the rational and free person, holo-somatic metaphysics argues that information is fundamentally essential in nature and thus depends upon a receptive interior mental space, physiological substrate, and neuronal architecture necessary for memory. Analogous to computer memory, human beings translate their sensual experiences or data into images, which are stored as memories.

Universal and fundamental laws demand one takes seriously physical forces and hence causality. Mechanical recursion models a subject's biological reflexive functions and anticipates her self-reflective capacities. Shannon's theory of information implicitly reworks previous hylomorphic theories of causality, insofar as an encoder serves the purposes of efficient causality, the material elements of information translate how the material cause relates to an essential or information matrix in the storage device. The information's structural language analogously mimics what past philosophical systems attributed to formal causality, and information's required destination in correspondence to its source gives us a new way to conceive of final causality.

Luciano Floridi explains, "the word information relates not so much to what you do say, as to what you could say. The mathematical theory of communication deals with the carriers of information, symbols and signals, not with information itself. That is, information is the measure of your freedom of choice when you select a message."[3] The mathematical theory of communication deals with the logical formalization of information by means of symbols and signals. [2] Information is adjudicated in symbols, images, and propositional statements that encode its semantic meaning and, in so doing, exhibit how the probable works with a specific functional architecture.[4] "Man never grasps the essence of things," Ulrich points out, "in the act of simple apprehension." [3] The binary set of 0 and 1, for example, identifies a QMO's informational state as a 50 percent chance that it will exist as a wave and a 50 percent chance that it will exist as a particle prior to its determination by observation.[5] Prior to this momentary removal of its potential identity, the QMO remains free to reveal as either state at each moment in time. Potency or its metaphysical rephrasing as ontologically differing finds its analogous manifestation in the QMO's probabilistic form. Beyond this, the two symbols of 0 and 1

[3] Luciano Floridi, C.E. Shannon and W. Weaver, The Mathematical Theory of Communication, (Urbana, IL: Illinois Press, 1949. Reprinted in 1998) quoted in Luciano Floridi, *Information: A Very Short Introduction* (Oxford, UK: Oxford University Press, 2010), 45. See also Seth Lloyd, "Computational Capacity of the universe," *Physical Review Letters* 88 (2002): 237901. Hereafter cited as *Computational Capacity*.

[4] See Arthur I. Miller, *Imagery in Scientific Thought* (Cambridge, MA: MIT Press, 1986), 225. See also Stephan M. Kosslyn, *Image and Mind* (Cambridge, MA: Harvard University Press, 1981).

[5] The so-called Pauli exclusion principle argues that no two electrons can be in the same stationary state. See Jeffrey Bub, Why the Quantum? *Studies in the History and Philosophy of Modern Physics* 35, no. 2 (2004): 241—266 and Rob Clifton, Jeffrey Bub and Hans Halvorson, Characterizing quantum theory in term of information-theoretic constraints, *Foundations of Physics*, 33 (2003): 1561–1591.

will be used as a shorthand symbol for all such philosophical conceptualizations of the extreme termini of complementary relationships such as being and nonbeing, truth or falsity.[6] These conceptual extremes are paradoxically unified by their shared identification with the concept of difference. As Ulrich explains, the free being's "appearance is negated in its multiplicity by the *indivisum*, as negation it is sublated and preserved in it."[7] [3] The holo-somatic principle argues that truth corresponds to relational differing just as material being can be codified given its relationship to causality or "motion."

Boolean algebra equips us with a mathematical notation to logically encode information (ideas). Mathematical operations are taken to be examples of specialized concepts of "motion" that are broadly directed by inferential and deductive differing. From basic logical propositions, a theoretical edifice can be built. There are two principles upon which such a mathematical structure is constructed, "on the one hand [for] symbolizing not only the notions used in the traditional branches of mathematics but those used in all deductive reasoning; and on the other hand [for] formulating explicitly the permissible rules of inference. This means that every inferential step can be (a) represented by the transformation of one or more symbolic expressions into another and (b) justified by appeal to clearly formulated rules."[8] [4]

Shannon mathematically defines information as a logarithmic distribution within a closed formal system, a "probability cloud" of sorts that gives one the option of seeing individual data as encoded "bits" that can be symbolically represented and physically stored. Theoretically, a single being or an infinite number of beings can be symbolically represented in hyper-dimensional geometries or in an infinite series. A symbol's ability to be represented and inhere in any number of material substrates gives information a variety of modes to communicate meaning. In each instantiation of information a new appropriation of relational differing is employed. Consequently, it is impossible to fully explicate the meaning of a being or information given its unique scientific, mathematical, or sacramental presentation and formalization. However, our ability to measure a being or informational state gives us a proportionate actualization of its meaning. A being and its truth is complementary to its informational transmission and conscious manipulation for: "if the object were indeed reducible to

[6] See *A Beautiful Question*, 180–185 and Nyss, *Against Eunomius*, *NPNF5*, Book 1, 22.
[7] *Homo Abyssus*, 401.
[8] *The Philosophy of Mathematics*, 34.

the representation, it would obviously not be subject to measurement; a mere model, after all, does not affect our instruments. Physical objects, on the other hand, evidently affect the appropriate instruments of measurement—by definition, if you like; this means that they have a certain existence of their own. The passage from representation to object, therefore, constitutes an intentional act no less enigmatic, certainly, than the humble act of sense perception."[9] [5]

The Heisenberg Uncertainty Principle argues that the experimenter's observations must be accounted for in an experiment, as they are influential to the final results. Experiments conducted with C.E.R.N.'s (*Centre Européenne de la Recherche Nucléaire*) particle accelerator (LEP, Large Electron-Positron Ring) show that how one represents information is critical. The success of C.E.R.N.'s computers in diagramming the hidden subatomic world depends upon the suitability of its algorithms. The collecting, measuring, and digitization of the information that results from particle-accelerator experiments must also be incredibly accurate and have minimal discrepancies. Conscious and instrumental measures remain impartial and probable in nature.[10]

A being never fully discloses its truth at any one time, hence every person remains partially a mystery. Knowledge is always in a state of revision. The hermeneutical approach one adapts must benefit as best as possible on the conceptual level with the specific manner of a being's symbolic representation. Heisenberg's insight is prescient when he states, "we have to remember that what we observe is not nature in itself but nature exposed to our method of questioning."[11] [6] Heisenberg reminds us, we must distinguish between a physical object, its manner of observation, and its formal explanation in a nondeterministic fashion.[12] QMT's explanation of being's relationship to probability brings these three elements together, as Heisenberg confided to Wolfgang Pauli when he stated, "I am truly convinced that interpreting [Bohr's theory] in terms of circular and elliptical orbits in classical geometry makes not the slightest physical sense, and my whole effort is to destroy without a trace the idea of orbits that cannot be

[9] Smith, 32.

[10] See Lee Smolin, *Three Roads to Quantum Gravity* (New York, NY: Basic Books, 2001), 146ff.

[11] Heisenberg, *Physics and Philosophy*, 58.

[12] See Niels Bohr, *Atomic Theory and the Description of Nature* (Cambridge, MA: Cambridge University Press, 1934), 116. See also Nicolas Gisin, "Non-Realism: Deep thought or a soft option?" *Foundations of Physics* 42 (2012): 80–85; and for a theological perspective, see *AE*, 168 and 192.

observed and to replace them with something better".[13] [7] The QMO's material state and formal identity is best described as a probability cloud or one might say, "chance" location.[14] The quantum particle's location prior to observation is uncertain and probably fluid afterward.[15] One can only understand the universe's informational state by accounting for the probable nature of the quantum structure, the QMO's location, and the laws of entropy.[16]

Neither Newtonian physics nor Daltonian atomism help us in regard to how the meaning of QMOs can change with regard to the alteration of their appearance or their non-local communication. Critics have also questioned whether postmodern systems such as Badiou's mathematical ontology gives attention to being's empirical conditions at the expense of promoting an unsupportable conception of being.[17] Mathematics and science use specific criteria to determine an object's empirical properties. The material and relational characteristics of a being are isomorphic to its logical structure and mathematical demonstration. In addition to a common sense of physicalism, Newtonian and Daltonian approaches share something that approaches a "no-gap" deterministic view of the physical world. This position echoes Pierre Laplace's understanding of the material and formal elements of causality. Specifically, his theory of the resultant predictability of physical reality leaves only a circuitous route for the transformations of the sort proposed by the miraculous transubstantiation and quantum physicists. Laplace views physical reality as a computing demon of sorts.[18] Among other difficulties that Laplace's hypothetical vision creates, it leaves little room for a divinely appointed Creation or those intri-

[13] Wolfgang Pauli, "*Scientific Correspondence with Bohr, Einstein, Heisenberg*," Edited by K. von Meyenn, et al., 2 vols., (Berlin, DE: Springer Verlag, 1979–), 231.

[14] See Peter Galison, "Images Scatter into Data, Data Gather into Images," in *Iconoclash*, Edited by Bruno Latour and Peter Weibel (Cambridge, MA: MIT Press, 2002), 306.

[15] See A. Hájek, *Interpretations of Probability*. Retrieved from the *Stanford Encyclopedia of Philosophy*: http://plato.standford.edu/archieves/win2012/entries/probability-interpret (2012), and T. Childers, *Philosophy and Probability* (Oxford, UK: Oxford University Press, 2013).

[16] See Caslav Brukner, "Questioning the rules of the game," *Physics* 4, no. 55 (2011).

[17] See Adrian Johnston, "What Matter(s) in Ontology: Alain Badiou, the Hebb—Event, and materialism split from within," in *Angelaki* 13, no.1 (2008): 28. See also Adrian Johnston, *Prolegomena to Any Future Metaphysic: vol. 1: The Outcome of Contemporary French Philosophy* (Evanston, IL: Northwestern University Press, 2013), 81–107.

[18] See Jeffrey Koperski, *The Physics of Theism: God, Physics, and the Philosophy of Science* (Singapore, SG: John Wiley Sons, 2015), 159–160. Hereafter cited as *The Physics of Theism*.

cate emergent informational webs that form the basis for random variation in Nature and human freedom. From my perspectives, I can say that such computational juggernauts as advocated by Laplace seem incompatible with the conceptual conveyance of the idea of Nothingness or the void that seem operative in QMT.

The proverbial issue of supernatural engagement with an ordered and semiautonomous universe traditionally finds traction in the theists' apologetic efforts and/or the religious philosophers' preview. In his letter to Samuel Clarke, the polymath Leibniz excoriates those scientific interpretations that allot space for divine miraculous action or what seems to amount to an imperfect evolutionary system.[19] Leibniz could not have imagined how GTR, STR, and QM would challenge Laplace's theoretical identification of the Divine with a hypothetical cosmic watchmaker. The mathematics employed by QMT, for example, gives us a way to move beyond the limitations of materialistic determinism, classical logic, and Euclidean mathematics. However, it t is possible to see how numbers in Laplace equations can be developed into holomorphic functions that metaphorically represent how beings like complex numbers holographically exist in multidimensional space.[20]

The quantum world of holographic realities, non-locality, action at a distance, entanglement, multidimensional realities, uncertainty, randomness, info-somatic or quantum logical states, and so on introduces us to a very different world than that proposed by classical physics. Mutuality between a physical entity's observation and its theoretical explanation is possible in part given the pervasive and multiple manner of information's virtual consignment as an event. With their unique perspectives and history, personal knowledge is a relative interpretation of being and truth, between what we know and the changing ontological state of affairs. This is not to suggest that the concept of absolute truth that exists for some philosophers and theologians is not related to those universal laws advocated by mathematicians and scientists. I believe that this alignment around truth and being follows from every methodological approach having either an implicit or explicit position on the concept of Transcendent. Predominantly, scientists do not recognize the Transcendent in their

[19] See G.W. Leibniz and S. Clarke, *The Leibnitz-Clarke Correspondence: Together with Extracts from Newton's Principia and Optics*, Edited by H.G. Alexander (Manchester, UK: Manchester University Press, 1956), 11–12. See also *The Physics of Theism*, 148–151.

[20] See Douglas Hofstadter and Emmanuel Sander, *Surfaces and Essence: Analogy as the Fuel and Fire of Thinking* (New York, NY: Basic Books/Perseus Books, 2013), 445–451. Hereafter cited as *Surfaces and Essence.*

theories, but this concept preserves in the idea that physical laws are immutable and universally hold for all physical reality in the universe. Likewise, the notion of Ur-Kenotic Divine Personhood gives us a way to understand simplicity just as non-locality and entanglement define the QMO's probabilistic identity. For "Entanglement is an especially damaging counterexample to reductionist views. Quantum mechanics is a fundamental theory. There is no scenario in which this quantum holism will be reduced away by some future theory."[21] [46]

We follow John Conway and Simon Kochen's lead in this regard, who state "we use the words 'properties,'' 'events' and 'information' almost interchangeably, whether an event has happened is a property, and whether a property obtains can be coded by an information-bit."[22] [8] While we can, as John Wheeler and K. Ford write, arrive at an "it from bit," the full expression of any informational or existential state is impossible.[23] The spatiotemporal explications of information deny its full observation and complete expression for at a given moment the "maximal knowledge of a total system does not necessarily include maximal knowledge of all its parts."[24] [9] The symbols 0 and 1 represent the subatomic particle's "ontological shape" or the truth of a quanta's interior structure. The accuracy of the QMT follows from the irreducible geometric properties of the QMO. This constant gives us a way to distinguish or "kenotically" differ between the QMO and its theoretical modeling and further between its symbolic representation and its empirical manifestation. As Niels Bohr writes, for "[t]he renunciation of pictorial representation involves only the state or atomic objects, while the foundation of the description of the experimental conditions is fully retained."[25] [10] Subjective observation, measurement, and knowledge mean that the quantum's spin, for example, results from a statistical averaging of the sum of the electron's distribution of energy with respect to the QMO's velocity. An analogy to listening to music on a subway car may clarify the issues at play at the subatomic level. Even with noise-canceling headphones, it is impossible to differentiate the

[21] *The Physics of Theism*, 231.

[22] See John Conway and Simon B. Kochen, *The Strong Free Will Theorem* (Princeton, NJ: Princeton University, 21 Jul., 2008), 5. Hereafter cited as *The Strong Free Will Theorem*.

[23] See John A. Wheeler, J. A. and K. Ford, "It from bit," in *Geons, Black Holes Quantum Foam: A Life in Physics* (New York, NY: W.W. Norton Company, 1998).

[24] Giulio Chiribella, Giacomo Mauro D'Ariano, and Pablo Perinotti, "Informational derivation of quantum theory," *Physical Review* A 84.1 (2011): 012311.

[25] Niels Bohr, *Atomic Physics and Human Knowledge* (New York, NY: Wiley, 1958), 90.

sound of recorded music fully from the surrounding environmental "noises." The quality (e.g., precision) and quantity of musical notes heard (e.g., data stored) and the manner of its retrieval and differentiation from "white noise" suggest something akin to a QMO's distribution of energy within a system.

The conscious subject is able to mentally locate or place a being or object given the senses' ability to see differences among beings. Memory makes it possible for the subject to conceive of these changes in appearance as a rudimentary temporal marker. Human imagination makes sense of individual differences in terms of various whole-part relationships. Memory holds a record of an object's accidental iterations over time as images—a *conversio ad phantasma* or conscious hologram. The object's ongoing development or "attunement" with internal and external influences plays a critical role in the grounding of human consciousness beyond its mere recognition of an object's physical nature.[26]

The virtual portrayal of physical beings plays a critical role in consciousness and hence in every theological and scientific theories. The reproduction of a signal, for example, involves more than the formal replication of its informational content. Knowledge transference includes an explicit or implicit reference to the efficient, material, formal, and final "identities" of the transmitted signal—for information and cognition have genealogies. I am not suggesting that a crude equivocation between reductive abstractions of the essential content of a being over and above its need to continually confirm its actual existence in the world is desirable or even possible. No known physical force can hold the universe's entropic tendency at bay. However, the holograph gives us a concrete example of how essential identity acts as an analogy for the permanence of individual holo-somatic identity. The holo-somatic principle integrates elements from both approaches, as the prefix "holo-" looks to a being's essential properties, and "somatic" incorporates the physical and material components of a being's identity. The concept of holo-somatic identity entails holding in conscious tension a being's ontological and essential properties. As a thought experiment, disciplinary articulations of either term can be unfolded.

It is possible to juxtapose theological and scientific representations of a being to draw out the preexisting proportion or symmetry between them.

[26] See Joseph Cardinal Ratzinger, "Message of His Holiness Benedict XVI to Archbishop Rino Fisichella".

One could take a theological interpretation of a being's non-subsistence relations (as represented in the qualitative "holo-properties" of being) and connect these to their scientific or physical quantification (e.g., their quantifiable somatic properties). In this way, it is possible to holo-somatically consider an object's essence from a philosophical and theological perspective while simultaneously explicating its physical properties as directed by QMT. Concurrently, the holo-somatic nature of being and information spawns a new metaphysics of being and its networked or collected influences and thereby revisits the concept of *natura naturata*.

On the surface, it is true that lasers (Light Amplification by Stimulated Emission of Radiation) can be used to produce what appears to be a "materially insubstantial" three-dimensional model of a phenomenal object. I place "material insubstantial" in quotation marks as light exists in physical reality as either a particle (photon) or a wave. However, light can take on a particular shape without a requisite mass. It is true that the holograph does little on first glance to represent the actual "density" of a material body. While this is correct given the complete translation possible between matter and energy, it is also true that a holograph can be seen to present the "energetic signature" of an object's external appearance and internal states. Laser light would be the envy of the Fauvist painter because it is coherent or "pure." An object's interior properties are understood as pure given they are seen as the delineation of coherent information represented by a series of numbers or points. This data can be translated into frequencies of light when patterned together as a holograph giving it dimensionality. The holo-somatic nature of identity fosters in the quantification and communication of information as a holograph and its symmetrical organic realization as mental images.

We epitomize our experience through those specialized "holographs" termed mental images, symbols, or signs. A sign, Origen explains, is something visible suggesting the idea of another visible thing (*signum dictum cum per hoc quod videtur aliud indicatur*).[27] The holographic measure is a suitable investigation of the strange worlds of transubstantiation and QMT because it allows us to conceive of formal explanatory systems that define a being's action in the universe. Thought of as a representational

[27] See Ambrose *On the Mysteries*, *NPNF2*, vol. 2, Book 10, Chapter 3, section 15, Translated and Introduced by Philip Schaff D.D., LL.D, and Henry Wace, D.D. (Grand Rapids, MI: Wm. B. Eerdmans Publishing Company, 1979) and Augustine *On Christian Doctrine*, *NPNF1*, Book 1, chapter 2, Translated and Introduced by Philip Schaff D.D., LL.D, and Henry Wace, D.D. (Grand Rapids, MI: Wm. B. Eerdmans Publishing Company, 1983), and *ST.* I, 85, 5.

being, the invisible life of the QMO is visible when understood as part of the larger system, just as the event of transubstantiation is only rational if placed within the larger schema of the Incarnation and salvation history. The communal nature of images follows, as no final definition has immunity from subjective interpretation or can escape its rolling non-subsistent or its probabilistic serialization by others. The universal meaning of a word or thing (*res*) partially disappears in each subjective interpretation. The proscribed meaning of a thing changes over time. Symbolic representations are intelligible when "flexible," that is, open to a meaningful agreement between a subjective being's conscious appropriation and its formal description.[28] [11] Meaning is generated in the subject's creative convolutions that regulate over time. The perceived object inspires an exotic response, but once placed in the existing effluvium of signification, its meaning "ages" and stabilizes. This process is part of the *ekphrasis* upon which civilization depends.[29]

A religious sign or "holograph" hopes to awaken an individual to the noumenon's relationship to the Transcendent.[30] The Eucharistic celebration is a dramatic retelling of a historical event open to subjective experience and is not, therefore, a simply repetitious chorus of sterile words and symbolic reifications of a deterministic interpretation of matter. The Eucharist is the real presence of the divine and when understood as a "real symbol," it is a representation that awakens the created order's spiritual relationship to the Transcendent, a sign or signal that pertains to the spiritual or sacred—a "*signum ad res divines pertinens.*"[31] The eternal marriage of the Transcendent and the immanent is satisfied in the person of Jesus Christ, the terminus and ontological unifier for all other Christian signs and signifiers. As Chrysostom states, "[f]or Christ hath given nothing sensible, but in things sensible all is intelligible." Holography facilitates an explanation of transubstantiation through the lens of QMT and Aristotelian

[28] See *The Pillar and Ground of the Truth*, 61, and John S. Bell, *Speakable and Unspeakable in Quantum Mechanics* (Cambridge, UK: Cambridge University Press, 1987), 152.

[29] See Friedrich Ohly, "On the Spiritual Sense of the Word," in *Sensus Spiritualis: Studies in Medieval Significs and the Philology of Culture*, Translated by Kenneth J. Northcott and Edited by Samuel P. Jaffe (Chicago, IL: The University of Chicago Press, 2005), 9.

[30] See Augustine, *The Letters of Saint Augustine*, "Letter 138: From Augustine to Marcellinus," *NPNF1*, vol.1, 7, Translated and Introduced by by Philip Schaff D.D., LL.D, (Grand Rapids, MI: Wm. B. Eerdmans Publishing Company, 1983). Revised and edited for New Advent by Kevin Knight. http://www.newadvent.org/fathers/1102.htm.

[31] See Gabriel Biel, *Canonis missae expositio, pars prima*, Lectio 27K, Edited by Heiko A. Oberman and William J. Courtney (Wiesbaden, DE: Franz Steiner, 1963), 265.

hylomorphism, that is, phenomenal reality's union of "form" and "matter."[32]

Sacramental theology has to argue the difficult point that at the consecration of the substances of the offertory species of bread and wine are replaced or become the divine substance without dislodging their accidental appearance in the believer's mind. The concept of "hologram" gives us a way to understand the tension of substantial matter being sacramentally "rewritten" and "perfected" at the consecration.[33] Past theologians' view that the Eucharist is a divinely instituted sacramental sign is commensurate with my hylomorphic and holo-somatic modeling of this mystery.[34] When the hylomorphic form-matter unity is rejected, one risks divorcing the natural order from its liturgical and eschatological conclusions. As Graham Ward comments, "Only God can make a copy of Godself, but human beings nevertheless are *homo symbolicus* because each is constituted as an image; each represents, incarnates, something of God."[35] Stated otherwise, where previous explanations argue that the underlying substances of the bread and wine are changed to signify sacramentally (*in signo*) the real presence of Jesus Christ, my approach argues for the holo-somatic image as a theologically and scientifically valid way to speak about being's representation or "measure" in reality.[36]

Matter's symbolic capacities drive my belief that a finite being can be analogously understood as a religious icon writ in the metaphoric "paint" of non-subsistent potentialities and relational differing. The person of faith intuits the essential identity of material beings and understands that this core reality emerges and is perfected in the human person. The way in which the artist and theologian approach the question of image is instructive to how they conceive of the nature of the essential. "Damascene calls the artist a portrayer of living forms [*zographon*] and associates him with a divine calling, apostleship [*aposteilanti*]. The artist is to make a portrait

[32] See Hampus Lyttkens, *The Analogy Between God and the World: An Investigation of Its Background and Interpretation of Its Use by Thomas of Aquino* (Uppsala, S: Almquist and Wiksells Boktryckeri AB, 1952), 170.

[33] See *Covenantal Theology*, 484.

[34] See *ST.* III, 75, 1–3 and Augustine, *The Exposition of the Psalms*, *NPNF8*, Psalm CVII, 1 and Psalm CXCIX, 1.

[35] Graham Ward, "The Beauty of God," in John Milbank, Graham Ward, and Edith Wyschogrod, *Theological Perspectives on God and Beauty* (Harrisburg, PA: Trinity Press International, 2003), 43. Hereafter cited as *The Beauty of God*.

[36] See *Sources*, *DS* 1651.

[*homoiographesai eikona*], not to make a copy [*poiesasthai mimema*]."[37] The icon materially actuates eternal prototypes as aesthetically pleasing forms. The writing of an icon does not look to be a photo-realistic representation—the spiritual image is always more. Florensky claims, for example, that the icon rejects arbitrary explanations in terms of becoming, change, innovation, or any attribution to the static properties of its two-dimensional transcription. Instead, as Florensky writes, "[t]he aim of art is the overcoming of sensual appearance, the naturalistic crust of the accidental, and the revealing of sensual appearance, the naturalistic crust of the accidental, and the revealing of that which is stable, unchanging, and has a general significance and general value in reality."[38] [12]

The religious icon testifies that it is not only possible but necessary that the spatiotemporal and material orders are spiritually infused in the "here and now of [the believer's] ontic disposability."[39] Catholic dogma states that the Eucharistic grace bestowed into the bread and wine at the consecration spiritually anticipates the eschatological sanctification of all created being. The essential truth of the Eucharist is referenced to in the religious icon and both are holo-somatic revelation of the material order's spiritual perfection. The heavenly banquet like Creation's present and future redemption is taken as a historical fact remembered at the liturgy. Participating in the liturgy allows us to share in the memory of the divine artistry that radiates and retroactively images the paradisal state in the eschatological reality of past, present, and future. This historical unity is made possible with the Incarnation that brings the Transcendent into history and testifies to the "all [is] in all" (1 Corinthians 15:28). The icon symbolically and aesthetically "recalls" the factual certainty of divine truth in the spatiotemporal realm. Consequently, perceiving and praying before an icon and the Eucharist are devotional experiences.[40] The priest and the congregation establish that the consecrated bread and wine "virtualize" the infinite graces radiating from Christ's assimilation of the sufferings in his sacrificial perfection on the Cross (Hebrews. 10:1–18). Christ's *pro nobis*

[37] Graham Ward, *The Beauty of God*, 45.

[38] Pavel Florensky, *Iconostasis*, Translated by Donald Sheehan and Olga Andrejev. Introduction by Donald Sheehan (Crestwood, NY: St. Vladimir's Seminary Press, 1996), 44. Hereafter cited as *Iconostasis*.

[39] See *God Without Being*, 170–171.

[40] See *Iconostasis*, 73.

(*ex opere operato*) holographically and historically presents the Transcendent for grace moves "from above downward."[41] The liturgy recalls for believers, Christ's oblation as a rememorative representation (*rememorativam repraesentationem*) and in so doing testifies to memory's endurance against all modes of sensual, cognitive and aspectual dissolutions.

Religious signs and ritual practices providentially guarantee that participants can virtually relive Christ's death in the liturgy. Relational differing and non-subsistence are transfigured in the material properties of the bread and wine to become part of reality's fundamental trans-annihilation or kenotic participation in Jesus Christ's person and mission. The statistical averaging employed by human measure is perfected as it now conforms to Jesus Christ's divine-human measure of the Transcendent. For "[b]y means of the Eucharist, he distributes his death, spilling it as life into the womb of the Church."[42] Repeating Christ's words, the priest consecrates the bread and wine and thus holo-somatically centers this dramatic replaying of final death into the divine substance of Jesus Christ's resurrected body.[43] The eternal echo of Christ's words is made possible given the proportionate and proximate power of the Spirit's safeguarding of human freedom as part of truth's universal and creative force. St. Ambrose rhetorically questions and answers "For that sacrament which you receive is made...by the words of Christ. But if the word of Elijah had such power as to bring down fire from heaven, shall not the word of Christ have the power to change the nature of the elements? You read concerning the making of the whole world: 'He spoke and they were made.'"[44] [13] The Eucharist embodies a virtual physiognomy (*rationes*) of pure Transcendent simplicity which is at its core "dimensionless" insofar as the Absolute Groundless Spirit brings into spatiotemporal and material existence the nonlocal communion of the Transcendent Son in the person of Jesus Christ—for "God is not a measure proportionate to the things measured."[45] [14] Similarly, mathematical truth does not merely describe a number's "meta-ontological" form but

[41] See Pseudo-Chrysostom, *Paschal Homilies*, *PG*, LIX, 735 and P. Eric Przywara, *Polarity: A German Catholic's Interpretation of Religion*, (London, UK: H. Milford Company, 1935), 34. Hereafter cited as *Polarity*.

[42] *TD* 3, 429–430. See also *TD* 4, 359.

[43] See Augustine, *On The Gospel of John*, *NPNF7*, Tractate, XCIX, 7.

[44] Ambrose, *On the Mysteries, NPNF2*, Chapter IX, 5. On the role of Christ speaking "through" the priest at the consecration, see also Gabriel Biel, *Canonis misse esposito*, Lectio 85 H (Oberman and Courtney, ads. pars quanta [Wiesbaden: Franz Steiner, 1967]), 103.

[45] *ST.* I, 13, 5, 3.

pronounces something about the event of a physical being actualization that fundamentally changes reality.[46]

It is possible to distinguish between a subatomic particle and its identity in the cosmos' "external" boundary or in a quantum wave's holographic pressuring of the edge of the universe with its presence. Thermodynamics is a suitable scientific quantification for how holo-somatic identity can be understood as local phenomenon or abstracted universal law. A being's dynamic behavior within a system gives theorists great predictive power given the theory of thermodynamics' ability to touch upon many of the hallmark characteristics of a metaphysical system. As Einstein says, "[a] theory is the more impressive the greater the simplicity of its premises is, the more different kinds of things it relates, and the more extended its area of applicability."[47] [15] The holographic principle follows the laws of thermodynamics, as the spatial metrics of the quantum particle includes an averaging of its subatomic motion.[48] Consequently, holo-somatic theory is logically consistent with the topological boundary of the quantum wave's (ψ) formal symmetry with the entire universe. Badiou's insights into the nature and metaphysical density of a number are applicable. This is mathematically shown by taking the number "1" as a vestige boundary for a complete being or the entire universe. With this move, it is possible to identify a discrete quantum particle's location in physical and complex space. The local quantification of a QMO's place in the universe is conventionally represented by the equation $\langle\psi|Pv|\psi\rangle \approx 1$. Congruent to the laws of thermodynamics or quantum electrical dynamics (QED), the independent quantum is nearly isomorphic to its collected identity within a field or system.[49] Consequently, over time, the microstates of a quantum wave

[46] See *Being and Event*, 8, *Briefings on Existence*, 40–41, *Theoretical Writings*, 4–14 and for an example of the functional role and applicable value of meta-principles for science, see Stephan Edelston Toulmin, *The Philosophy of Science* (New York, NY: Harper Row, Harper Torchbooks, 1953), C. Liu. "Gauge invariance, Cauchy problem, indeterminism, and symmetry breaking", *Philosophy of Science 63 (Supplement), Proceedings of the 1996 biennial meetings of the philosophy of science association, Part I: Contributed papers* (September, 1996): 571–579.

[47] Albert Einstein, *Autobiographical Notes*, Edited by P.A. Schlipp, *Albert Einstein: Philosopher-scientist* (Evanston, IL: Library of Living Philosophers, Inc., 1949), 32.

[48] See N. Sieroka and E.W. Mielke, "Holography as a Principle in Quantum Gravity? Some Historical and Systematic Observations," *Studies in History and Philosophy of Science Part B: Studies in the History and Philosophy of Modern Physics* 46 (2014): 175, Allori et al. (2008), 363 and Herman Weyl, *Space, Time, Matter* (London, UK: Methuen Press, 1922).

[49] What follows develops upon the seminal work, "Holography and Emergence," 300–311. See Appendix B.

works toward nearly duplicating the macro-state of its thermodynamic, topological, or holographic boundaries and ultimately mirrors the universe as a whole. Russell compares the fulfillment of this potential with the collapse of the quantum wave.[50] Thus, holo-somatic identity acts as a being's essential and existential differing as an equilibrium point in the physical universe as an image of what lies beyond it.

The hologram here is likened to the quanta's mathematical equivalence as a dense matrix within the system as a whole. Thus, the QMO's probability cloud equates to the state's equilibrium ($\hat{W}$) or the informational status of a collection of atoms within a thermodynamic system. This density field can be understood as the quantum wave's condensation prior to its collapse as a particle when observed in three-dimensional spatiotemporal reality. Immediately prior to its collapse, one can argue that the wave would ripple forever mirroring physical space—to the end of the universe, if such an end exists.[51] This thought experiment gives us a way to see in the quantum wave a retroactive image of the identity of the universe as a whole and as a multidimensional medium for the communication of motion (energy). Something akin to this genealogical placing of a particle's identity is co-opted in the theory of the Big Bang. What is important is that the holographic principle gives us an accurate way to approximate a QMO at a particular location at a stated time; in short, the mystery of being is reduced to its best resolution. Thus, one can see in a holo-somatic metaphysics based on a physical being's motion a way to theologically and scientifically resolve the intractable mereological problem.

Truth is always "dilated, dispersed and distributed" in various representations and idioms."[52] [16] The temporal response to space's contraction by proportional frames of reference (e.g., distending) and natural human limitations, Ulrich points out, becomes the source of one's encounter with Jesus Christ and the personal growth that this engenders.[53] Physical reality hyper and supernatural dimensions promise that there is more to every immediate perception. Believing that the universe objectively exists prior to its observation or its conditional quantification inherently endorses a

[50] See *Quantum Shift*, 55.

[51] See John C. Polkinghorne, *Quantum Physics and Theology: An Unexpected Kingship* (New Haven, CT: Yale University Press, 2007), 25–26, 36–38, and 44–46.

[52] Barbara Maria Stafford, *Echo Objects: The Cognitive Work of Images* (Chicago, IL and London, UK: The University of Chicago, 2007), 93. Hereafter cited as *Echo Objects.*

[53] See *Homo Abyssus*, 175–176, 228 and 357.

form of metaphysical realism that does not collapse being along radically deterministic lines of inquiry. Furthermore, all beings mentally catalog existing physical and mathematical events in the universe as an antidote to radical subjectivism. The internal essential identity of a being and its representation is not realized in immediate perception, for to do so would make subjective truth a Promethean absolute that succumbs to postmodern philosophy's own correlational theory of being.

My concept of holo-somatic structure develops theologically in light of science's own use of holograph. Theoretical physicists see holography as a way to model physical reality's primitive structure. The holograph gives us a way to discuss matter's quantification as an image without completely jettisoning what is theologically fundamental to hylomorphism. Many advanced physical theories argue that holographs are equivalent to physical reality's informational identity. Black hole entropy shows just this—that a holograph embodies a being's "essential" information as a pattern, shape, topology, or configuration space. The laws that describe its behavior in ideal conditions elucidate information about a being or object. The Law of the Conservation of Energy works in tandem with the logic of thermodynamics if one agrees that all cosmic energy existed prior to the Big Bang or the totality of its potential derives from an irreducible cosmic constant (e.g., Nothingness).[54] This gives us at the very least a theoretical basis for the maximum motion (speed of light or c) of a being in a physical system.

The laws of electromagnetism and quantum metaphysics verify that the universe exists in relationship to nothingness via its progressive disorder and entropy.[55] An electromagnetic force has theoretically unlimited degrees of freedom and thus suggests that this property is shared with QMT's probability calculus and mathematical arrogation.[56] In simple terms, the spatiotemporal movements of the QMO and its electromagnetic descriptions and algebraic identity are balanced. In this spirit, Hermann Weyl (1885–1955) recognizes that a compass needle's electromagnetism is "a reflection with respect to the plane in which current and needle lie, maps the current into itself, but interchanges the north and

[54] See *Quantum Shift*, 94.

[55] A primitive principle of emergence operates throughout reality. For example, in String Theory, the most primitive symmetries in physical reality are generated from Nothingness.

[56] See Max Tegmark, "The Mathematical Universe," Found. Physics (Sept. 2007):1–13. Revised, see p.9, from: arXiv:0704.0646v2 [gr-qc]. 8 Oct 2007. Hereafter cited as *The Mathematical Universe*.

south poles of the magnet."[57] [17] The quantum particle's spatiotemporal momentum and its relative energy circumscribe physical matter. The Law of the Conservation of Energy's dialectical correspondence to the Second Law of Thermodynamics formally accounts for another description of matter's directional motion and energy's dimensional progress.[58] Physical matter manifests in the universe under the influences of disorder, chaos, order, and entropy. Paul Davies simplifies the Second Law of Thermodynamics to say that the universe has a "natural tendency for order to give way to chaos."[59] [18] Ludwig Boltzmann's insights into thermodynamics give us an important example of this process.[60] His discoveries about the nature of entropy are key to my own theory of holo-cryptic metaphysics and transubstantiation. Planck's constant defines the vacillation of the QMO as its essential measure or quantification. Boltzmann notes, "I cannot really imagine any other law of thought than that our pictures should be clearly and unambiguously imaginable." Students are often introduced to the counterintuitive relationship between the world of quantum action and sensual experience by considering thermodynamic states. They are often asked to imagine gas particles as marbles bouncing around inside a balloon. My thought experiment can be extended to include the identity of the human person or the Eucharistic species, insofar as "all physical phenomena can be completely reduced to movements of invariable and similar particles or elements of mass."[61] [19] The nature of the internal motion of the consecrated species is not immediately

[57] Herman Weyl, *Philosophy of Mathematics and Natural Science* (Princeton, NJ: Princeton University Press, 1949), 160.

[58] This formulation of conservation is a gross simplification but creates a working context for what follows. It does, however, presume that high-energy and low-energy states are mutually determined. See, for example, R. Shankar, "Effective theory in condensed matter physics," in: *Conceptual Foundations of Quantum Field Theory*, Edited by Tian Yu Cao (Cambridge, UK: Cambridge University Press, 2004), 47–55.

[59] Paul Davies, "Complexity and the Arrow of Time," in *From Complexity to Life: On the Emergence of Life and Meaning*, Edited by Niels Henrik Gregersen (New York, NY: Oxford University Press, 2003), 73. See also Dean Rickles, Penelope Hawe, and Alan Shiell, "A Simple Guide to Chaos and Complexity," *Journal of Epidemiology and Community Health* 61 (2007): 934.

[60] See Richard A. Muller, *Now: The Physics of Time* (New York, NY: W&W Norton Company, 1923), 111–112.

[61] Max Planck, "The Place of Modern Physics in the Mathematical View of Nature" (1910), *in A Survey of Physical Theory*, Translated by R. Jones & D.H. Williams (New York, NY: Dover Books, 1960), 28.

apparent. To understand how the outer appearance of bread and wine manifests the informational oscillations of its interior state, I must continue to clarify the sacramental efficacy of sign and signification.

Traditional theological explanations offer an explanation for how the words of consecration transform or effect the substantial properties of the pre-sanctified bread and wine. Based on the theological and metaphysical properties of angelic being, Peter Abelard saw in the "invisible" air the perfect medium to convey the transfigurative power of the Eternal Word. For his part, Thomas focuses on the formalization of such actions with regard to the concept of quantity. Thomistic metaphysics prioritizes an actual existing being and retroactively premises its interior essential ordering. This change within physical reality is an interior movement and thus can be spoken of as an informational or essential modification. Holocryptic metaphysics argues that a parallel movement can be seen in the interior lives of rational believers. A believer and a nonbeliever may both see no physical difference in the bread and wine after their consecration. However, the baptized Christian is holo-somatically ordered to resurrection, so for them, the ecclesial communion mirrors the eternal kingdom as part of the Eucharistic feast. The holo-somatic concept of being theologically revises the essence's non-subsistent relationship to an actual existing being in the same vein as QMT retools past deterministic and materialistic paradigms of physical reality by means of probability. Through these theoretical developments, it becomes possible to define the physics and metaphysics of transubstantiation in a theoretically and scientifically mature way. Taking on a Eucharistic personality means one lives as a Christian and thus one's being already inheres as part of Jesus Christ's resurrected state. The holo-somatic principle allows us to conceptualize this eschatological state within the interpretative structure of one's unique spatiotemporal observations as a physically and historically embodied actor.[62] Through the concept of temporal succession, we are able to comprehend and symbolize eternity analogously, albeit impartially.

If our universe is fundamentally informational in nature, it is also symbolic and holographic for "symbols of various aspects of an object, of various world perceptions, of various levels of synthesis... [for although] methods of representation differ from each other, not as the object differs from its representation, but on the symbolic plane... all are symbolic by

[62] See *Quantum Shift*, 51.

nature."[63] [20] Symbols are only possible when their information is meaningfully "incarnated" or "embodied" in some material medium. Following this logic Origen calls the Eucharistic bread the "typical" and "symbolic body" of Christ.[64] The use of linguistic and mathematical symbols in ritual or in the lab exemplifies the power of the image to explain how even non-sentient matter systematically conforms to an ultimate meaning within the observational confines of its physical boundaries. Exploring the nature of the subject-object relationship gives us insights into how conscious subjectivity engages physical reality, whereas the formal and methodological use of symbols and signs (*signum-signatum*) communicate a material being's formal properties.

Previous scholarly work in theological hermeneutics, anthropology, material culture, philosophy, and aesthetics ground my holo-cryptic metaphysic's theory of a material being's holo-somatic truth. Here, medieval theories of signification are important. Scholars such as Aden Kumler and Christopher R. Lakey argue that signification inaugurates the personal and social meaning of material, cultural, and religious artifacts.[65] No being lacks a signatory significance and this fact is especially relevant for spiritual art. For "'all things, herbs, seeds, stones, and roots reveal in their qualities, forms, and figures [*gestalt*] that which is in them.' If 'they become known through their *signatum*,' then '*signatura* is the science by which everything that is hidden is found, and without this art nothing of any profundity can be done."[66] [21] Theologians and believers argue that the in-eradicable stain of original sin or misunderstood notions of negativity perverts and transforms all significations of physical beings.[67]

[63] *Pavel Florensky, "Reverse Perspective"*, in *Beyond Vision: Essays on the Perception of Art*, Edited by Nicoletta Misler and Translated by Wendy Salmon (London, UK: Reaktion Books LTD., 2002), 254. Hereafter cited as *Reverse Perspective.*

[64] See Origen, *Commentary on the Gospel According to John Books 13-23*, Translated by Ronald E. Heine (Washington, D.C.,: The Catholic University of America Press, 1993), 111–116 and *In Mt.* 11:14: *PG* 13, 948d and 949b.

[65] See Aden Kumler and Christopher R. Lakey, "Res et signification: The Material Sense of Things in the Middle Ages," *Gesta: International Center of Medieval Art*, vol. 51, no.1 (2012): 1–2.

[66] Paracelsus, "Von den natürlichen Dingen," in "De Liber Prodigiis," in *Paracelsus*, "Concerning the Signature of Natural Things," in *The Hermetic and Writings*: Vol. 2.4, Edited by Arthur Edward Waite (London, UK: James Elliott, 1894), 259.

[67] See Richard Wolin, *Walter Benjamin: An Aesthetic of Redemption* (Berkeley, CA: University of California Press, 1994), 130.

The seductive power of signification within a non-theistic ideological formula is also discerned in techno-scientific worldview. Material bodies and symbolic forms are no longer perceived as necessarily passive entities but reveal as possessing at their core a transformative and anagogic force. These forces spawn when the instrumental and efficient causes of technological manipulation appeal and give meaning to the material order. It is beyond the scope of this book to undertake an in-depth analysis of this trajectory.[68] Thus, it will undoubtedly become a more pressing concern in the coming days of artificial intelligence (A.I.). Indeed, when A.I. systems are creatively writing code as an extension of their own algorithmic programming, the question of the material signification of the designed and produced product will need to be revisited along the principles set out in holo-cryptic metaphysics.

The geometric, formal, and logical ordering of the QMO and its place within a larger system requires complex or hyper-dimensions that are analogous to what has long been seen as imperative to spiritual life. The fact that QMOs mutually define one another even when at distances that preclude direct physical contact they give us an experiential verifiable example of what the Church believes in regard to the nonphysical presence of the Spirit that exists in the interior state of a believer. These distances are possible because Jesus Christ is at once fully human and divine in the world and simultaneously the Transcendent Eternal Word. The geometric configuration of the QMO and that of the conscious partaker of the Eucharist share an analogous formal or symbolic ordering. The impossibly difficult internal geography of being is fully revealed from the perspective of a Christological hermeneutic. Balthasar opines that "if the form of Christ itself is what it shows itself to be of itself, then no particular age or culture can of *itself* be privileged in respect of this phenomenon. The decisively illuminating factor must be in the phenomenon itself, and this is two senses. First, in the sense that the figure which Christ forms has in itself an interior rightness and evidential power such as we find--in another, wholly worldly realm—in a work of art or in a mathematical principle. And second, in the sense that this rightness, ...possesses the power to illumine the perceiving person by its own radiating light."[69] [22] The holo-somatic principle supports the Church's position that the efficacy of receiving the

[68] See Martin Heidegger, *Being and Time*, Translated by John Macquarrie and Edward Robbins (New York, NY: Harper and Row, 1962), 195.

[69] *GL* 1, 465–466.

Eucharist depends, in part, on the manner of a person's reception. The Transcendent Spirit transforms bread and wine into the body, blood, soul, and divinity of Jesus Christ with the free consent of the priest and the faithful. The historical event of the mass demands the metaphysical interior spaces of the believer's heart and mind holographically replay the Passion of Jesus Christ. The free reception of the Eucharist does not require faithful absolute certainty as to how transubstantiation occurs materially in time. A believer understands the Blood of the Lamb has redeemed her and holo-somatically lives in the hope that the resurrected embodiment may be finally granted her at the far side of end times. Holo-somatic identity like all reality is conformed to the truth of resurrection. The pure simplicity of the resurrected Christ is unbounded by any material or dimensional limitations or counter-causal forces.

Subjective perception and cognition take place within a spatiotemporal framework but nonetheless aim for that total freedom evident in the Ur-Kenotic Spirit. The human person's essential state proportionally reflects the created order as a whole as all find common identity as being. The imprinting of information into some material component and its later transmission or forecast reveals in the rational mind's "essential" and free reception of Being's infinite modalities. Reality is fundamentally free, and the concepts of multiplicity and infinity have a role in clarifying how uncertainty, chaos, and probability relate to a metaphysics of liberty. The human person prospers by entering into relationships with other beings and by engaging new ideas and taking unexpected paths one experiences growth. These journeys into the unknown reveal Creation's inherent goodness—a work that ideally theological and scientific endeavors strive to participate in. The mystery of being rejects surrendering its full truth to anyone system or type of analysis, for "the 'kenotic' disposition of the image of deliberate wandering in the borderlands of mystery ought not lead us to underestimate its epistemic capacities."[70] [23]

The names or exemplars that God gives in the act of creating are eternal essential realities that physically and metaphysically appear.[71] The essential or informational status of created beings or processes is not and cannot finally be completed outside of its transcendent measure found in its source as an exemplar. Alternatively, as stated in the disciplinary language of QMT, all scientific laws depend upon the normalization of scalars and statistical

[70] Trevor Hart, "Lectio Divina," in *Theology, Aesthetics, & Culture*, Edited by Robert McSwain and Taylor Worley (Oxford, UK: Oxford University Press, 2012), 231.

[71] See Walter Benjamin, "Awakening" (*Arcades*, 462; n2a, 3).

quantification of empirical data (e.g., *adaequatio intellectus ad rem*). Theories and results are, more often than not, idealized by convention or necessity—precision must be weighed against immediate need and one's available resources. Often, the most advanced theological or scientific models and systems are the most abstract; thus, simplicity is ideal. Physical theories of reality, for instance, rarely consider "friction and dissipative effects, hierarchical structures, feedback effects, or the causal efficacy of information, and they do not take quantum uncertainty into account."[72] [24] The probable nature of our universe is not a nod to the nihilistic any more than classical laws idealization of measure is a quantifiable argument against material entities' underlining non-subsisting freedom.[73]

The interior geometry of the QMO holographically rendered shows an orthogonal structure that is determined yet opens to free and developing relationships. Further, the theory of relativity demands we consider much that is counterintuitive to immediate experience. Hermann Minkowski argued that these events result from our experience of a four-dimensional universe from the perspective of our three-dimensional experience.[74] Max Weber's artistic vision is instructive. Weber sees in consciousness' ability to receive and metaphorically "enfold" information as proof of the mind's essential four-dimensional nature. Human consciousness can conceive of how all-material beings coexist in relationship to the fourth dimension, not as time *per se* but rather as "spirit." The fourth dimension acts as an analogy, a theologian may argue, for what was previously defined by grace's operations in physical and supernatural spaces. The four-dimensional properties of the beautiful object inspire a visceral response, as this reality "is real, and can be perceived and felt. It exists outside and in the presence of objects and is the space that envelops a tree, a tower, a mountain, or any solid or the interval between objects or volumes of matter if receptively beheld. ... It arouses imagination and stirs emotion."[75]

[72] George F.R. Ellis, "Physics in the Real Universe: Time and Space—Time," in *Relativity and the Dimensionality of the Word*, Edited by Vesselin Petkov (Montreal, Canada: Springer, 2007), 74.

[73] Timothy McGrew, Lydia McGrew, and Eric Vestrup, "Probabilities and the Fine-Tuning Argument: A Skeptical View," *Mind* 110 (440): 1030.

[74] See Vesselin Petkov, *Relativity and the Nature of Spacetime*, 2nd ed. (Berlin, DE: Springer, 2009), 163.

[75] Max Weber, "Fourth Dimension from a Plastic Point of View," *Camera Work* 31 (July 1910): 25. Cited in Linda Dalrymple Henderson, "Mysticism, Romanticism, and the Fourth Dimension," in Maurice Tuchman and Judi Freeman, Edited by et al., *The Spiritual in Art: Abstract Painting: 1890–1985* (New York, NY: Abbeville Press, 1986), 224.

[[25] cited in [26]]. Rather than focusing on a unifying higher dimensional presence in reality, most postmodern proponents resolve to interpret the material order only by its objective three-dimensional or deterministic deconstructed or hedonistic measure. The quantification of being does not logically necessitate the complete reduction of a being or its encompassing verification and description by a conscious observer.[76] Minkowski's spacetime is thought of as lines within a light cone.[77] Minkowski's work does not explore the role of grace but asks us to consider the independent role that light plays in any historical observation. The four-dimensional spacetimes of GTR and QMT are presented as a natural development of three-dimensional spacetime.

My views on the holographic informational structure of being align with Minkowski's exploration of the essential role light plays in four-dimensional spacetime. A key difference between Minkowski's spacetime and my holo-somatic spacetime is the latter's dependence on an Ur-Kenotic theory of being and understanding of light's translations as a physical manifestation of non-subsistent differing.

Scholars have noted that the light-cone model of reality is encompassing and important for sensual and conscious operations. To this end it promotes many questions for traditional theories of free will. Vesselin Petkov remarks, "In the Minkowski four-dimensional world…there is no free will, since the entire history of every object is realized and given once and for all as the object's world tube."[78] [27] However, as Howard Stein notes, it takes time for light to travel, and each person's perspective is variable or free in relationship to others within a given system. A subject has a personal and restrictive experience of time and space given their exclusive position within a Minkowski light cone.[79] Each of these approaches claims that an interior freedom of a given being or material object conveys how light functions unique to one's dimensional identity. A mathematical metaphor for the theological concept of interiority is possible because the QMO and its system are formally distinct from our experienced three-dimensional experience. The quantum world and the embodied soul exist within multidimensional spaces.

[76] See *The Physics of Theism*, 111.

[77] See, 253.

[78] Vesselin Petkov, *Relativity and the Nature of Spacetime*, 2nd ed. (Berlin, DE: Springer, 2009), 172.

[79] See Howard Stein, "On Einstein-Minkowski Space-Time," *Journal of Philosophy* 65, no. 1 (Jan. 11, 1968): 5–23.

Non-Euclidean geometry gives us a way to understand the interior spaces of the human heart relationship to the transcendent and Jesus Christ's supernatural presence in the Eucharist. The absence of light is also a powerful expression for how the supernatural apophatically reveals itself in physical, reduced, and complex dimensions: "Then the cloud covered the tent of meeting, and the glory of the Lord filled the tabernacle. And Moses was not able to enter the tent of meeting, because the cloud abode upon it, and the glory of the Lord filled the tabernacle" (Exodus 40: 34–35). The complete subtraction of light (darkness) reduces a person's perceptual field. Advanced computer graphics programs are able to represent complex geometries and spaces in two dimensions and thus mimic previous advances in this area made by abstract artists such as Ad Reinhardt. In his "black paintings," Martin E. Marty notes that Reinhardt "[uses] nearly [all] black pigments to achieve absolutely flat, matte surfaces, which would not reflect and therefore did not seem to extend further than the confines of the two-dimensionality of the canvas and its handmade frame."[80] [28] The dialectical relationship that exists between light and darkness in modern art is often used symbolically to denote how absence hierarchically correlates to absolute flourishing.

In his work, *Monadology*, Gottfried W. Leibniz (1646–1716) sets out the metaphysical principles that drive his philosophical approach. At the heart of this system is the concept of a monad. A "monad" is the term Leibniz ascribes to a simple "substance" (*in median res*) that is, the "real atoms of nature." Monads have no qualities of their own, as they lack size or shape and are indivisible—like the mathematical point—and consequently, do not properly occupy space or time. Leibniz's concept of the monad shares much of the same theoretical space or "geography" as was previously dominated by Greek philosophy's theories of essence and Medieval Theologian's interpretations of the nature of the rational and non-rational souls.[81] Monads are indivisible but rational and therefore can combine to create compound substances if such harmonies are based on sufficient reason and are not contradictory. By means of rational principles, monads combine in new "fulgurations" that ultimately manifest a

[80] Martin E. Marty, "Tibetan Mandala: Ad Reinhardt, Thomas Merton, John Cage," in *Negotiating Rupture: The Power of Art to Transform Lives*, Edited by Richard Francis (Chicago, IL: The Chicago Museum of Contemporary Art Chicago: 1996), 78.

[81] See Leibniz, *Discourse on Metaphysics/Correspondence with Arnauld/Monadology*, Introduction by Paul Janet, Translated by Dr. George R. Montgomery (LaSalle, IL; The Open Court Publishing Company, 1980) 256 (*Monadology*, paragraph 22) and Ruth Lydia Saw, *Leibniz* (London, UK: William Clowes and Sons, Limited, 1954), 32.

divinely purposed universe.[82] A monad's identity remains unchanged even though it aids in the new composite being. The interactions of monads that lead to new configurations ground Leibniz's model for perception.[83] This modeling of perception inspired later thinkers' understanding of how an unchanging entity makes succession a key to conscious observation.

Leibniz's theory of monads theoretically rejects the notion of the pure void or the ether and thus, in some ways, anticipates the later discovery of physical fields. Leibniz's philosophy works to mathematically define being insofar as each monad, like a number, is immutable. The concept of monad welcomes a parallel to the QMO's "self-reflexive" identity as quantified by its orthogonal structure and the state of its internal spin. Furthermore, if Henry Stapp is correct in stating that Leibniz "argued for the relational view that space is naught but relations among actually existing entities: completely empty space is a nonsensical idea," then the central question is this: Does the idea of a monad successfully account for quantum relationships? The hyper-dimensional spaces that a physical being appears to occupy are impossible to represent fully in the monochromatic spaces of the artist's canvas, the monad, or the mathematical diagram. However, it is also the case that these attempts continue to inspire our work toward achieving such an end.

It is not surprising that so many physicists continue to find inspiration in Leibniz's philosophy. The mathematician Hermann Weyl attributes the philosopher's work as influencing his own investigations of the quantum world. Weyl sees in his abstract conceptualizations of the hidden world of the QMO a motif commonly associated with the artist. "My work always tried to unite the true with the beautiful" Weyl remarks, "but when I had to choose one or the other, I usually chose the beautiful." The holosomatic theory of identity explains how a QMO or any being's holographic representation presents its identity in the universe. The image's dimensional "skin" takes the "possession" of physical and hyper-dimensional spaces within the universe and a system. The probability motion of a QMO models in its density pattern as a holographic projection. On February 22, 1909, the Italian poet F. T. Marinetti published the "Futurist Manifesto" in *Le Figaro*. In this work, the modern art world finds the principles that articulate the principles and sensibility for an increasingly technocratic world. At the heart of this new school was the attempt to produce art that bids the viewer to consider how reality is

[82] Ibid., 33.
[83] Ibid., 30.

unceasingly in motion.[84] A being's beauty and meaning (these artists argue) follows when "lines of force" emerge from an image as kinetic motion in lines or paint on the canvas.

The QMO's identity relays itself in complex geometric patterns and hyper-dimensional spaces. Remarkably, these abstractions avail themselves of the human imagination. This is important because a similar experience is required to better understand the supernatural or "hyper-dimensional structures" of sanctified bread and wine. Modern and postmodern artistic movements distinguish themselves from my approach when they reject the theological values of nothingness and the void. The sacramental sign holo-somatically interpreted defines the soul as at once fully "incarnate" in every element of one's physical embodiment and as a vertical and hyper-dimensional emblematic thread—a single geographic zone or "weave" that is part of the Absolutely Groundless Spirit's presence in reality. This is a particular challenge with QMT and GRT, as the philosopher Tim Maudlin states: "The moral of the theories of relativity is not that classical spatiotemporal notions are rendered merely relative, but that they are expunged from physics all together."[85] Finite freedom requires that the mutually informing moments of transcendence and autonomy coexist. Without a transcendental trajectory, the finite ordering of the infinite and void is impossible, and risks becoming radicalized in one of three extreme trajectories.[86]

3.2 Holo-somatic Metaphysics and Infinite

One of the most important ancient thinkers around the question of the nature of infinity was Aristotle. He argues that the infinite is characteristically formless or undetermined. For Aristotle, infinity is closely aligned to the question of potency's ability to inform the material order. The potential prefigures in the universe's motion or "telos," but, insists Aristotle, does not warrant belief in a Creator or an actual existing Infinity. The mathematics of infinity raises profound metaphysical issues when one

[84] See Sarah Newmeyer, *Enjoying Modern Art* (New York, NY: Mentor Book, 1962), 150.

[85] Tim Maudlin, "Non-Local Correlations in Quantum Theory: How the Trick Might Be Done," in *Einstein, Relativity and Absolute Simultaneity*, Edited by William Lane Craig and Quentin Smith (London, UK: Routledge, 2008), 156.

[86] Ibid., 244, 246, and 258. See also Linda Dalrymple Henderson, "Mysticism, Romanticism, and the Fourth Dimension," in Maurice Tuchman and Judi Freeman Edited by et al., *The Spiritual in Art: Abstract Painting: 1890–1985* (New York, NY: Abbeville Press, 1986), 220.

accepts only relative infinity because it risks unchecked relativism. On the other hand, if one accepts only the existence of an Actual Infinity, what potential infinity affords for finite freedom is undermined. The existence of potential infinities or relative actual infinities inspires Thomas' own dialogical ordering of ontological differing to an Actual Infinite Being.[87] The answer to this balancing act, Etienne Gilson argues, rests in attempting to know precisely "if existence can be nothing [or]... an empty logical concept in the mind or a relation in the thing."[88] [29] Aristotle's argument that only relative infinities exist supports his belief that the cosmos exists in an unending cycle. This so-called eternal return is problematic for the Christian, as it does not fully explain a Transcendent Creator. The difficult issue of how relative infinities work causally in the created order as a metaphoric organic whole demands greater study than can be provided here. We must, however, focus briefly on the critical nature of an individual being's understanding of their hypothetical relative infinity.

Thomistic metaphysics defines finite self-reflexivity as a participatory aspect of the good of the greater whole because it models the divine-human nature of the incarnate person Jesus Christ. Jesus' obedience to the Father exemplifies the finite nature of the holo-somatic entrance into the eternal space of the Transcendent through non-subsistent relationships. There is wide-ranging consensus within the scientific community that time is not to be imagined as a "snake eating its tail" but rather as "the flight of an arrow." The modern physicist's deciphering of time is often concurrent with science's wholehearted rejection of material, formal, and final causes and perhaps even that of the efficient cause. Philosopher-scientist Francis Bacon rejects the idea of final causes, seeing them as a dissembling force against scientific research and therefore as requiring a "severe and diligent inquiry of all real and physical causes."[89][30] However, with the theories of relativity and quantum physics, it is necessary to question the wisdom of fully discarding past philosophical and religious commentaries on the nature of causality, infinity, and spacetime. Mathematics conceives of infinity as the sequential computation of "transfinite" or "supra-finite" numbers.[90] These numbers stretch our rational and imaginative capacities, but

[87] See Aquinas, *On Being and Essence*, Chapter 4.

[88] Etienne Gilson, *Being and Some Philosophers*, 2nd ed. (Toronto, ON: Pontifical Institute of Medieval Studies, 1952), 177. Hereafter cited as *Being and Some Philosophers.*

[89] Francis Bacon, "Advancement of Learning," Book 2, III, 350 in *The Works of Francis Bacon* new ed., Edited by J. Spedding, R.L. Ellis, and D.D. Heath (London, UK: 1870).

[90] See *Being and Event*, 142–149.

through their unique virtual representations, some insights are possible. As David Foster Wallace notes, "[t]he fundamental flaw of all so-called proofs of the impossibility of infinite numbers is that they attribute to these numbers all the properties of finite numbers, whereas infinite numbers... constitute an entirely new type of number, a type whose nature should be an object of research instead of arbitrary prejudice."[91] The concept of the infinite qualifies how finite holo-somatic identity images the Pure Spirit, that is, the Divine Lord.

3.2.1 *Identity as Closure*

The limitations of running an experiment a finite number of times with a non-infinite measuring device will fail to fully account for the complexity of either an evolving system or being.[92] Even though the QMO's structure is invariable, the complete and simultaneous measurement of all its properties is impossible. However, "the mathematical formalism of the quantum theory is capable of yielding its own interpretation."[93] Relative to the *z*-axis, the two states of the QMO are represented as spin-up (↑|z⟩) = (1, 0) and spin-down (↓|z⟩) = (0,1). The spin-up orientation portrays space along the horizontal axis, whereas the spin-down measures placement on the vertical axis. Because the spin between particles is common in all quantum systems, it instigates a coherent or ideal form (εἶδος) for the QMO. The QMO's wave or particle's internal state reveals itself in one of four possible combinations (e.g., 00, 01, 10, 11). One can imagine that the internal architecture of a being is equally externalized in and contained by the "crust" or "lines" of its projected image.[94] The constellation of the quantum particle's Janus-like existence as wave and particle are part of its single metaphysical economy. Even with only four possible states to draw from, one cannot conclude that the quanta's observation is predetermined. As there is a 50 percent chance that one of the QMO's two possible states of wave or particle will display within the previous parameters, the concept of non-subsistence gives us a metaphysical process that directly relates to the material

[91] David Foster Wallace, *Everything and More: A Compact History of Infinity* (New York, NY: Atlas Books/W.W. Norton & Company, 2003), 40.

[92] See Steven Weinberg, "The Trouble with Quantum Mechanics," in *The New York Review*, January 19, 2017, http://www.nybooks.com/articles/2017/01/19/trouble-with-quantum-mechanics/.

[93] See B.S. DeWitt, "Quantum Mechanics and Reality," *Physics Today*, 23, no. 9 (1970): 33.

[94] For a traditional theological perspective on the issue of essence-existence relation as an informational construct see *ST.* I, 50, 2, 3, *SCG*, 1, 38, 52–54 and *De ver.* 27, resp. 1.

nature of a spatiotemporal being. The proposed symmetry existing between the non(One) and its image in Badiou's philosophy finitely models infinity's authentication of being via Nothingness and the void. As Franz Kafka explains, "It is up to us to accomplish the negative. The positive is given."

The formlessness of potential infinity disposes being by means of its logical antipode of the void. Metaphorically these concepts function in a similar way to how color and darkness function within a painting. Colors are differentiated given the physical properties of the paint and how its spatial placement reflects light.[95] The infinite like color delineates forms and the dark line like the void defines a particular being, for: "every multiple being in its turn [is] nothing other than a multiple of multiples—is the law of being. The only stopping point is the void."[96] [31] Rejecting the Christian understanding of the Transcendent and hierarchy, Badiou's mathematical quantification of being includes his own views on "higher" levels of formalization. He defines a physical being in terms of its relationship to the infinite and the void, and its serialization and network are limitless. A horizontal genealogy partially drives Badiou's rationale for non-subsistence and physical, historical, and conceptual boundaries. Of the infinite and void's proactive correspondence to mathematical beings, he writes, "In set theory, the primitive name of Being is the void, the empty set. The whole hierarchy takes root in it. In a certain sense, it alone "is." And the logic of difference implies that the void is unique. Indeed, it cannot differ from another since it contains no element (no local point) that can aver this difference."[97] A careful reading of Plato reveals how he influences Badiou's understanding of the Transcendent and his singular and realistic approach to mathematics. Ricardo L. Nirenberg and David Nirenberg comment that "Unlike mathematical 'formalists' like David Hilbert, for whom truth or falsity depends on the axiom system one chooses to deal with, just as, in a game, a certain move is allowed if it does not violate the rules; unlike 'intuitionists' such as Henri Poincaré, L. E. J. Brouwer, or Herman Weyl, who handle infinity with pincers and refrain from proofs by contradiction; unlike post-Tractatus Ludwig

[95] See Friedrich Ohly, *Sensus Spiritualis: Studies in Medieval Significs and the Philology of Culture*, Translated by K.J. Northcott and Edited by Samuel P. Jae (Chicago, IL: The University of Chicago Press, 2005), 102.

[96] Alain Badiou, *Ethics: An Essay on the Understanding of Evil*, Translated by Peter Hallward (London, UK: Verso, 2001), 25.

[97] *Briefings on Existence*, 98.

Wittgenstein, Badiou is a Platonist for whom the huge universe of set-theoretical objects is real."[98] [32]

The void and the infinite are perhaps best thought of as Platonic Forms. So understood, the mystery of these forms will always remain, but they also possess a high degree of interpretative latitude. If one conjectures that an all-encompassing infinite number exists (ω), it is possible to conceive of it as being both primitive and transcendent to all other infinities. For this organization of various types of infinity, it is necessary to assume that they are distinct, that is, separated by some "space" or "void." The so-called Actual Transcendent Infinite Number guarantees the existence of all other potential infinities and the void. This claim follows from earlier philosophical advances. Thomistic metaphysics argues that a being's actual existence conceptually transcends or is in some ways prior to its essential truth in the same way that the whole is assumed before all its component parts are known. All mereological questions around the nature of part and whole counter-intuitively assume some actualization of nothingness or the void as a relational or differing principle. This truth seems operative in Gribbin's explanation that "The pieces of the system interact with one another to produce something that is greater than the sum of its parts. In addition, that's complexity founded upon deep simplicity."[99] [33] Consequently, the Platonic idea or form of an infinite set exists prior to all other infinities. Such a schema gives us a rationale for how infinite sets can contain different members. The existence of a transcendent infinite set makes the individual parts or members of all other sets freely configured. Unlimited multiplicities logically depend on an actual transcendent and pure being. From this perspective, the infinite form and void work together to circumscribe a margin, restriction, or boundary. With these holographic delimitations, individual beings and multiplicities are possible and further make possible community.

The love that integrates a person and grounds community must positively engage all being, forces, and modes of negativity. Nothingness, void, and the infinite are used to re-evaluate traditional understandings of being and it relationship to non-subsistence. Holo-cryptic metaphysics states that a physical body's holographic actualization implicitly takes into

[98] Ricardo L. Nirenberg and David Nirenberg, "Badiou's Number: A Critique of Mathematics as Ontology," *Critical Inquiry* 37 (Summer 2011): 584.

[99] John Gribbin, *Deep Simplicity: Bringing Order to Chaos and Complexity* (New York, NY: Random House, 2004), 147.

account the void and Nothingness given their necessary correspondence to creation and the Transcendent. In finite being's correspondence of material reality, void, and Nothingness, its infinite "transcendental" purposes emerge. The post-Newtonian revision of physics gives us a new way to speak about how nonbeing acts as a handmaiden to being. Thus, it is essential that we critique how the concepts of infinity and different modes of Nothingness are philosophically and theologically explicated and unified in Being, for "when Being is confronted as love the threat which infinity poses to finitude vanishes."[100] [22]

When Badiou likens the void to the "presentational occurrence of inconsistency as such, or the ruin of the One," we do not see a decisive argument for community's necessary destruction or an intractable statement of God's nonexistence given the reality of multiplicity. Rather, we see a coded defense of the unification that is the Absolute Difference of the Spirit's infinite nonmaterial presence. Mathematical subtraction, nonsubsistence, the transubstantiation of matter-energy fluctuations, inconsistency, nothingness, and the void, holo-cryptic metaphysics argues, do not overthrow the existence of an Actual Transcendent Infinity but confirm the Spirit's mysterious work from a finite perspective.[101]

The act of counting and subtracting under infinity's "gaze," the propositional statement, or the philosophical distinction all give rise to how one rationally and morally conceives the nature of formal and final causes. An Actual Transcendent's infinite manifestations as finite and punctuated multiplicity repackage past notions of the Creator-created covenant. The infinite generative capacities of Nature virtually image the conceptual distinction between the Divine Persons. Created diversity is a sign of Creation's moral goodness when witnessing the absolute difference between created and Transcendent Spirit. Believing that God alone is absolutely good, it is therefore the case that no finite being can claim to achieve self-completion or moral perfection on his own. Unlike the abyss that separates the Ur-Kenotic Spirit from finite reality, the created entity is never in relationship to "simple" or "pure" difference. However, without non-subsistent differing relationships, the human person would never be able to freely choose a task or ultimately commit to a meaningful life path.

[100] *GL* 1, 159.

[101] See *Being and Event*, 93 on Pure Being's reflexive relationship to Nothingness and chaos and on the role of downward causation and the divine, see Arthur Peacocke, *Theology for a Scientific Age: Being and Becoming—Natural, Divine and Human*, 2nd ed. (London, UK: SCM Press, 1993), 53–55.

The human person must, Ulrich argues, be able to rationally and freely self-differ or choose. The monotony of undifferentiated Nothingness leaves little room for human reason or choice. Relational differing makes possible the full declaration of a being's essential truth. Lacking the modulations of negativities such as the void or non-subsistence, Ulrich, means a being could never "enter into a difference from itself, because it would no longer come about 'over against' its transnihilation through the essence as a result of the principles of the essence, and thus the evolution of the cosmos ('from below') toward the 'being that is given as a task' would either go astray into a direction of the absolute has-been, or would degenerate into a completely indeterminate process that intends 'nothing.' In the enactment of the transnihilation, being would have 'in view' not itself as concrete, finite substance (which grounds in truth the fact that God creates by arming himself...but only itself in pseudo-subsistence."[102] [3] Ulrich's position develops upon Thomas' theory of the philosophical and spiritual implications of ontological differing.

A being's essence must be modulated and transmitted in a discrete or spatiotemporal manner in the universe. For the rational person, this unceasing process of differing or moral perfecting safeguards one's irreducible meaning or signification. Thomas' insight into the *a priori* import of actual existence over hidden essential difference and Badiou's defense of the historical realization of the event are critical elements of this discussion. The Thomistic category of relative infinity actualizes physical structure through the differing of a being's form: "outside the pure act of existing, if it exists, nothing can exist save as a limited act-of-being. It is therefore the hierarchy of the essences which establishes and governs that of beings, each of which expresses only the proper are of a certain act-of-being."[103] [34] To deny the meta-leveling properties of the historical realization of the ontological differing of essential and existential inevitably ends in human reason and freedom's evisceration. As Balthasar notes, "becoming... does not mean that the object's existence has a temporal form, while its essence, unperturbed by coming to be and passing away in time, supposedly represents its supra-temporal truth. For its essence is

[102] *Homo Abyssus*, 464–465.

[103] Etienne Gilson, *The Christian Philosophy of St. Thomas Aquinas*, Translated by L.K. Shook (New York, NY: Random House, 1966), 36. The actual is always taken as more perfect than the potential. See *ST.* I, 8, 4.

precisely what exists, but what exists cannot be unaffected by alteration."[104] [35] The vacillations of interior ontological differing are structured to harmonize with exterior differing.

The interplay of the interior structure and external appearance of the QMO gives us a way to understand the nature of ontological differing. The granular or orthogonal nature of the quanta is hidden when the QMO exists as a quantum wave (ψ). Quantum oscillation makes direct observation of the wave difficult but not impossible, as quantum indeterminacy can be statistically normalized.[105] Cultivating the possible theological implications of the nature of measure in QMT, Robert J. Russell remarks that "If we adopt the interpretation that these quantum statistics reflect ontological in-determination, then I may argue that God can act together with nature to bring about all events at the quantum level, and that these events give rise to the classical world."[106] [36] The QMO's admission of identity is complementary to its probable measure. Truth domiciles in being and is open to being described by domain-specific languages—whether those favored by scientists or as part of theologian's views of a divine providential plan. [40]

The "perichoretic" quality of (ψ) includes its physical appearance as a particle. The QMO, similar to a gas particle, exists in a system and can thus be scaled given its motion. Classical and quantum systems elucidate the nature of fundamental particles as a function of their statistical norm in their thermodynamic or entropic state. Outside of the Trinity, the plurality of "things" exists in a graded placement based on their relative superiority and subordination to a presumed all-encompassing and Transcendent Infinite Point.[107] Even though one may reject outright all theological configuring of the QMO to an Actual Eternal Transcendent Being, nonetheless the "transcendental" inclination to a definitive and immutable point is not easily discarded. Mathematical, philosophical, and theological descriptions of a Platonic form of transcendent infinity, for example, find their analogous depiction in the QMO's place in the potentially infinitely divisive hyperspaces and the celebration of emergent multiplicity writ into the natural order.

[104] *TL* 1, 59.

[105] See *TL* 1, 192 on essence-esse relationship.

[106] Robert J. Russell, "God's Providence and Quantum Mechanics," *Counterbalance.* http://www.Counterbalance.net/physics/qmprovid-fram.html. Accessed April 11, 2013. See *The Physics of Theism*, 161–162.

[107] *ST.* I, 42, 2 and *SCG* III, 2.

3.2.2 Holo-somatic Identity and Infinite

A physical object and its holographic equivalent are co-constitutive when projected into higher or multidimensional spaces. A being's correspondence to a particular spatiotemporal location is analogously mimicked in a number's inclusion or exclusion within an infinite set. Consider how strict inclusion functions in set theory.[108] By definition, set *P* is contained in set *R* if and only if set *R* contains all the members of set *P*, as well as at least one additional term that distinguishes set *P* from set *R*. Restricting *P* to be a finite set, it is strictly bounded in set *R*, and *R* has more elements than set *P*. For example, all loafs of bread in set *P* are unleavened, and there is some bread that is not unleavened in set *R*. In our speculative schema, we may further add that this action is "kenotic" in nature as identity "produces" by means of a "subtractive relationship." The infinite set correlates the order of all members (e.g., loafs of bread) of the set to each other and to the set as a whole. The coincident ordering of one-to-one relationships and one-to-the infinite requires a liberal interpretation of the notion of absence or "virtual negative" spaces. The one-to-one correspondence of numbers assumes an irreducible (formal) identity for each number. However, such a concession can only be made abstractly when one looks at an infinite series or more directly at the nature of our emergent universe. A being's formal identity can be analogously modified within an infinite series, as is proper for any other finite-infinite corroboration.[109]

A being or number's placement in a particular grouping or set is distinct from the final overarching logic driving its relational identity to all other present or relevant beings or numbers. This raises the second important aspect of the finite-infinite relationship that must be clarified. As previously noted, the infinite inward division of a being is untenable when it exculpates a being from its generative external relationships. As a model for infinity, consider, for instance, *N*—the set of Natural numbers and $2N$ the set of all even numbers of *N*. Thus, *N* and $2N$ exist in a specific one-to-one (bijective) relation. This correspondence (e.g., a mathematical function) creates a kind of "inverse mimesis," as every natural number is only associated with one other even number that is its double. Each of these even numbers correlates to a single natural number with half its numeric value. The individual identity of each number follows from a

[108] My theory of infinity was guided in part from several conversations with Fady Chidiac, S. J.

[109] See *The Philosophy of Mathematics: An Introduction*, 128, 146 and 174–187.

direct relationship with one other number and its "embryonic" identity as part of a potential infinite series of numbers. This mathematical function can serve as a prototype for how a finite being experiences infinity in her daily life. Similarly, Thomas summarizes this complex and unlikely union in terms of an existing being's heterogeneous identity by means of comparison to all other beings: "now being, as being, cannot be diverse; but it can be diversified by something besides itself." Infinity isolated from all external correspondences duplicates the meaninglessness of inescapable repetition, whereas succession ordered to the other (continuum) inspires a creative response.

Cantor's work proves that not all infinities are equal. Membership, structure, and topology may vary among infinite sets. The numbers of members within a set determine its structure, whereas the value of each number establishes its topology in relationship to other sets (e.g., $A \cap B$). A number's identity is disclosed, in part, because of its serial position within a set. This position or "in-differing" gives beings or mathematical terms the freedom to relate "reflexively" and promulgate correspondences to other members of a given Set. W. D. Hart explains, "if any infinite set had as many members as any other, the arithmetic of infinity would be boring. Cantor's proof that there are more reals than naturals hints that it is not. However, first we should explain 'more.' A set *A* has fewer members than or as many members as *B* just in case *A* has many members as a subset of *B*, or equivalently, some function maps *A* one-one into *B*."[110] [37] The issue at hand is this: Does the unique identity of numbers only reflect their place within a set? The set's identity warrants or qualifies whether a number or a being meets the criteria for its membership in a set by implicitly defining each member of the set. An infinite series such as a finite being's perfection is met when its properties create a series of lateral relationships between the members within and outside the set with which it identifies—the infinite demands we consciously hold that a being can be continually "fractured" or divided and subdivided without ever reaching Nothingness or its symbolic representation as zero. These correspondences are virtually finite in regard to each other but actually infinite when taken as a whole. Consequently, I argue that a finite being's ideas and desires when taken together can be similarly understood as infinitely self-reflexive.[111]

[110] W. D. Hart, *The Evolution of Logic* (Cambridge, UK: Cambridge University Press, 2010), 15.

[111] See *Emblems of Mind*, 58.

Internal freedom is only possible if the universe as a whole is itself free, that is, if its finitude is infinitely open-ended. The meeting of the finite (numbers or ideas) and infinite (series or desire) in a mathematical set or being serves to locate their relative place within the universe and point to their means to transcend their limitations.[112] From a religious perspective, a being's infinite perfection involves an interior transformation that requires more than an endless increase in relationships or the metaphoric addition of new members to an unending numerical sequence. Bulgakov sees in finite measure the infinite's spiritual presence: "a quantitative scale, the fullness of their number, cannot be applied directly here in the sense of 'the more, the better'; for besides the *opus operatum* ['the work wrought' or spiritual effect] there is also the *opus operantis* [the efficacy of the agent], different forms of the reception and appropriation of the gift of grace."[113] 47 It is no more possible to quantify spiritual grace than it is to finally quantify the divine chosen medium of Nothingness to create. However, in the mathematical expression of the void or the concept of Nothingness, a preliminary definition is given in the empty set. In comparison, Thomas is cautious to attribute a philosophical or theological boundary to the Transcendent, the void, or its metaphoric demarcation in infinity.[114]

It was not until Cantor's famous *Grundlagen* that the infinite was mathematically analyzed and given an interpretation radically different from classical positions such as forwarded by Aristotle. In distinguishing between relative infinities and an actual infinity, Cantor illustrates for Badiou how it is possible to see the void's imaging and possible grouping as multiplicity. The postmodern position is correct when stating the unpleasant truth that the concepts of void and Nothingness are unavoidable, and theologians should not undervalue Badiou's insights in looking for answers to questions about the mathematical concept of the empty set.

Finite and non-subsistent replications of Ur-Kenotic relationships require some conceptual "buy-in" to the idea that members of an infinite set can be formally determined. The "subtractive" mathematical action is a putative "transcendental" example of the finitization of the infinite. Badiou categorizes the immeasurable continuum of the infinite as a

[112] See ibid., 11 and 21.

[113] *The Bride of the Lamb*, 290.

[114] See *ST*. I, 7, 4. See also Joseph Warren Dauben, *Georg Cantor: His Mathematics and Philosophy of the Infinite* (Princeton, NJ: Princeton University Press, 1990), 120–125.

deconstruction of the number One. Pre-Socratic thinkers such as Anaximander (610–547 B.C.) employed a similar hermeneutic when he sought the infinite's theoretic first principle (Ἀρχή, *arché*). So imagined, there is no unary concept that *a priori* limits the universe's "eternal motion" or limits an infinite set's ongoing potential reduction or growth.[115] The ONE and the essential formlessness of the infinite (apeiron) reveal itself in the binary disclosure of finite beings' latent and infinite natures.

Thomas proposes that an analogous formal ratio exists between the pure actualization of the Divine and the metaphysical state of finite beings, as "the Transcendent's infinite simplicity and eternity are evident in the formal ratio that finite beings participate."[116] 14 Matter's potential changes and infinity's transpositions in reality make it difficult for the subjective observer to be objective about what they perceive. Cantor's theories do not oblige a single manifestation of the infinite.

One must be cautious when attempting to determine a final meaning for multiplicity, the void, or the infinite. One must take conscious responsibility for their actions, but it is important to note that one can never fully account for their hidden subliminal bio-survival urges or external influences.[117] [39] Disciplines such as psychology and anthropology suggest that a finite being subconsciously experiences the infinite. There is an element of pseudo-artificial ordering to one's experiences of reality and intuitions about the infinite or the void. The subject's grasp of being and truth is never perfectly resolved by human perception or measurement. The description of a being's relationship to the true often blurs the unique identity of each and this leads to denoting meaning in a circular manner. For "we know the object by way of the principle, and the principle by way of the representation, which in turn is arrived at by way of the data ensemble."[118] 5

We compensate for the loss of observational certainty about a particular object when insights are generated about the nature of its geometric and spatiotemporal elements. These images serve as a means to place it in a system or universe as a whole. In these cognitive and artistic gestures, a

[115] See Aristotle, *Phys.* 3, 254–255a. This question infers if one argues both for a minimal length in reality, anticipating Planck's later work, and for a space-time continuum that Einstein defined mathematically. While beyond the scope of this work, this paradox underlines later discussions on quantum gravity.

[116] On the simplicity and power of the Divine, see *ST.* I, 3, 7, and the Divine's creation of time, see *ST.* I, 46, 1.

[117] See Aristotle, *Ethic Nic.*, 1140a7–1140a17.

[118] See *Smith*, 44.

person symbolically and actually participates in the infinite exemplarity that is the created order.[119] The mathematics of infinity thus provides us with an example of how the location of physical objects—say bread and wine—determines their symbolic, creative, and holo-somatic ordering of the eternal hypostatic identity of Jesus Christ as fully human and divine in the Eucharist. The Eternal Logos never "leaves" the hypostatic union of the Trinity, but in the Spirit, the person of Jesus Christ is infinitely actualized in the Eucharist.

3.2.3 *Infinity and Holo-somatic Non-subsistence*

Subject-object relations are fundamental to the conscious ordering of the created order and anticipate one's spiritual interpretation of the Transcendent's correspondence to the immanent. A holo-somatic account of information can be interpreted in a manner that is favorable to the principles of hylomorphic theory and thereby our own Ur-Kenotic interpretations of the Creator's role in the act of transubstantiation. This relationship reveals itself to the believer through the use of sacred representation. As Florensky notes "as for the content of this symbol, it cannot be a rational content. It can only be a content that is immediately experienced in the experience of self-creativity, in the active self-building of the person, in the identity of spiritual consciousness. That is why the term 'numerical identity' is only a symbol, not a concept has an inner connection with that which it symbolizes; it is endowed, if only partially, with the spiritual power of the signified."[120] [11] Our ability to judge, to take a stand in the world, follows from the ineffability of what we image. A being's essential identity is rarely conventional or linear in its expression or easily equated with or concomitant with accepted empirical "facts," language use, or subjective perception.[121] Transubstantiation is a physical and supernatural event that appropriates matter in its entirety at the essential level. The believer participates in their own life this moment of transubstantiate disclosure. Christ's identity in the Eucharist is a viable reordering of matter's identification of nothingness, the void, and the infinite. Florensky presents us with a way to understand how space is simultaneously unlimited, empty, and contained.

[119] See Thorleif Boman, *Hebrew Thought Compared with Greek*, Translated by Jules L. Moreau (Philadelphia, PA: The Westminster Press, 1960), 129–130.

[120] *The Pillar and Ground of the Truth, 61.*

[121] See *The Pillar and the Ground of the Truth*, 40. See also *ST*. I, 81, 2.

An unconventional understanding of the spatiotemporal is needed if we are to re-reposition how physical matter brackets divine substance within the consecrated species. Florensky's tri-part division of space argues that (1) geometric space is infinite, (2) the bounding of this visual space is possible, as evident in representational art, and (3) the non-homogeneous or isotropic geometric space finds its organic parallel in the physiological space of human consciousness.[122] [75] The phenomenon of transubstantiation is never limited to the particular consecration of bread and wine, as the grace it radiates generates throughout the universe. This ritualistic engagement spiritualizes the underlining non-subsistence of the material order because the incarnation promises the renewal and ultimate resurrection of the material and finite. Transubstantiation supremely exemplifies how the principle of non-subsistence divulges in a being's accidental properties. The objective nature of transubstantiation is needed, as this act is irreducibly tied to the redemption of our habitual sinful nature. Thus, the structure, operative efficacy, and communal nature of the sacrament of Reconciliation share much with the Eucharist (Matthew 18:15–20). No sacrament is reducible to objective principles because all theological claims are verified by the faithful's free and conscious commitment to God.[123]

As immanent beings only enter into non-subsistent relationships, all truth is experienced as a "floating middle," an ideal whose teleological aims mutually define divine life and human fallibility as part of a single whole. Various interpretative strategies are used to establish truth in the face of a being's ongoing change and motion. This act demands holding on to an image of the "*verbum mentis*" by constantly turning back or remembering an image of a being's original appearance.[124] The state of the human soul immaterially manifests nature's becoming as an interior restlessness. Gilson states, "[t]he question never is for a soul to become *what* it is (it is such *qua* form) but *to become that which it is.* In other words, a human soul has more and more to actualize its very definition."[125] [29] The Christian covenantal relationship to Jesus Christ is lived in the present but future and virtually lived as part of the heart's desire. For the Christian, the future is not a blind flaying against the unknown but a reflection of

[122] See *Reverse Perspective*, 267.

[123] See *TD* 4, 394–399 and *TD* 5, 115.

[124] See *TL* 2, 254. See also *TL* 1, 157 and Aristotle, *Prior Analytics*, 1-4-11 at classics/mit.edu/Aristotle/prior.1.i.html.

[125] *Being and Some Philosophers*, 181.

one's intrinsic freedom in an uncertain universe. Truth sustains us with meaning, all the while inviting us into deeper exploration of its mysteries.[126]

The desire for ultimate truth never ceases to demonstrate its source in nonbeing's various "intonations" such as change. In commenting on an inspiring passage of the *Book of Zechariah*, Didymus the Blind reflects, "the soul has within itself [a]... principal of activity [that is]... in constant perpetual activity."[127] The person, who desires, Nyssa reflects, is without rest because "such a one has soared up into the indeterminate and the infinite."[128] The human person is metaphysically structured for change given consciousness' ability to understand the "space" or potential "informational ratio" of the indeterminate and the infinite within the created order. The experience of sensing and understanding motion is reenacted in one's personal life, salvation history and liturgical celebration.[129]

The unbounded and limited coexist because non-subsistence and the conceived infinite series are both sourced and given their quota and definition by the Divine. Sacred scripture brings these themes and principles together in the aesthetic and metaphysical description of angelic being. Angels are often symbolized by light and color. The angels' white wings and robes call to mind their pure spiritual being and purposes to history's eschatological fulfillment. Thus, David Tracy states that in "the New Testament," angels are always described as dressed in white, according to Jewish tradition. White is prominent in the *Book of the Apocalypse*, identified as the color of the celestial creatures and consequently as that of Jesus' robe—a connection notable in the account of his Transfiguration: "And he was transfigured before them, and his clothes became dazzling white, such as no one on earth could bleach them" (Mark. 9:2–30).[130] [41] This life-giving tension reveals in the light of Christ at Mount Tabor.[131] [42]

[126] See *End of Modernity*, 175.

[127] Didymus the Blind, *In Zachariam*, *SC*, I, 43. 5–6.

[128] Nyssa, *De Vita Moysis*. I, 301 AB. Cited in *Presence*, 44.

[129] Didymus the Blind, *De Trinitate*; *PG* XXXIX, 721 A and Clement of Alexandria, *The Fathers of the Catholic Church*, vol. 85, *Stromateis*, Book Two, 17–18, Translated by John Ferguson, (Washington, D.C.,: The Catholic University of America Press, 1991), 209–216,

[130] David Tracy, "The Catholic Imagination: The Example of Michelangelo," in *Heavenly Bodies: Fashion and the Catholic Imagination*, vol. 1, Andrew Bolton, et al. (New Haven, CT and London, UK: Yale University Press, 2018), 18.

[131] Norman Russell, *The Doctrine of Deification in the Greek Patristic Tradition* (New York, NY: Oxford University Press, 2006), 306.

The light of Christ's transfiguration perfectly actualizes the properties of what we intend in our own concept of holo-somatic interior freedom. The differences existing between beings is not simply a matter of accidental differences but points to a being's interior metaphysical life and thereby the capacity of the virtual or essential to communicate one's actual existence. The essence-esse aspect of one's transactional or intentional life assimilates into the cosmos' own growth without loss of individual identity. Reinhardt argues that the essence of non-sentient beings, great art images, and the soul are properly described as irreducible, unreproducible, imperceptible, and thus defy characterization as "usable," "manipulable," "salable," "dealable," "collectible," "graspable" a "commodity" or a "jobbery." He explains, "Art is not the spiritual side of business."[132] The essential never reduces to transactional calculations or final quantification but opens windows into the infinite powers of conversion and transformation granted by the Transcendent.[133] The concept of the holo-somatic alliance between the human person and Jesus Christ and infinity's reflective proper ties conceives and models how two apparent contraries can be productively ordered.[134]

3.2.4 Holo-somatic Self-consciousness

The conscious subject must account for the mutual reciprocity existing between her and universal truth. On some level she confirms an object's reflexive interior mathematical properties as an image (*phantasm*) or conscious holograph. An unavoidable paradox defines conscious life, that is, "things are unveiled in my own presence-to-myself: namely, the absolute incommunicability of my own being (as 'I') and the unlimited communicability of being as such (which is not 'used up' by the fullness of all the worldly existence in which it subsists)."[135] Subjective perspicuity and reciprocity among beings is always positive, as it draws upon that inborn goodness that is realized when individual identity and the other is apprehended. To participate in this insight, the conscious observer must overcome egotistical tendencies—those conscious manipulations that disavow mutual disclosure of truth. The negation of these "bad" infinities requires

[132] Ad Reinhardt, "Art-as-Art," *Art International* 6:10 (December 1962), reprinted in *Art in Theory, 1900–1990: An Anthology of Changing Ideas*, Edited by Charles Harrison and Paul Wood (Oxford, UK: Blackwell, 1992), 809.

[133] See *ST.* I, 7, 2.

[134] See Rudy Rucker, *Infinity and the Mind* (Boston, MA: Birkhäuser, 1982), 50.

[135] *TD* 2, 208.

that one attest to the positivity of difference. This process is part of the "higher, spiritual law of identity, [that] self-affirmation lies in self-negation, whereas according to the lower fleshy law of identity, self-affirmation lies in self-negation."[136] [11] The rational subject can reject Creation's goodness and move to unreservedly see reality as a prefiguring of Hell. The conscious choice to work against Nature's telos or necessity (ἀνἀγκη) without which Aristotle reasons, "A thing cannot live… [For] the conditions without which the good cannot be or come to be, or without which we cannot get rid or be freed from evil [–such] compulsion is a form of *ananke*."[137] The egotistical consciousness depends on an anti-creation vision that is blind to the gift of Nature's causal and emergent forces—a vision that serves the ego's Promethean will.

Past theological visions suggest that Hell follows from the results of Lucifer's personal rejection of the Divine. Lucifer's act of defiance determines his entire existence toward an inescapable end. By his disobedience, Lucifer "the Light Bearer" isolates "his" immaterial identity from the goodness of finite dominion. Further, Lucifer is powerless to create or even reduce his informational ground state to Nothingness. However, "his" non-dimensional nature holographically exudes the sense of Hell's infinite move toward a null state. The freely rewritten non-composite infinite angelic nature source and grounds the anti-creative geography of Hell. In comparison to Hell, the created order exists in creative correspondence with Nothingness—the moment of Creation, and this correspondence reveals who without the Divine, non-subsistent relationships necessarily become dissipated. In comparison, truth directs finite testimony to reality's holographic imaging of Ur-Kenosis.

The distinctions existing between subject and object depend on proposing an actual or virtual "transcendent" point that encompasses both parties and reality as a whole. Without that freedom made possible by the Transcendent, a being exists in Hell in a qualifying and quantitatively discontinuous state. Grouping an infinite sequence of numbers in a mathematical set or the phenomenal entity in complex hyper-dimensional spaces gives the postmodern thinker new images for the mystery of the soul's nature and means of existence in the supernatural realm. Each of these examples has to be explored on its own merit with special reference to human consciousness. The essential or core identity of a being remains

[136] *The Pillar and Ground of the Truth*, 224–225.
[137] Aristotle, *Metaph.* 5, 4, 1015a, 20–34.

regardless of its spatiotemporal placement or Hell's deconstruction of this universal experience of embodiment and dimensionality.

Removing a number from its position in an existing mathematical set and placing elsewhere among the collected members, for example, does not totally "subtract" or "remove" its ongoing potency or original identity within a set. Badiou stresses that an individual being or number's organization never fully eviscerates its original identity. The guarantee of a number's identity follows from a subject's "buy-in." However, Badiou rightly sees all such "subtractions" or "displacements" of a number in a set as a move that devalues or weakens its potential efficacy.[138] The context in which truth derives or the manner of its conscious arrival in the world is important. The mathematical set makes the actualization of a number's existence possible by formalizing its identity in terms of a mathematical operation or function. [38] This placement like historical existence defines a number and a being against relational non-subsistence and infinity's formless "soup" of multiplicity and potentiality.[139]

The mathematical set holographically simulates the identity of innumerable infinities in a manner to how letters and words make possible an unlimited collection of sentences. Meaning is denoted in symbols, and this process makes possible conscious self-reflexivity. Comparatively, "set theory ontology may be said to be an ontology of immanence, retaining being within its inscriptions."[140] [43] The idea of the set parrots the bracketing of a being, as it gives us a way to speak of the reflexive identity of each unique infinite sequence. The logic of bracketing can be extended to any being or system that is open to its own growth and critique. This follows when the various elements of a group or set are placed in mutually productive differing correspondences, as Ray Brassier explains: "[w]hat renders formal systems productive is their stratification understood as the differential network through which mathematical symbols and operators are assigned a signifying function, and variously combined, produce distinct statements."[141] Two numbers can be seen to be compatible regard-

[138] See Peter Hallward, Badiou: *A Subject to Truth*, (Minnesota, MN: University of Minnesota Press, 2003) 328ff.

[139] See Ibid., 61.

[140] Oliver Feltham, "Translator's Preface", in Alain Badiou, *Being and Event* (New York, NY: Continuum International Publishing Group, 2005), xxiv.

[141] *Key Concepts*, 69. See also Alain Badiou, *Theory of the Subject*, Translated and Introduced by Bruno Bosteels (New York, NY: Continuum, 2009), 68–69. Hereafter cited as *Theory of the Subject*.

less of their accidental properties or value if they belong to the same set or exist in a shared mathematical equation. The metaphoric "topological" structures of all numbers are theoretically commensurate if the number 1 is taken as the fundamental number. Some sense of shared identity is immediately evident when the locations of two numbers intersect on a graph or a conjunction exists when collected and defined by the mathematical set.[142] In this regard, I believe that mathematical philosophy shares a logical genealogy with scholastic theology's axioms concerning a being's correspondence to nonbeing. The philosophical concept of genus and the act of mathematical bracketing are indispensable to any categorization but uniquely detailed in every grouping. Thomas writes: "Therefore, when something belongs to a genus, its nature, or what it is, must differ from its existence. As I have shown, although, this difference does not exist in God. God, therefore, clearly cannot be a species within a genus."[143] [14] The theory of mathematical classes brings to the fore several points about identity, reflexivity, relationship, and transcendence that I develop in later chapters around the themes of quantum superposition and holographic information. I do so indirectly by focusing on the logical antimony written into a traditional theory of classes.[144] Such moves are only possible by showing how apparently contradictory particular classes are in fact logically and consistently part of the study of mathematics in total.

Zermelo-Fraenkel's axiomatic theory assists Badiou in his desire to quantify a being's self-reflectivity, freedom of choice, and evolution. Badiou suggests that the study of mathematical probability engages the question of being's in-differing. The Axioms of Extension and Choice give us a way to formulate mathematical rules for a number's inclusion and exclusion within a set, and this metaphorically models how any free being can intentionally exist in an uncertain universe. Badiou argues that choice is never simply an isolated subjective construct as all potential infinities or information is only meaningful when consciously ordered. One understands data by structuring it, that is, discovering a result or meaningful pattern that gives one a sense of closure. By ordering information, one repulses or undermines chance's universal influence within a system. The clinamen is "a specific, beyond necessity, absolutely out-of-place,

[142] See Jason Barker, *Alain Badiou: A Critical Introduction* (Sterling, UK: Pluto Press, 2002), 154.

[143] *ST. I, 3,5.*

[144] See *The Philosophy of Mathematics*, 45 and *Being and Event*, 130–132.

unplaceable, unfigurable: chance"[145] that must be addressed. Badiou states that being is always undergoing change and as such exist in relationships with other beings. Hallward argues that Badiou reduces uncertainty or the potential when he validates and synthesizes two ostensibly contradictory notions by means of the Axiom of Choice.

Probability must hold universally; otherwise, the specification of being or their actions remains determined by the manipulations of exterior forces. The risk of unchecked relativism like uncertainty increases if one rejects choice under the force of determinism. The subject must accept that their freedom is complementary to what determinism enforces. Through choice the subject must move beyond accepting unbridled freedom or its opposite of determinism. The conscious subject balances these two undesirable options when a subject follows the Transcendent Being as part of our postmodern conversation. A coherent and random reality for the believer, Badiou states, forces the subject to choose between becoming subordinate to "existential judgments" and "finite and controllable linguistic protocols" or make admissible an appraisal of being that posits the Transcendent's existence.[146] Badiou's mathematical philosophy offers one way to bridge these two untenable possibilities. The ONE like the Transcendent is retrospective to a finite being's awareness of his situation in the world. Badiou conceives of this action in terms of probability and counting and the believer in terms of spiritually discerned actions.[147] Although the non(ONE) and nothingness make possible the mind's "structural retro-action" of meaningful appropriation of multiplicity and potentiality, it remains an open question whether such a philosophical schema fully addresses what is entailed in the idea of Christian desire. For example, Badiou likens computation to the act of gambling: it is not simply a notional or relational act but one that always exists under the influence of chance.[148] The random gives a being the opportunity to make clear her subconscious and premeditated desires.

[145] *Theory of the Subject*, 59. On the influence of the work of the French poet Stéphane Mallarmé (1842–1898) on Badiou's understanding of chance, see *Theory of Subject*, 92–97.

[146] See *Briefings on Existence*, 55. I take Badiou's rejection of constructive mathematics to point to the fact that he sees such approaches as inadequate as linguistic descriptions of being.

[147] See *Being and Event*, 197–198, where Badiou reduces this complex undertaking to the imperative "Decide from the standpoint of the undecidable."

[148] Badiou's understanding of chance originates in the concept of the "*clinamen*," proposed by Lucretius as the random process of atoms falling into the void.

The subjective being creates a "world" that confirms her choices in an infinite drama of probability.[149] As some of a being's contacts are accidental, they are set in reserve in one's subconscious. This information is conserved for observations of future appearances. The vaulting of subconscious information is possible, Badiou argues, as its "interred assignation" is insured by the void's negation and non-existence's interminable arrival in conscious operations.[150] Our own interpretation argues otherwise, with its defense of an actual Transcendent Spirit that human freedom is viable and redeemable in the face of an uncertain future—an eschatological force drives reality and conscious operations toward fulfilling God's will.

All beings have unique essential, conceptual, or informational properties that are shared or naturally reveals in the universe, and these mereological "edges" stimulate personal freedom and motion. A person's memory, reason, and sensual faculties are pneumatically structured. Human perception, whether in the church or the lab, is never reduced to a single practical, ritualistic, or aesthetic gesture. Desire and sensibility exist in a sort of motion between the two theoretical brackets that the ONE and the non(ONE) represent or the faithful and the Transcendent. For "[b]y virtue of this speculative presupposition, the spirit has already begun to 'materialize' itself and matter has begun to become 'spirit.' As the non-essential dimension of pure potentiality. But what opens it to all things is the pseudo-subsisting being as such, its ***activum*** is the Anti-God (***quo est omnia facere***), while it becomes the maternal principle. This is an abyssal event! Receptive reason has generated into pure sensibility."[151] [3] Individual consciousness grounds abstract ideas and thereby holographically situate non-subsistent relationships into a meaningful perception of the sensual and the conceptual as part of the human subject's existence as part of the larger humanity. The cosmos not only makes us who we are: in a type of "fractal" reversal, we, likewise, participate with the Creator to make the world meaningful. Finite freedom and creativity initiates in "spaces" made possible and structured by the "absolutely other" of Trinitarian Ur-Kenosis.

[149] Dialectics always remains a "program or initiation" that is qualified by mathematics' "preexisting procedure." On the dialectical creation of "worlds," see *Theoretical Writings*, 29 and *Logics of Worlds*, 306–308.

[150] *Logics of Worlds*, 104–111 and 302. See also ibid., 223, 311 and 363, where Badiou says that "a site is a multiple which happens to behave in the world in the same way with regard to itself as it does with regard to its elements [e.g., its materiality], so that it is the ontological support of its own appearance."

[151] *Homo Abyssus*, 450.

3.3 Infinities and Providential Design

Even though I reject the atheist's claim that the Transcendent does not exist, I still see the concept of divine nonexistence (non(ONE)) as a useful construct, as it helps us understand Hell and the mystery of Jesus Christ's death on the Cross. Moreover, the proposed holo-somatic relationship between Jesus Christ as the ONE and the hypothetical non(One) parses our interpretation of Ur-Kenosis' quantification and qualification of Nothingness and the void's conceptualization in created existence. Mystics, philosophers, and theologians have long seen an irreducible relationship between physical reality and the void or its semantic equivalent.[152]

Conscious subjects understand finite reality through one's relationships with others and the symbolic manipulations of knowledge by using interior and exterior interfaces. What separates Badiou's interpretation of reality from that of most theologians is the role he gives to the void and infinity for engendering violence and revolt.[153] Badiou's revolutionary Maoism associates violence with social progress. Badiou's philosophical orchestration of revolutionary-inspired social justice cultivates his atheism as a negative arrogation of the infinite and ONE as non(ONE).[154] The extremes of this correspondence can be associated with the brackets used in set theory or the boundary limits of a wave's motion in a quantum system and the universe. Each of these models supports practical judgments and philosophical decisions, as they are open to various denominations, nominations, or interpretations of relational differences that lead to multiplicities, potentialities, and the infinite that symbolize the sequential count or providential unfolding. In comparison, the Christian economic order follows from the eternal harmony and supra-emanation of the hypostatic identity of each of the Divine Persons—movements grounded and illumine the hoped for realities of peace and justice.

The paradoxical union of eternal divine relations and historically nonsubsistent correspondences creates a space for my understanding of tripart monotheistic identity within the world's unjust structures. I risk overstating the truth if I hypothesize that Badiou's philosophical approach to social change directly engages a traditional Trinitarian theology; likewise, I undermine the truth to state that no points of contact are

[152] Anonymous, *Meditations on the Tarot: A Journey into Christian Hermeticism* (New York, NY: Penguin Putnam, Inc., 2002), 85. Hereafter cited as *Meditations on the Tarot.*

[153] See *Being and Event*, 163–164, 168 and 169–170.

[154] See *Being and Event*, 188.

detectable. Badiou's atheistic philosophy informs his political theory of revolutionary action. The disruptive act, Badiou avers, arms individual identity, but I believe, at the cost of rejecting a personal relationship with God. Thus, "[w]e find ourselves on the brink of a decision, a decision to break with the arcana of the one and the multiple in which philosophy is born and buried, a phoenix of its own sophistic consumption. The decision can take no other form than the following: the one is *not*."[155] The atheist's rallying cry is one that has echoed through the ages. It sets human freedom in a necessary and violent relationship to claims of Transcendent monotheistic unity. Three Divine Persons must be understood this line of argument suggests, at best as an anthropological fiction that must be nullified from human consciousness for reality's liberation. One cannot avoid making a decision about the Transcendent, as human freedom and consciousness are creatively bound to the realities of the Transcendent and infinite. I assert that Jesus Christ's nonviolence and willing acceptance of redemptive death motivates a course of radical reception of reality. Christ's actions inverts and rebuffs the Maoist or any other schools of thought that correlate violence, human action, and the disclosure of fundamental identities. It is an open question for each person to decide for himself or herself if this dilemma is resolved by means of prayer or revolutionary praxis.

The mathematical normalization of a being is not a purely reductive act in Badiou's system insofar as his philosophical formalization(s) do not undercut the ordering of physical reality or the immaterial realm that consciously fosters subjective freedom. However, chaos and violence's extreme recounting in history leaves little conceptual "space" for any "thick" description of divine providence to take hold of the human spirit. The closest analogy for theological providence in Badiou's system rests in the axiomatic communication of mathematical truth. The axiom of choice gives us a way to ensure a mathematical set is logically structured. However, this strategy of illuminating the abstract and formal can in no way be equated to the Divine's inspiring the hearts and minds of the faithful. A being's strict representation by mathematical theorems is not discounted with my own holo-somatic theory of being for a weak comparison can be made with Badiou's understanding of chance and freedom's relationship to the non(ONE), the void, and the mathematical complication in axiomatic theory. However, my position moves beyond these considerations and rejects explanations that derive solely from a purely empirical, formal, or idealistic

[155] *Being and Event*, 23.

methodology. Unlike Badiou and other atheists, the Christian perceives chance's spatiotemporal arrival to be holographically part of the hyper-dimensional Transcendent life as revealed in the humanity of Jesus Christ. This mystery results when a likeness between the Transcendent and finite is not founded on the mutual predication of a single essence of formalization but rather when a similarity of relations (*similitudo habitudinum*) or a similarity of proportions (*similitudo proportionum*) exists as part of the human person's subordination to an actual Transcendent Creator.[156]

As the Passion is a definitive event, any and all violence, negativity, and chance in history exist for the Christian as part of the Divine's plan to have free creatures. The Providential gesture actualizes all of Nothingness' harbingering of nonbeing, multiplicity, infinity, potency, and the void as an apophatic revelation of Jesus Christ's freely undergoing in his Passion. The Spirit makes possible our participation in Christ's redemption of Creation. All modalities of Nothingness—from nonbeing to its abstracted spatiotemporal manifestation as deferral, omission, and non-subsistence—are necessary if the holo-somatic is to be a free reflection of the Ur-Kenotic Divine. This event moves beyond, reorders, and perfects the postmodern philosopher's litanies of deconstructed exuberance.[157] An individual being's placement within an infinite relational series or its immediate individuation with reference to the extreme termini of Nothingness and Pure Being, freedom and determinism, does not image meaningless dispersal but rather supports pyscho-somatic identity and the stability of one's conscious grasp of the unfolding salvation history.[158]

3.4 Angelic Natures and Holo-somatic Embodiment of the Infinite

The angelic order gives us a way to conceive of how a spiritual being is uniquely created and part of a community. The angelic being gives us an analogy of how created freedom and irreducible identity metaphysically stand between the theoretical extremes of absolute infinity and potential

[156] See *De ver.*, 2, 2, resp. 11.

[157] See *The Bride of the Lamb*, 54, 60–61 and 261.

[158] See Alain Badiou, *Infinite Thought: Truth and the Return of Philosophy*, Translated and edited by Oliver Feltham and Justin Clemens (London, UK: Continuum International Publishing Group, 2011), 58–68. Hereafter cited as *Infinite Thought*. See also Alain Badiou, *Manifesto for Philosophy*, Translated by Norman Madarasz (Albany, NY: SUNY Press, 1999), 135. On the material being seen as a determination in the domain of potential infinite possible presentations, see *Being and Event*, 127.

nonexistence.[159] John Damascene says of Adam that "[t]his man He set in paradise which was both of the mind and of the sense. Thus, while in his body he lived on earth in the world of the sense, in his spirit he dwelt among the angels, cultivating thoughts of God as being nurtured by these."[160] An angel's ontology is comparable to the dimensionless state prior to the constructions of space or time, that is, the state where nothingness reflects infinity's formlessness prior to creation. Angels are metaphysically structured around this point prior to divine providence's historical realization of immaterial and matter's potential place within the economic order.[161] Physical matter's *metanoia* is only spiritually pliable if one accepts both Creation's source in the Father and the Son's kenotic embrace of the Spirit's ongoing transformation of Nothingness and the full gamut of material deprivations and entropy.

The angles are hierarchically ordered as Angels, Archangels, Principalities, Powers, Virtues, Domains, Thrones, Cherubim, and Seraph. Each of these nine groups represents a unique genus and is taken as an analogy of nine penultimate subsets of potential or created infinities directly under the actual infinity—the Transcendent—I AM WHO AM. Infinity does not necessitate universal formlessness or chaos, as various magnitudes of infinities when understood as generated from a common source. Moreover, one need not presume that a dichotomous relationship exists between the infinite and the finite. The infinite dimensions of human desire and the unlimited potential existence of created multiplicities in reality leave room for autonomous finite freedom. As nonmaterial beings, angels are in no way spatiotemporally limited as physical agents. The angel is not ontologically above the created order, but the angel's non-bounded existence can be taken as an analogy for the Spirit. Angels exemplify how a finite being participates in reality as an embodied spiritual being. It is perhaps these themes that also implicitly drive Thomas' intuition that the angelic order is numerically greater than the sum of all multiplicities found in Creation.[162]

Cantor gives us the mathematical tools to understand transfinite sets, and Thomas gives us a theological vocabulary to speak of potential

[159] See *TL* 1, 103–105.

[160] *John Damascene, An Exact Exposition of the Orthodox Faith, NPNF2*, vol. 9, Book 2, chapter 11, Translated and Introduced by Philip Schaff, D.D., LL.D., and Henry Wace, D.D. (Grand Rapids, MI; Wm. B. Eerdmans Publishing Company, 1983).

[161] See *Presence*, 47–55.

[162] See *ST*. I, 47, 1.

infinities in the angelic and spiritually embodied realms. Angelic perfection or absence models how a created being exists in relationship to the actual infinity of the eternal Transcendent life. Thus, "since … the divine being is not a being received in anything, but He is His own subsistent being as was shown above, [therefore] it is clear that God Himself is infinite and perfect."[163] The eternal Transcendent mind is proportionally reflected in the spiritual "ever actual intelligence" of angelic consciousness. Indeed, "[f]rom the moment of their creation,…the intellect of angels is perfected by innate forms giving them all the natural knowledge to which their intellectual powers extend, just as the matter of celestial bodies is completely terminated by its form."[164] The angel's conscious apprehension of a created being inverts human consciousness' dependence on the material properties, as angelic intellect immediately understands by infused knowledge what must be unveiled for us—the buried essential truth of a being.

Lacking material limitations, the angelic body holographically spreads across the physical universe like a drop of water's union to the distant shore of the ocean or the QMO's propagation across the cosmos. The angel's spiritual body experiences material reality in the manner in which light reflects a physical being. "The angelic nature," Thomas asserts, "is itself a kind of mirror representing the Divine image."[165] Most obviously, angelic being has traditionally been depicted in Christian theology as spiritual or immaterial. Thus, Thomas reasons of angels that angelic nature exists as a simple spiritual body. [166]

Angelic nature or quiddity is not received in any matter but is a self-reflective spiritual or essential holo-somatic embodiment. It is said that intelligence is unlimited in regard to one's capacity to receive but limited from above or in terms of one's actual existence. [44] Angelic spiritual embodiment is a holo-somatically looking glass for the essential truth of every being and the simplicity of the divine essence. This metaphysic affords me with how mental images can be comprehended as holographic. Although angels do not know God perfectly, their spiritual nature means they infinitely know the divine, since "an angel's intellect and essence are infinitely remote from God, it follows that [he] cannot comprehend Him;

[163] *ST.* I, 7, 1.
[164] *De ver.*, 8,4. See also *ST.* I, 55, 1.
[165] *ST.* I, 56, 3.
[166] See Aquinas, *On Being and Essence*, chapter 4.

nor can he see God's essence through his own nature."[167][14] The Divine holo-somatically manifests the pure divine essence nominates the Spirit qua Spirit. The materially boundless angelic nature virtually embodies a created analogy for divine thought's eternal self-reflectivity in the created order. Relational differing is the operative force that makes the relative self-subsistence of created identity an endless project—no one engagement of divine exemplarity exhausts its ability to inform and reveal the Transcendent. Thomas reasons that "If we speak of the infinite in reference to form, it is manifest that those things, the forms of which are in matter, are absolutely finite, and in no way infinite [as Aristotle would argue]. If, however, any created forms are not received into matter, but are self-subsisting, as some think is the case with angels, these will be relatively infinite, inasmuch as such kinds of forms are not terminated, nor contracted by any matter."[168] Angelic identity spiritually proportions the essential as nonmaterial creation. Thomas' theological and philosophically develops how the spiritual nature of angelic embodiment and knowledge reflects in human existence, thought, and judgment.

The union of the angel's spiritual form and holo-somatic essence means that their identity is fundamentally decided by a single choice—whether to be obedient or not to the Transcendent. All future choices and actions made by an angel conform and support this decision—without loss of their angelic nature. Furthermore, given the nonmaterial angelic embodiment, this choice for or against God and their future actions is nonmaterially present in all their subsequent decisions. A trace of their relationship to the Divine "diffuses" throughout the physical universe. Once an angel has defined their orientation to or against the Transcendent, they, like all other human beings, engage and associate the relational tensions emerging from the unending iterations born in created hylomorphic non-subsistence. Unlike the angels, the future is more open-ended for free finite beings that make sense of their lives as grounded in possible conversion in an uncertain universe. The inimitable and evolving nature of a created beings' unlimited potency and ability to enter into non-subsistent relationships also shares something of angelic actualization of the infinite. The angelic being confirms subtraction's paradoxical engendering of the positive. In the ideal, angelic nature directly aligns with infinite perfecting, for "[t]he fact of its not extending itself to its own proper limit renders the

[167] *ST*. I, 56, 3.
[168] *ST*. I, 7, 2.

spirit limitless and infinite in a certain way: the nature of souls and of angels does not know limits, and nothing stops their respective natures from progressing to the infinite."[169] [45] The infinite signifies here that the Passion of Jesus Christ perfectly ripples throughout the metaphysical order—there are no conceptual or ontological obstructions or dimensional hierarchies that keep the Spirit from revealing its holo-somatic truth. Where Badiou speaks of how a being parallels the cardinality of the infinite mathematical set, Ulrich transposes the Ur-Kenotic nature in our lives as he sees it to be key to the faithful person's ability to know. Ulrich writes that for "the angel, the decision concerning obedience with respect to the necessary sense of being is made in the moment of creation with a view to man, not in the successiveness of physical spatiotemporality. But, man, as embodied spirit, so long as he is in this body, has to undergo the temptation of the negation of the kenosis of being in its super-essentiality."[170] [3]

The positive core of the negative testifies to the same mystery found in Ur-kenosis and holographically in the Passion and Eucharist. The angelic order "infinitely" testifies or relates to the Eucharist, as every angel is a singular species and consequently instantiates a singular intellect and observational perspective of the formal or sacramental principles in the cosmos. Choice for the hylomorphic creature is essential, as it moves finitely toward inevitable death—but one that models the person of Jesus Christ. Like information, the human soul depends upon the body to serve as a material "transducer" to encode, "record," and "transmit" its infinite essential movements. The finitization of spatiotemporal being is lived daily at the physical level given the inordinate role memory plays in our day-to-day encounters. The biased and truncated nature of our experiences means that a being never fully reveals in the moment its hidden resurrected "corporeity" in the physical universe. In the Old Testament, the angels at Passover prefigure the Spirit's power to transform and reorder death to redemption and resurrection. Blood biologically and symbolically refers to a being's "life-force." Thus, it is "perfectly obvious why the event of the destroying angel sparing the first-born of the houses marked with the blood of the lamb should be an outstanding symbol of redemption. We

[169] Nyssa, *Answer to Eunomius' Second Book*, *NPNF1*, vol. 5, Translated by M. Day, Revised and Edited for New Advent by Kevin Knight, <http://www.newadvent.org/fathers/2902.htm>.

[170] *Homo Abyssus*, 321.

have here one of the essential theologoumenon of Scripture: the world is under the dominion of sin, judgment is to strike the guilty world, but God spares those who are marked with the Blood of Christ."[171]

Mary experiences this life-giving contradiction as she holds her dead bloodied son symbolically disgorged from all evil and sin when taken down from the Cross. Mary's identity becomes Eucharistic at the foot of the Cross: she is the first to experience Creation's anticipation of conversion that is made possible in the ecclesial and sacramental celebration of the Risen Lord. Mary experiences the Eucharist as part of her ontological state as human mother and proleptically as Theotokos, the first spiritual intercessor for the future Church.[172] Mary gives ecclesial existence its underlying personal dimension because she is physically, spiritually, metaphysically, historically, and holo-somatically unified with Christ. Mary is assumed into heaven without dying outside of her experience of her son's death. She freely and infinitely radiates the resurrected state in her life. The Eternal Logos' Incarnation as a person with divine and human natures gives us a way to imagine the created universe's holo-somatic ordering. In the sinless person of Jesus' mother Mary and in angelic nature, we find exemplified the theological basis for the sacramental altar of sacrifice and its fulfillment in resurrected citizenship in New Jerusalem.

In the holo-cryptic event of the meeting of the Logos' "non-causal" divine nature and Mary's human nature, Nothingness' critique of the infinite is realized. The mysticism of infinity is Christological since the eternal irrupts into the finite as an endless moment that simultaneously witnesses in Jesus Christ's Incarnation, Passion, and the Eucharist.[173] Angels can "contract" their infinite immaterial form to reveal themselves in the spatiotemporal order. In this way, the angel analogously lives what QMT has experimentally shown with the QMO's collapse and what the Eucharist asks of the faithful to "see spiritually." The essence of a material or immaterial being spatiotemporally reveals impartially in a being's appearance. A physical artifact's meaning depends upon the conscious state of its observer. To state otherwise risks initiating a new idolatry around matter's commodification. Paradoxically, the object can only become an idol when

[171] *The Bible and The Liturgy, 163.*

[172] See John Damascene, *An Exact Exposition of the Orthodox Faith*, *NPNF2*, vol. 9, Book 5, 13.

[173] Hans Urs von Balthasar, *Elizabeth of Dijon: An Interpretation of her Spiritual Mission* (London, UK: ET,1956), 64–65.

the Divine perceives it as such. The fabricated object "exhausts the gaze presupposes that this thing is itself exhausted in the gazeable. The decisive moment in the erection of an idol stems not from its fabrication, but from its investment as gazeable, as that which will fill a gaze."[174] One can never immediately judge that one's perception of a universe-created good is suspect. However, every re-representation of the universe as a mental image involves human freedom and is therefore by definition a moral act. A being's holo-somatic disclosure argues for a being or image's trans-historical and supernatural dimensions. The spatiotemporal truth of a being at any "one time" is at once a metaphysical anticipation of the future (advent), a memorial of past revelations of ontological differing or holo-cryptic identity (Eucharistic), and an immediate turning and attempt to hold on to the present needs of one's situation (pneumatic). The chronology of past, present, and future outside of its theological and holo-somatic truth solidifies the spatiotemporal in its autonomous finitization in unending succession.[175] In contrast, when infinity is taken as a holographic image of eternity, finitization transmutes into Ur-Kenotic created fecundity. To this end, Balthasar seems to look to Mary and the angelic order, for the angels were not created in time but "in eternity to accompany the lamb of self-sacrifice… the somatic presence of the Eucharistic Son and the woman clothed with the sun."[176] Angels can be ontologically quantified in spatio-temporal reality because they are not constrained by physical laws or material conditions but the Transcendent. The angel's spiritual identity foresees the metaphysical freedom that will be accorded to the resurrected body—a somatic state that believers share when they participate in the Eucharist with the hope and the faith that "we shall be like him" (1 John 3:2).

The Eucharist is a part or a "seed" of a larger cosmos eschatologically working toward its own "resurrected" or converted state. The role the holographic representation plays within a holo-cryptic metaphysics gives us an ontic and noetic language to speak of this mystery and be sensitive to a scientific worldview. The distance separating an electron and its holographic display, for example, is prescient of the material body's holo-somatic annunciation of resurrected flesh and immaterial embodiment. We see a similar holographic calculation when the relevant theological, sacramental, and metaphysical roles are accorded to the angels at Christ's

[174] Jean-Luc Marion, *God Without Being*, Translated by Thomas A. Carlson (Chicago, IL: University of Chicago Press, 1991), 10. Hereafter cited as *God Without Being*.

[175] See *Homo Abyssus*, 325–26.

[176] *GL* 3, 382.

tomb and at the liturgy. Christ's historical return preambles the world's "final harvest" by the angelic reapers—an end that returns Creation to its original holographic imaging of Paradise. The angelic body ontologically portends the metaphysical freedom and the effected differing of bringing in the harvest of human souls—the metaphoric infinite wheat—the spiritual partner to the eternal bread of the Eucharistic sacrifice. The angelic person and mission define heavenly existence and simultaneously participate in the Father's eternal plan for Creation. As spiritual beings, angels' nonlocal nature is localized when observed. This ontology finds it material parallel in the QMO that can exist as a wave or particle. As wave, the QMO manifests at the theoretical edge or at the universe's "end." One of the hallmarks of the quantum system is its indeterminacy, and as Ulrich argues, "uncertainty has its analogous philosophical explanation in terms of the essence's spatiotemporal correspondence to materiality."[177]

3.5 Holo-somatic Personhood

The human person cannot fully comprehend an actual infinity but can abstract from created reality its image. Impervious to measure, divine life is without latency, it is continuous and immediate as "*Deus est actus purus non habens aliquid de potentialitate.*"[178] God's self-sufficiency premises divine perfection and station our own limited conscious capacities.[179] These representations of pure beings are taken as a holo-somatic simulacrum for created intelligence and are foundational for our identity as human persons.

The Person of the Spirit's absolute freedom defines the original foundation of Creation's ultimate end. The Spirit's revelation of Creation's final and first cause allows for penultimate and secondary causal expressions of the same divine ground, for we act "In things willed for the sake of an end, the whole reason for our being moved is the end, and it is that, that moves the will, as most clearly appears in things willed only for the sake of an end."[180] Nothingness becomes a positive factor when informed by Divine thought. Various modalities of Nothingness fence in and serve to

[177] See *Homo Abyssus*, 243.

[178] See *ST.* I, 3, 4 and *ST* I. 7,4.

[179] See Nyssa, *On the Holy Spirit: Against Macedonius*, *NPNF2* vol. 5, Translated and Introduced by Philip Schaff, D.D., LL.D., and Henry Wace, D.D. (Grand Rapids, MI; Wm. B. Eerdmans Publishing Company, 1983).p. 316 and 318.

[180] *ST.* I, 19, 2, resp. 2.

individuate beings when bounded; they arrange those places or metaphysical transitions necessary for individual identity without providing any access to the conclusive meaning that an actual Transcendent Spiritual Being provides. The concept of nothingness purposes various conscious notes in existential life. Most notably, nothingness serves to virtually deconstruct the very being made possible by its metaphysical framing to engender new and emergent images of being and thus creative spaces for novel perspectives. Postmodern thought advances a philosophical, theological, and scientific basis for attributing to the void and nothing the ultimate means to promote the truth's multidimensional reality without any material limitations. All finite displacements of nothingness and nonbeing are virtual and never challenge the extreme ideality of Trinitarian identity at "[t]he moment that the essences no longer have their place (in the mode of nonbeing) in God's ideality as a *nihil intellectum*, but instead the nothing's internal contradiction drives it from the divine's intellect, God necessarily gets determined by an otherness to which he is subject…the attempted dislodgment of the nothing 'shrinks' God into an abstract absolute identity, afflicts the Logos with the other to himself, depotentiates the Father, and introduces the negation of negation that belongs to the metaphysics of Spirit."[181] [3] By rejecting Nothingness' ability to medium cognition, one simultaneously denies a personal Creator in the person of the Spirit.

The Trinity's ingress into finitude with the Incarnation contravenes neither divine unity nor the human being's natural desire to know and act.[182] Jesus Christ is the eternal Son whose human supereminence unlocks the way for all other human beings to participate in created Nothingness as the Transcendent's pressure under judgment on the Cross. Christ's divine-human person corresponds perfectly to the mission the Father has given him. For, "it is only possible to apply qualities and attributes of the one nature to the other because both are united in the one person of the logos—not by way of nature, but by way of person… the properties of each remain unimpaired [*sozömenes*]."[183] The absolute difference

[181] *Homo Abyssus*, 182.

[182] See *TD* 2, 87–88 and *TD* 4, 81–84.

[183] *TD* 3, 222. The Council of Chalcedon (451) teaches that Jesus Christ's union of the divine and human represents the assumption but not absorption of human nature. It is in the dramatic integration of the divine and human that gives Balthasar his argument against

separating the states of divine and finite consciousness is paradoxically and flawlessly intertwined in Trinitarian life and in the consciousness of Jesus Christ.[184] The inscrutability of the divine-human union persists beyond all attempted explanations and in the theological categories of person and mission when ordained to the Transcendent. Jesus Christ never loses sight of Satan's attempt to disentangle God's will from Creation—evil desires an autonomous Nothingness.[185]

Christ's humiliation is possible because he freely renounced his transcendent state even to the point of being "obedient unto death"[186] (Philippians 2:8). The transcendental orientation of his person is realized in his prayer and obedience. He makes possible ecclesial existence that stands against a deterministic universe. God will not let the creature unknowingly fall into a false sense of Nothingness.[187] The nature of the Spirit's consciousness remains a mystery, but one that increasingly reveals both Jesus' consciousness when facing the Cross and later as the first resurrected human being. All human beings are asked to struggle with this mystery. For, we have "faces unveiled, reflecting as in a mirror the glory of the Lord is being transformed into his very image from glory to glory, as through the Spirit of the Lord" (2 Corinthians 3:18). God's eternity is not a matter of endless time but gives divine tri-personal reflectivity an endless life. The Divine life is simple and eternal and is not bounded or limited by the infinite. [45] The Spirit draws out the eternal implications of Jesus' finite actions for the Trinity and future generations (Proverbs 16:4).[188] The Spirit's work is a requisite part of Christ's Parousia and is tethered to the Spirit's own sense of self.[189] Divine glory (*doxa*) is to be seen in Jesus' unmerited abuse, his final failed gasp for air, as well as his death and descent into Hell. All are part of his and the Trinity's engagement with reordering sin and evil to their proper metaphysical state of Nothingness. This can only be the case if Jesus' degradation is holo-somatically part of

Nestorianism. With the resurrection, the fully human Jesus Christ enters fully into the Trinity an offers a new vision of human embodiment. See also *TD* 5, 474.

[184] See *TD* 3, 230 and Augustine, *On the Holy Trinity*, *NPNF1*, vol 3., Book 9, chapter 4. Translated and Introduced by Philip Schaff, D.D., LL.D., (Grand Rapids, MI; Wm. B. Eerdmans Publishing Company, 1980).

[185] *See TD* 3, 230–231.

[186] *TD* 3, 521.

[187] See *GL* 1, 628.

[188] See *TL* 3, 310–334.

[189] See *The Bride of the Lamb*, 396. See also *TL* 3, 205–217.

the Son's eternally generative relationship with the Father as Pure Actual Being.[190]

The Eternal Groundless Spirit does not overthrow the natural order of Jesus Christ's human consciousness but ensures its perfected state.[191] The death of Jesus Christ and his descent into the presumed non-redemptive Nothingness of Hell precedes and inspires the Spirit's own mission consciousness. Most postmodern thinkers would likely reject the idea that the Transcendent's inventory of Hell and Nothingness are generative principles of human consciousness. Directed solely by the dictates of a materialistic philosophy, human cognition never exceeds what can be achieved by the creative cloning, ordering, and conscious representation of various physical beings. François Laruelle's work, for example, highlights how self-reflection leads to an actual, and perhaps total, foreclosure of being. Applying this interpretative approach to the question of the Passion, most postmodern thinkers hesitate to take Jesus' understanding of his own death and the possibility of nonbeing as positive influences on divine consciousness and instead see them as the rhetorical ornament of deluded gospel writers. The introduction of any finite property—even that of immaterial thought—is regarded as a different metaphysical or trans-metaphysical order when applied to the Divine. It is nonetheless not ironic to attribute characteristics and properties to Jesus Christ's holo-somatic reflection of the impassible Transcendent. Laruelle's philosophy suggests otherwise if one accepts claims that finite thought is determinate and results in mutual foreclosure.[192]

The occlusion of being, Laruelle argues, is tied to a non-existing Transcendent. Laruelle philosophically explores the nature of negation—an avenue important for mathematics and theology alike—it is not simply academic. The faithful believe that Jesus Christ converts the created order back to its original divinely appointed end by negating its most glaring negation—death. Through prayers and symbolic and sacramental action (*vi verborum*), the faithful celebrate reality's redemption from death and its effects (Revelations 7:9–17, 8:1–4; 14:1–4). The spatiotemporal nature of history and finite being's natural path of ongoing perfection support the Divine's plan because we are all infinite insofar as we exist in a state of potency. Thus, the Roman Catholic Church does not frown upon the multiplication of masses, there being no limit to how much prayer and

[190] See *ST.* I, 2, 1.
[191] See *TL* 2, 78.
[192] See *Dict. Non-Phil.*, 97.

grace can be applied to any particular issue, nor any limit to the "speed" or "distance" that this specific form of information (prayer) can travel. Even though death destroys our three-dimensional means of framing the spatiotemporal, the essential truth of this activity remains after death.[193] As Jesus states, "not one of them you gave me have I lost'" (John 18:9). Fully divine and human, Jesus' person and mission confer upon believers the sanctifying grace found in the Eucharist (1 Corinthians 11:26). The Mass draws its efficacy from the living God, the eternal mystery who is partly comprehended, by observing the memorial of the Cross and resurrection. As he says, "I am with you always, even unto the end of the world" (Matthew 28:20). The Eucharist proclaims in words and remembers in deeds the historical death and resurrection of Jesus Christ (1 Corinthians 5:7). Within the context of the liturgy, the memorial elucidates a theological interpretation of the spatiotemporal.[194] In the reciting of sacred scripture, the Church reveals its trans-historical character in the Eucharistic anamnesis of the liturgy. Christ reminds us that he will bring "all things to your remembrance, whatsoever I have said unto you" (John 14:26).[195] Death is the natural end of an organic biological process and consequently the means to sin in the world. In part, it defines one's "immaterial or spiritual" moral order.[196] While the dead no longer have the ability to pray, Bulgakov remarks, they remain part of Providence's unfolding.[197] Death is part of salvation's economy, for although death "is the supreme punishment for man's original fault, it is death, nonetheless, that detaches us from the world and purifies us of all concupiscence."[198] Evil is not to be equated with matter, but rather with that lack that results from willfully perverting the goodness in the created order and turning away from the Divine.[199] Immorality's holo-somatic mode manifests the existence of the nonbeing of the good. It is part of Nothingness' reiterative ordering of the

[193] *Mystery of the Eucharist*, 300–301.

[194] See *God Without Being*, 172-173.

[195] See *GL* 1, 577.

[196] See Nyssa, *On the Soul and the Resurrection*, *NPNF2*, vol. 5, Translated and Introduced by Philip Schaff, D.D., LL.D., and Henry Wace, D.D. (Grand Rapids, MI; Wm. B. Eerdmans Publishing Company, 1983), 447.

[197] See *The Bride of the Lamb*, 365.

[198] See Nyssa, *On Making Men Moral*, *NPNF2*, vol. 5, chapter XVII, Translated and Introduced by Philip Schaff, D.D., LL.D., and Henry Wace, D.D. (Grand Rapids, MI; Wm. B. Eerdmans Publishing Company, 1983).

[199] See Nyssa, *On Making Men Moral*, *NPNF2*, vol. 5, chapter XVIII, Translated and Introduced by Philip Schaff, D.D., LL.D., and Henry Wace, D.D. (Grand Rapids, MI; Wm. B. Eerdmans Publishing Company, 1983).

material domain that it does not eradicate the essential positivity of Creation. In the following chapter, we turn to the study of QMT to prove our concept of holo-somatic identity.

References

1. Mark C. Taylor, *After God* (Chicago, IL: The University of Chicago Press, 2009).
2. Luciano Floridi, *Information: A Very Short Introduction* (Oxford, UK: Oxford University Press, 2010), 45.
3. Ferdinand Ulrich. *Homo Abyssus: The Drama of the Question of Being*, Translated by D.C. Schindler (Washington, DC: Humanum Academic Press, 2018), 401.
4. Stephan Körner, *The Philosophy of Mathematics: An Introduction* (New York, NY: Harper Torchbooks, 1960), 34.
5. Wolfgang Smith, *The Quantum Enigma: Finding the Hidden Key*, Forward by Seyyed Hossein Nasar (San Rafael, CA: Angelico Press: Sophia Perennis, 2005).
6. Werner Heisenberg, *Physics and Philosophy: The Revolution in Modern Science* (New York, NY: Harper Row, 1962), 58.
7. Wolfgang Pauli, *Scientific Correspondence with Bohr, Einstein*, Heisenberg," Edited by K. von Meyenn, et al., 2 Vols. (Berlin, DE: Springer Verlag, 1979–), 231.
8. John Conway and Simon Kochen, *The Strong Free Will Theorem* (Princeton, NJ: Princeton University Press, 21 July 2008), 5.
9. Giulio Chiribella, Giacomo Mauro D'Ariano, and Pablo Perinotti, "Informational derivation of quantum theory," *Physical Review* A 84, no.1 (2011): 012311.
10. Niels Bohr, *Atomic Physics and Human Knowledge* (New York, NY: Wiley, 1958), 90.
11. Pavel Florensky, *The Pillar and Ground of the Truth: An Essay in Orthodox Theodicy in Twelve Letters*, Translated by Boris Jakim. Introduced by Richard F. Gustafson (Princeton, NJ: Princeton University Press, 1997).
12. Pavel Florensky, *Iconostasis*, Translated by Donald Sheehan and Olga Andrejev. Introduction by Donald Sheehan (Crestwood, NY: St. Vladimir's Seminary Press, 1996), 44.
13. Ambrose, "On the Mysteries," *NFPF2*, vol. 10, 9, 52.
14. Thomas Aquinas, *Summa Theologiae* I, (Garden City, New York: Doubleday Company, 1964–1976).
15. Albert Einstein, *Autobiographical Notes*, in *Einstein: Philosopher—scientist* Edited by P.A. Schlipp, (Evanston, IL: Library of Living Philosophers, Inc., 1949), 32.

16. Barbara Maria Stafford, *Echo Objects: The Cognitive Work of Images* (Chicago, IL: The University of Chicago, 2007).
17. Herman Weyl, *Philosophy of Mathematics and Natural Science* (Princeton, NJ: Princeton University Press, 1949), 160.
18. Paul Davies, "Complexity and the Arrow of Time," in *From Complexity to Life: On the Emergence of Life and Meaning*. Edited by Niels Henrik Gregersen (New York, NY: Oxford University Press, 2003), 73.
19. Max Planck, "The Place of Modern Physics in the Mathematical View of Nature," (1910), in *A Survey of Physical Theory*. Translated by R. Jones D.H. Williams (New York, NY: Dover Books, 1960), 28.
20. Pavel Florensky, "Reverse Perspectives," in *Beyond Vision: Essays on the Perception of Art*, Edited by Nicoletta Misler and Translated by Wendy Salmond (London, UK: Reaktion Books LTD., 2002).
21. Paracelsus, "Concerning the Signature of Natural Things," in *The Hermetic and Writings*: vol. 2, no. 4, ed. Arthur Edward Waite (London, UK: James Elliott, 1894), 259.
22. Hans Urs von Balthasar, *Glory of the Lord: A Theological Aesthetic*: vol.1, Edited by Joseph Fessio, S.J and John Riches. Translated by Erasmo Leiva-Merikakis and John Riches (Edinburgh, UK: T&T Clarke, 1982).
23. Trevor Hart, "Lectio Divina," in *Theology, Aesthetics, Culture*, Edited by Robert McSwain and Taylor Worley (Oxford, UK: Oxford University Press, 2012), 231.
24. George F.R. Ellis, "Physics in the Real Universe: Time and Space—Time," in *Relativity and the Dimensionality of the Word*, Edited by Vesselin Petkov (Montreal, QC: Springer, 2007), 74.
25. Max Weber, "Fourth Dimension from a Plastic Point of View," Camera Work 31 (July 1910): 25.
26. Linda Dalrymple Henderson, "Mysticism, Romanticism, and the Fourth Dimension," in Maurice Tuchman and Judi Freeman Edited by et al., *The Spiritual in Art: Abstract Painting: 1890–1985* (New York, NY: Abbeville Press, 1986), 224.
27. Vesselin Petkov, *Relativity and the Nature of Spacetime*, 2nd ed. (Berlin, DE: Springer, 2009), 172.
28. Martin E. Marty, "Tibetan Mandala: Ad Reinhardt, Thomas Merton, John Cage," in *Negotiating Rupture: The Power of Art to Transform Lives*, Edited by Richard Francis (Chicago, IL: The Chicago Museum of Contemporary Art Chicago: 1996), 78.
29. Etienne Gilson, *Being and Some Philosophers*, 2nd ed. (Toronto, ON: Pontifical Institute of Medieval Studies, 1952).
30. Francis Bacon, "Advancement of Learning," Book 2, III, 350 in *The Works of Francis Bacon*, Edited by J. Spedding, R.L. Ellis, and D.D. Heath (new ed.; London: 1870).

31. Alain Badiou, *Ethics: An Essay on the Understanding of Evil.* Translated by Peter Hallward, (London, UK: Verso, 2001), 25.
32. Ricardo L. Nirenberg and David Nirenberg, "Badiou's Number: A Critique of Mathematics as Ontology," *Critical Inquiry* 37 (Summer 2011): 584.
33. John Gribbin, *Deep Simplicity: Bringing Order to Chaos and Complexity*, (New York, NY: Random House, 2004), 147.
34. Etienne Gilson, *The Christian Philosophy of St. Thomas Aquinas.* Translated by L.K. Shook (New York, NY: Random House, 1966), 36.
35. Hans Urs von Balthasar, *Truth of the World* vol.1. Translated by Adrian J. Walker (San Francisco, CA: Ignatius Press, 2000).
36. Robert J. Russell, "God's Providence and Quantum Mechanics," Counter balance, http://www.Counterbalance.net/physics/qmprovid-fram.html. Accessed April 11, 2013.
37. W.D. Hart, *The Evolution of Logic* (Cambridge, UK: Cambridge University Press, 2010), 15.
38. Joseph Warren Dauben, *Georg Cantor: His Mathematics and Philosophy of the Infinite* (Princeton, NJ: Princeton University Press, 1990).
39. Aristotle, *The Nicomachean Ethics.* Translated with an Introduction. Re-issued 1980, revised by J. L. Ackrill and J. O. Urmson (Oxford, UK: Oxford University Press), 1140a7–17.
40. Thorleif Boman, *Hebrew Thought Compared with Greek.* Translated by Jules L. Moreau (Philadelphia, PA: The Westminster Press, 1960), 129–130.
41. David Tracy, "The Catholic Imagination: The Example of Michelangelo," in *Heavenly Bodies: Fashion and the Catholic Imagination*, Vol. 1, Edited by Andrew Bolton, et al., (New Haven, CT: Yale University Press, 2018), 18.
42. Norman Russell, *The Doctrine of Deification in the Greek Patristic Tradition* (New York, NY: Oxford University Press, 2006), 306.
43. Oliver Feltham, "Translator's Preface," in *Alain Badiou, Being and Event* (New York, NY: Continuum International Publishing Group, 2005), xxiv.
44. Thomas Aquinas, *De Ente et Essentia* 4, 92. Translated as *Being and Essence.* Translated, interpreted and adapted and html-Edited by Joseph Kenny, O.P. https://isidore.co/aquinas/DeEnteEssentia.htm.
45. Nyssa, "Against Eunomius," in *NPNF1*, vol. 5.
46. Jeffrey Koperski, *The Physics of Theism: God, Physics and the Philosophy of Science* (Singapore, S.G., : John Wiley Sons, 2015).
47. Sergius Bulgakov, T*he Bride of the Lamb*, Translated by Boris Jakim (Grand Rapids, MI: Wm. B. Eerdmans Publishing Co., 2002).

CHAPTER 4

Holo-cryptic Metaphysics in an Entropic Universe

4.1 Toward a Quantized Metaphysics

Classical physics claims that it is possible to understand a system if one determines its initial state and all forces acting upon it. By identifying all causal forces in a particular physical system, this line of inquiry proposes, one can place each of its parts into a model of the larger whole. Newton's clarifications of a particle's position and momentum assume such knowledge.[1] The possible extent of our knowledge is limited, as Nicolas Gisin explains, "Raising these theories to the status of ultimate truths in a dogmatic, almost religious way amounts to a straight error of logic, since they are contradicted by our existence of free will."[2] [1] Quantum probability can be taken as a physical analogate for free will.[3] The physical actualization of a QMO's position and momentum, its spin, its energy, relative duration, the value of a particle's field, etc., all determine its formal arrangement and probable measure. In observing a QMO, the subject allocates its potential or probable physical states in a physical measurement and conscious images. The correlation of essential or formal explanations

[1] See David Z. Albert, *Time and Chance* (Boston, MA: Harvard University Press, 2000).

[2] Nicolas Gisin, *Quantum Chance: Non-locality, Teleportation and Other Quantum Marvels*, forward by Alain Aspect (Cham, Heidelberg: Springer, 2014), 90–91. Hereafter cited as *Quantum Chance.*

[3] See Henry Stapp, *Mind, Matter, and Quantum Mechanics* (Berlin, DE: Springer Verlag, 1993), 16–20. Hereafter cited as *Mind, Matter and Quantum Mechanics.*

M. P. Fusco, *The Physics and Metaphysics of Transubstantiation*,
https://doi.org/10.1007/978-3-031-34640-8_4

of the QMO as conscious image is a critical addendum to post-Newtonian research in light of my own holo-cryptic metaphysics.

The modern story of QMT in many ways begins when Johann Jakob Balmer (1825–1898) devised an experiment to test the properties of the hydrogen atom. Balmer's charting of the hydrogen atom's wave-like structural topology is preliminary to later conclusions on the nature of the QMO. Building upon Balmer's work, James Clerk Maxwell (1831–1879) distinguished a subatomic particle's orbit and its specific electrical charge. Specifying the material properties of an atom's charge and its associated geometries, Maxwell formulates his laws of electromagnetism. Thus, a material object is associated with its own image as a field of energy. The uniform electromagnetic radiation emitted from a black box inspired Max Planck's assertions about light and quantum mechanics. Previous experiments show that light filtered through a prism breaks down into its component parts—a continuum of distinguishable colors. Planck experimentally verified and mathematically defined that these colors are discrete packages of energy—again, the physical object correlates to its energetic image. The electron's orbital spin radiates continuously, and this movement distributes as an electromagnetic field.

Planck mathematically determined the relationship between a QMO and its presence as a field or "image." Furthermore, he shows how an object corresponds to its spatiotemporal dimensions as the QMO's spin and angular movement creates a wave whose frequency mimics its momentum (energy). The quantum's angular momentum is $\hbar/2$, where $\hbar$ equals Planck's constant (approximately 6.626176 G10 joule-seconds) divided by 2p. Thus, QMO's energy (E) in a system is equal to the frequency of its electromagnetic wave (v) multiplied by a unique constant ($\hbar$ that is E = v). Light's propagation at a constant speed is, at a deeper level, a collective effort of various frequencies or "speeds" of radiation (X-rays, gamma rays, etc.) within an observed frame of reference. The formal level of description of subatomic realities given by Planck's constant $\hbar$ gives us a way to grasp how a QMO's different frequencies engage spatiotemporal and higher dimensions. An overarching harmony prevails as various frequencies work together to give light to its constant speed. I see in this physical quantification of light an example of the metaphysical principle of "unity-in-difference." In his *Atoms and Human Knowledge*, Bohr applauds Planck's explanation of the relational or "subtractive" dimensions of a QMO. Planck's work, I suggest, presents an interpretation of physical matter that is commensurate with my view of reality that is radically relational or "kenotic."

Einstein advanced the study of QMT with his work on the photoelectric effect, that is, the effects of a photon when making contact with a metal plate. David Park explains that "doubling the intensity of the light should double the number of electrons without changing their average energy, whereas since the energy of a photon is proportional to the frequency of the light, increasing the frequency should increase the energies of the electrons that get knocked loose. This, it turned out, is exactly what happens."[4] [3] The hidden motion or temperatures of the metal plate increases as electrons are emitted after photons strike it. In this causal interaction the atom's chromatic properties are confirmed and anticipates Einstein's later insight into the reciprocal relationship between mass and energy ($E = mc^2$). Counter intuitively; given sufficient momentum (energy), a massless particle (photon) changes physical reality.[5]

Louis de Broglie (1892–1987) applied Einstein's formula for the photoelectric effect and Arthur Compton's findings concerning the collisions of photons and electrons to arrive at his theory of quantum waves.[6] De Broglie defined a particle's momentum (p) and wavelength (λ) to arrive at his formula: $\lambda = \hbar/p = \hbar/mv$.[7] Where the photoelectric effect conceptualizes the QMO's momentum in relationship to two-dimensional space (e.g., a metal plate), de Broglie highlights the quantum wave as a material particle in three-dimensional space-time. Holding electrons are to waves as photons are to particles, de Broglie discovered "matter—waves." The quantum particle translates as a wave's amplitude (height), wavelength (l), and the distance between their adjacent crests of the wave's summit. The

[4] David Park, *The How and the Why: An Essay on the Origins and Development of Physical Theory* (Princeton, NJ: Princeton University Press, 1988), 308. Hereafter cited as *Park*.

[5] See Albert Einstein, "*Uber einen die Erzeugung und Verwandlung des Lichtes betreenden heuristisches Gesichtspunkt,*" *Ann. Physik* 17:132–148 (1905), Translated by H.A. Bootes and Edited by L. Motz *The World of the Atom*: Vol. 1 (New York, NY: Basic Books, 1966), 544. See Niels Bohr, *Atomic Theory and the Description of Nature* (Cambridge, UK: Cambridge University Press, 1961), 4 and 34, and Frank Jackson, *From Metaphysics to Ethics. A Defense of Conceptual Analysis.* (Oxford, NY: Oxford University Press, 1998), 23–24. Energy's proportionate relation to its mass informs my claims about holographic identity. Planck's insights into the Scylla of the quantum wave and Einstein's grasp of the Charybdis of the quantum particle established the two pillars upon which QMT builds its theoretical edifice.

[6] See *Park*, 315.

[7] A complex number takes two real numbers a, b under the form: $a + b\sqrt{-1}$.

"horizontal" identity of particles and waves includes a "vertical" ordering.[8] Of de Broglie's imaging of the QMO, Park writes that "[t]he idea that struck him was so simple that nobody had thought of it: instead of worrying about the pathological behavior of light, why not, nature being consonant to itself, assume that the behavior is not pathological at all? What if every particle is also a wave? The scope of the problem is enlarged and the problem certainly is not solved, but it is transformed."[9] [3]

Erwin Schrödinger (1887–1961) developed the concept of "matter waves" by mathematically describing QMO algebraically in terms of matrix mechanics.[10] Schrödinger's explanation involves a thought experiment about a cat being imprisoned in a box that when unobserved simultaneously exists in a state of death and life. This state of existential "superposition" gives us an example of the QMO prior to its determination as a wave or a particle. The life or death of Schrödinger's cat or, by extension, the QMO's state is determined by subjective observation and judgment.

Schrödinger favors the formalization of the QMO as a complex wave of interacting fields, rather than its more traditional atomic representation as a central core of mass with electrons held in satellite. Schrödinger's equation defines a quantum particle according to its imaging within a quantum system, that is, its "holographic" and "dimensional" measurement. Crediting Einstein's earlier work and adapting Neo-Platonic theory, Schrödinger distances himself from Heisenberg's methodology and explanations of uncertainty.[11] In an interview with *Time* magazine, Heisenberg's commitment to a philosophical perspective is made apparent. He states that "[t]he future theory of matter will probably contain, as conceived in Plato's philosophy, only assumption of symmetry can be stated to a large extent; they seem to show that the future will be very simple and concise in its fundamentals, despite all complications of its inferences."[12] Quantum

[8] See Niels Bohr, *Atomic Physics and Human Knowledge* (New York, NY: Wiley, 1958), 74.

[9] *Park*, 315.

[10] On Schrödinger's equation ($i\hbar\ \partial\psi/\partial\psi = \mathrm{H}\ \psi$), see Bruce R. Wheaton, *The Tiger and the Shark* (Cambridge, UK: Cambridge University Press, 1983) and Jürgen Renn, "Schrödinger and the Genesis of Wave Mechanics," in Wolfgang I. Reiter and Jakob Yngvason, *Erwin Schrödinger—50 Years After* (Zürich, Switzerland: European Mathematical Society, 2013), 9–36.

[11] See Walter J. Moore, *Schrödinger: Life and Thought* (Cambridge, UK: Cambridge University Press, 1989), 228.

[12] *Time* Magazine, May 5, 1958, p. 53. See also *Erwin Schrödinger,* "My theory was inspired by L. de Broglie," *Ann. de Physique* 10 (3), (1925): 22 (Thèses, Paris, 1924).

probability develops the classical conception of reality's description of actually existing objects, whose identity is revealed as waves in localized fields. Schrödinger's equation helps as it records the observed values of the quantum wave and consequently "[t]here is... no conflict between determinism and indeterminacy; and as a matter of fact, quantum theory insists upon both. To be precise, ... the Schrödinger equation reveals quantum determinism, even as the Heisenberg principle describes irreducible indeterminacy."[13] [4]

The mathematics of probability locates the QMO's relative position against the potential array of infinite possibilities. In his *Atomic Physics and Human Knowledge*, Bohr writes that the quantum's stochastic unfolding conceptually revives the insinuation of the uncertain as "the appropriate physical interpretation of the symbolic quantum—mechanical formalism amounts only to predictions, of determinate or statistical character, pertaining to individual phenomena appearing under conditions defined by classical physical concepts."[14] The augmentation of the particle "rests" in a range of possibilities—the factual given of the particle is always a fleeting point in the pattern of the wave's greater harmonious fluctuations. Arthur I. Miller notes that it was Max Born (1882–1970) who first proposed that Schrödinger's wave function does not visualize the electron's charge as an orbital path around an atom but rather as a probability field represented as a mathematical pattern.[15] Of the QMO's perplexing marriage of certainty and the probabilistic, Born assumes a " 'complete analogy' between a light quantum and an electron in order to postulate the interpretation the 'de Broglie-Schrödinger waves' (the wave function in a three-dimensional space) [acts as] the 'guiding field' for the electron. Owing to the mathematical character of the wave function, Born attributed physical reality to its magnitude squared... although the carrier of probability, that is, the

[13] See also Timothy Maudlin, "Nonlocal correlations in quantum theory: How the trick might be done," in W.L. Craig and Q. Smith, Editors, *Einstein, relativity, and absolute simultaneity* (London, UK: Routledge, 2008), 156–179. A further obstacle arises, as *GTR* and the *QMT* mechanics exist in an unsettling relationship, as they appear to contradict each other. In summary, both theories prioritize the subjective perspective of the observer, a move that may challenge the further claim that truth is inviolable and universal.

[14] See Werner Heisenberg, *Physics and Philosophy* (New York, NY: Harper and Row, 1958), chap. III and Henry P. Stapp, "The Copenhagen Interpretation," *American Journal of Physics* 40, 1098 (1972): 64.

[15] Arthur I. Miller, "Erotica Aesthetics and Schrödinger's Wave Equation," in *It Must Be Beautiful: Great Equations of Modern Science*, Edited by Graham Farmelo (London, UK: Granta Books, 2003), 122.

wave-function, developed causally, all final states were probable."[16] [5] The quantum particle is conceived as an ordering of *a priori* probable states with a defined structure.[17] [6] There is no way to further simplify this state, as Born notes: "[n]o concealed parameters can be introduced with the help of which the indeterministic description could be transformed into a deterministic one. Hence, if a future theory should be deterministic, it cannot be a modification of the present one but must be essentially different."[18] [7] There are only three ways to ascertain the QMO. One can understand the wave function as an object or a physical manifestation of a universal law or as an accidental property.[19] These three possibilities are empirically established in 1803 in Thomas Young's so-called double-slit experiment.[20]

In Young's experiment, light (from a monochromatic source) is directed through two separated slits (*S*1 and *S*2) onto a screen that records the light's relative concentration after passing through respective openings. As one might expect given theories advocating light's identity as a particle, when Young covered one of the slits (*S*1) and shone light through the remaining opening (*S*2), a band was recorded on the screen behind that was a near-exact copy of the opening through which the light passed. However, when focusing a beam of light with the same amplitude or phase through both openings (*S*1 and *S*2) simultaneously, a "splatter" pattern appears. The fact that the density of this pattern decreases as one moves away from the opening supports light's identity as a propagating wave.[21] Young's experiment proves that light promulgates as either a wave or a particle prior to its observation (measure).[22] To test his hypothesis, Young shot out of phase light through two openings (*S*1 and *S*2). As might be

[16] Arthur I. Miller, *Imagery in Scientific Thought*, (Cambridge, MA: MIT Press, 1986), 146.

[17] See Valia Allori, "On the Metaphysics of Quantum Mechanics," in *La philosophie de la physique: d'ajourd'hui a demain*, Edited by Soazig LeBihan (Devon: Editions Vuibert, 2013). The probable states are finally derivable as the wave function's square such that: $|\psi|$ [1].

[18] Max Born, *Natural Philosophy of Cause and Chance* (Oxford, UK: Oxford University Press, 1949), 109.

[19] See Gordon Belot, "Quantum States for primitive ontologists: A Case Study," *European Journal of the Philosophy of Science* vol. 2, (1) (2012): 71. Hereafter cited as *Primitive Ontologies.*

[20] J. de Barros Acacio, Gary Oas, and Patrick Suppes, "Negative probabilities and Counterfactual Reasoning on the double-slit Experiment," arXiv preprint arXiv:1302.3465 (2013).

[21] See *Smith*, 130.

[22] For pre-modern thinkers, a subject exists in perceptual harmony with another object when account can be made of one's contact with said object. However, if an observer is to be in harmony with a physical object, she must grasp its immaterial and prescribed properties.

expected, when the two phases of light waves intersect, they cancel each other out. Young's experiments prove the binary identity of light.[23]

Over the last century or so, the double-slit experiment has been repeated and refined without overturning Young's original paradoxical results. The photons' impact points on the screen differ when (1) only one opening is available for light to travel through—either *S*1 or *S*2 is closed—but (2) when both slits are open (*S*1 and *S*2), the light can theoretically "choose" which path to travel. These emblematic results suggest that some *a priori communiqué* exists between photons. The discrepancies between classical and QMT concerns not only the attributes each accords a being but also how each is observed.[24] Young's emblematic experiment suggests that a non-subsistent relation exists between a subjective observer and the QMO as "[t]he only intrinsic properties of the entities are those embodied by those relations between them."[25] [8]

The results of Young's experiment suggest that a review of the presumptions of classical physics and metaphysics is necessary.[26] It is important for a theorist to consider whether his or her methodological approach is instrumental, positivist, realist or looks to some other metaphysical or epistemological school.[27] As we delve deeper into the unexpected geometries and topologies of a quantum system, it is helpful to remind ourselves that scientists and theologians can and must make reasonable claims about crucial questions concerning the nature of immanent and supernatural realities.[28] Martin Heidegger argues that meaning is tied to the nature of measure, and he thus finds a natural ally in art's proclivity to engage proportion and perspective. The act of quantifying or measuring sets the

[23] See John Gribbin, *In Search of Schrödinger Cat: Quantum Physics and Reality* (New York, NY: Bantam Books, 1984), 171.

[24] See John Gribbin, *Schrödinger's Kittens and the Search for Reality: Solving the Quantum Mysteries* (Boston, MA: Little, Brown Co., 1995), 12.

[25] *The Mathematical Universe*, 2. See also Frank Wilczek, "A Piece of Magic: The Dirac Equation," in *It Must Be Beautiful: Great Equations of Modern Science*, Edited by Graham Farmelo, (London, UK: Granta Books, 2003), 136–137.

[26] See Valia Allori, "Primitive Ontology in a Nutshell", *International Journal of Quantum Foundations* 1(2): 107–122, https://philarchieve.org/archieve/ALLPOIv1., accessed: March 21, 2017, at Research Gate.

[27] See J. Faye, "Niels Bohr and the Vienna Circle," J. Manning and F. Stadler, Edited in *The Vienna Circle Institute Yearbook* 14 (Dordrecht, NL: Springer, 2010), 33–45.

[28] See J.A. Barrett, "Are our best physical theories (probably and/or approximately) true?," *Philosophy of Science*, 70 (5) (2003): 1206–1218.

work of the scientist and artists on a similar path, as both bring an order to what our subconscious experience registers as disorder or uncertainty.[29]

The quanta's ontological status, axiomatic description, and systematic modeling are receptive to a wide range of influences and consents to plurality of surveying. Irregular experimental results must ultimately answer to an existing or yet to be articulated mathematical or physical law.[30] The unexpected imbrication of the QMO as wave and particle can be statistically normed when the dual manifestation of the QMO is accounted for from a variety of vantages. The quantum wave models the infinite as its continuous modulations ripple throughout the universe in all directions, whereas the quantum particle takes a particular path. Consequently, the QMO's "self-reflexive" identity as wave and particle is only made sensible by locating it in physical and complex or hyper-dimensional spaces. The identification of the QMO in higher dimensions finds its analogous version in the theologian's association of human self-consciousness and aseity in supernatural dimensions.[31]

Physicists map out the QMO's wave-particle dyad by employing the complex space aptly named "Hilbert space" in honor of the great mathematician whose equations helped define it. Hilbert space has become the orthodox theoretical foundation for representing the byzantine dimensions of the quantum world. Not only does QMT allow us to conceive of a material object as virtual, as C. N. Villars posits it furthermore allows us to conceive of immaterial Hilbert space as physical.[32] According to Hilbert space, the quantum exists in six dimensions (6n). This space is obtained by adding the three values for the quantum's position and the three values for its momentum in a demarcated location.[33] These hyper-geometric topologies also owe a debt to Georg Cantor, as one can represent a vector line from any single point of origin indefinitely. In Hilbert space, the QMO

[29] See Martin Heidegger, "The Question Concerning Technology," in *The Question Concerning Technology and Other Essays*, Translated by William Lovitt (New York, NY: Harper and Row, 1977), 4.

[30] On the relationship between irregular events and their axiomatically defined regularity, see Ludwig Boltzmann, *Lectures on Gas Theory* (Berkeley, CA: University of California Press, 2012). For a theological perspective on information, see *TL* 1, 44, and 248–249.

[31] We return to the quantum object and human person's mutual configuration in the hyper-dimensional in Chaps. 5 and 6.

[32] See C.N. Villars, "Observables, States and Measurements in Quantum Physics," *European Journal of Physics* 8, no. 2 (April 1987): 181.

[33] Park, 321–322.

relates to a set of eigenvectors ($\alpha\,|\psi\rangle$, $\beta\,|X\rangle$) taken as the weighted sum of a complex number (e.g., α). These hierarchic dimensional addenda are philosophically justified if we take seriously that the quantum's non-subsisting relationships exist in correspondence to the system they inhabit. The values of the physical QMO can be conceptually separated from their supervening basis as "higher-level laws and phenomena supervene on lower levels, but cannot be reduced to those lower levels… supervenience is something weaker than causation."[34]

Outside influences of decoherence effect the measurement of the QMO. The probable identity of the quantum wave is theoretically infinite, as they can exist anywhere in the universe. Consequently, the quantum wave's collapse involves more than its immediate spatiotemporal position.[35] Experimental measurements disrupt a physical entity's identity within a given system.[36] Nima Arkani-Hamed cautions that any quantification of the quantum wave is problematic, as any instrumental device introduces into the process what might be called a "false flag," as a scientific instrument is engineered of a material substance similar to that which it attempts to measure. Consequently, the instrument that measures and the human subject that interprets its results are never fully separate from the physical world both seek to objectify. The exact measurement of a QMO theoretically requires two unattainable ideals: (1) an infinite measuring device that can instantly record all quantum interactions under observation and thereby overcome one's limited scope (e.g., the measurement problem); and further, (2) to observe the probability of quantum system demands, the ability to infinitely observe a QMO.

The quantum's dual existence as particle and wave both weigh in or are significant to its phase state.[37] In a similar manner, a new artistic creation communicates the artist's inspired vision within existing materials, but its meaning changes with every viewing. Concurrently, the unknown future arrival of the aesthetic rapture is analogous to the near ineffable arrival of

[34] *The Physics of Theism*, 232.

[35] See Brian Greene, *The Fabric of the Cosmos: Space, Time and the Texture of Reality* (New York, NY: Alfred A. Knopf, 2004), 94. Hereafter cited as *The Fabric of the Cosmos.*

[36] See *Smith*, 60–61. For a full explanation to why the quantum wave ontologically defines in relationship to its measure remains elusive, see *Quantum Theory and Measurement*, Wojciech Hubert Zurek and John Wheeler, Editors, (Princeton, NJ: Princeton University Press, 1989), 152–167.

[37] See Gordon Belot, *Primitive Ontologies*, 77.

the QMO. The identity of the QMO is consciously determined, just as faith is personal event or an original art work must be sensually understood: "the continuance and influence of rapture depends upon an attenuation and relocation of its 'being' as both event and significance. In the very act of mediation or representation, there is a necessary threat to art's originary or essentializing presence. This is artistic rapture's most difficult negotiation with the agora or interpretation for in order to ensure the survival of art, its authenticity or autonomy has to be partially erased."[38] [9] The subject uniquely participates in the arrival of an artwork or QMO hylomorphic communication and meaning. Classical physics presumes that the identity of a physical object concludes in its exterior topological shape as "each point of the system's phase space corresponds to a position-and-momentum property the system might have. We can think of the phase space as a whole as representing a determinable family of properties, with each point of space corresponding to an individual determinate property."[39] [10] However, the subject's grasp of the complexity of the QMO's physical and dimensional identity requires a non-classical approach—indeed, one may say, a theological and artistic approach.

4.2 The Interior Structure of the Quantum Object

The QMO's internal geometry and its external spatiotemporal appearance find its crude presentation via mathematical vectors. An arrow with a definitive point of origin exemplifies these attributes. A vertical bar followed by a Greek letter and a bracket (e.g., $|\psi\rangle$ or $|X\rangle$) symbolically represent the two underlying states of the QMO as a vector and, by extension, an eigenvector. In QMT, three mathematical operations visualize the interaction of eigenvectors: (1) vector addition, (2) scalar multiplication, and (3) "inner product" multiplication. The mathematics of the inner product of vectors allows us to obtain an analogous sense of the "interior state" of a QMO. Given this, we can say that the informational or essential identity of the QMO finds its metonymic movement by calculating the

[38] Homi K. Bhabha, "Aura and Agora: On Negotiating Rapture and Speaking Between," in Richard Francis, *Negotiating Rapture: The Power of Art to Transform Lives* (Chicago, IL: Museum of Contemporary Art, 1996), 12.

[39] Gordon Belot, *Primitive Ontologies*, 69. https://doi.org/10.1007/S13194-011-0024-8.

inner product (e.g., $OU \times OU$) of its mathematical properties.[40] We can imagine that the interior properties of two vectors portray a metaphoric "inherent freedom" in relation to each other and hence the environment—a strange claim when reality is taken as a materially deterministic closed system. A quantum eigenvector states something of the QMO's intrinsic movement as either spin-up ($|\uparrow_z\rangle$) or spin-down ($|\downarrow_z\rangle$). It is impossible to measure two Eigen states (U and V) simultaneously.[41] Consequently, in quantum systems, the following convention applies: the horizontal axis nominates the vector space where the QMO exists in a state of "spin-up," whereas the vertical axis bounds the arena of the QMO's status as "spin-down." When plotted, the QMO's spin-up can be represented by its placement (1, 0) and the QMO's state of spin-down with the point (0, 1). This description does not make the internal interminable motion of a single QMO any less problematic.[42]

From a certain point of view, Schrödinger's equation provides one way to conceptualize quantum geometry in a deterministic manner. As Zecevic explains, "[t]o get a better sense for what this truly means, we first must recognize that the 'state' of a quantum particle is generally described in terms of a wave function, this complex-valued function is usually denoted by the Greek letter ψ, and its temporal evolution is governed by Schrödinger's equation. Given the set of initial conditions, this equation wave Schrödinger's equation allows us to compute at any point x and at any time t."[43] [11] Thus, Schrödinger's equation ideally preserves the interior identity of the quantum wave as the "angle" and as a function of its internal states of U and V at rest (e.g., time is 0 or $t(0)$). This relationship holds over time, as the causality influencing the QMO is evident in its effective agency (*causa latet in effectu*). Observation of the QMO's temporal advancement (t) through space is limited to its observed local

[40] See Appendixes for further background.

[41] The measure between U and V is mathematically calculated by taking the weighted sum of the differences influencing U and V ($\Delta U \Delta V$). This mathematical operation is always positive, but in a state of ongoing change (e.g., $\Delta U \Delta \geq \{U, V\}$.

[42] See: Valia Allori, Sheldon Goldstein, Roderich Tumulka, Nino Zanghi, "On the Common Structure of Bohmian Mechanics and the Ghirardi-Rimini-Weber Theory," *The British Journal for the Philosophy of Science* vol. 59, (3) (2008): 335–389.

[43] Aleksandra I. Zecevic, *Truth, Beauty, and the Limits of Knowledge: A Path from Science to Religion* (San Diego, CA: University Readers, 2012), 52.

manifestation.[44] The ideal presentation of the QMO as holographic downplays the complication of the interior nature of the QMO.

The wave period or the quantum particle's momentum remains unknown prior to its surveillance, but its interior geometric ordering determines how subject observation will bring both into existence. The identification of the quanta, like the consecrated species, works in conjunction with the prescribed narration of its secondary or accidental properties—momentum, location, sacramental identity, ritualistic space, etc. The Eigen state can be thought of as the ontological and logical conveyance of the QMO's probable physical properties.[45] When the ensemble of potential Eigen states is reduced to a single eigenvector, a mathematical model for the collapse of the quantum wave is given. The formal description of the QMO gives us a conceptual way to "see" what remains concealed from direct observation. The Eigen state—Eigenvalue Rule (*EER*) describes the rapport between an Eigen state, its eigenvalue, and subjective conceptualization or its phenomenal observation. The *EER* confirms the role played by conscious observation insofar as it maintains that "an observable (i.e., any genuine property) has a well-defined value for system *S* when and only when *S*'s quantum state is an Eigen state of that observable."[46] [6] The ascertainment of the two QMO vectors (*U* and *V*) depends upon its distinct physical and conceptual formalizations. The metaphoric "total energy" or mathematical "identity" of the Eigen states does not change, whereas the "energy" or spatial orientation of the eigenvectors varies when multiplied by a nonzero complex number. In other words, something of the internal movements of a being and its external motion can be abstractly adjudicated. The formal mathematical delineation of the QMO's quantification recounts how the conscious subject imagines higher dimensions and observational points of views. Gordon Belot explains of the quanta's relational nature that its "material degrees of freedom posited by (extant) primitive ontology approaches [that] are not dynamically autonomous: in order to write down the equations

[44] See Peter Forrest, *Quantum Metaphysics* (Oxford, NY: Basil Blackwell, 1988), 19. Hereafter cited as *Forrest*.

[45] See *Smith*, 147.

[46] Valia Allori, *On the Metaphysics of Quantum Mechanics*, http://philsci-archive.pitt.edu/9343/1Allori-LeBihan-On the Metaphysics of Quantum Mechanics-finale.pdf, p. 4 Accessed March 22, 2017.

governing their evolution, one has to assign each system a wave-function evolving according to the Schrödinger equation or one of its relatives."[47][10]

The description of the QMO remains partial given its future emergence as its complementary state. The quantum state gives us the best scientific analogy for the unexplained substantial change that occurs at the moment the subject becomes aware of transubstantiation at a particular historical moment. For "the intervention of the experimental process that causes the physical system to jump, to change instantly from one state to another, without passing through a continuous array of intermediate states (in accordance with the Schrödinger's equation)."[48] [4] Schrödinger's equation describes a quantum wave's oscillating behavior and energy over a period of time.[49] In the state of superposition, the QMO appertains along the *z*-axis such that its Eigen state evidences both a spin-up ($|\uparrow\rangle$) and spin-down ($|\downarrow\rangle$) orientation.[50] The coincidence of both options can be symbolized as "YNO" (YES & NO, "unity-in-difference"). In quantum superposition, *U* and *V* exist simultaneously, which seems to logically reject the law of contradiction.[51]

It is difficult to isolate entangled electrons and their magnetic field. Paul Dirac (1902–1984) pioneered the study of quantum electrodynamics (*QED*). Dirac's *QED* equations illustrate how a balance is struck between an individual QMO's Eigen spinors and its boundary conditions.[52] The importance of Dirac's insight and its application cannot be overestimated, as Frank Wilczek says, for "[t]he Dirac equation became the fulcrum on which fundamental physics pivoted."[53] [12] The *QED* confirms the holographic principle whose hermeneutical presupposition is that a being's mathematical representation virtually details a QMO and its respective system.[54] [15] Dirac discovers the mathematical formulas that give us provisions to explain how an electron's gyrations create an energy signature

[47] Gordon Belot, *Primitive Ontologies*, 69.

[48] *Smith*, 146.

[49] See R. Tumulka, "The 'unromantic pictures' of quantum theory," *Journal of Physics* A 40 (2007): 3245–3273.

[50] See *Forrest*, 19.

[51] *The Philosophy of Mathematics*, 164.

[52] Sean A. Hartnoll, Andrew Lucas, and Subir Sachdev, *Holographic Quantum Matter*, (Cambridge, MA: The MIT Press, 2018), 158–160.

[53] *A Piece of Magic*, 133.

[54] See P.A. M. Dirac, *The Principles of Quantum Mechanics*, 4th ed., (Oxford, UK: Clarendon Press, 2004), 15.

within the larger electromagnetic field.[55] A QMO never permanently domiciles at any particular place but exists in a neighborhood of interacting fields. If all the observables can exist in nonlocal states, then the QMO's location becomes a key way to understand its emergent identity.[56] Dirac explains superimposed subatomic particles by contracting their area of movement. Cooperative forces define the QMO and its place in reality for "[t]he approximate separation between things, which is brought about by the 'locality of the fields of force' that give rise to approximate autonomy and identification of things (physical phenomena, properties, etc.), must in some deep sense be reconciled with the universal relationality that underscores the wholeness of our world."[57] [14]

The QMO is only theoretically separate from the larger quantum system because a quantum particle's spin orientation partly creates its own subspace.[58] The beauty of *QED*, Close muses, follows from its ability to unite the comparative properties of the quantum's informational or holographic signature as energy or "light" (e.g., quantum electrodynamics *QED*), and from it, material identity manifests at the local to cosmic scales.[59] Dirac discovered the key or alpha number and its explanatory equation in light of the electromagnetic force. Mlodinow contextualizes this imaginative breakthrough by writing that "according to the mathematics of the time, the delta function was, simply, equal to zero. According to Dirac, the delta function was zero everywhere except at one point, where its value was infinite, and, when used in conjunction with certain operations of calculus, yielded answers that were both finite and (typically)

[55] On Dirac's explanation for the electron's spin, see Paul. A. M. Dirac, *Proc. R. Soc.* A., 133, 60 (1931) and A.I. Miller, *Early Quantum Electrodynamics: A Source Book* (Cambridge, UK: Cambridge University Press, 1994).

[56] See Steven Carlip, "Challenges for emergent gravity," *Studies in History and Philosophy of Science Part B: Studies in History and Philosophy of Modern Physics* 46 (1) (2014): 200.

[57] P.A. Ligomenides, "Scientific Knowledge as a Bridge to the Mind of God," Edited by John Polkinghorne in *The Trinity and an Entangled World: Relationality in Physical Science and Theology* (Grand Rapids, MI; Cambridge, UK: William B. Eerdmans Publishing Company, 2010), 89.

[58] Alexandru Baltag and Sonja Smets, "Quantum Logic as a Dynamic Logic," *Synthese* 179 (2) (2011): 296. https://doi.org/10.1007s/1129-010-9783-6. Hereafter cited as *Dynamic Logic*.

[59] See Frank Close, *The Infinity Puzzle: Quantum Field Theory and the Hunt for an Orderly Universe* (New York, NY: Basic Books, 2011), 25–26. Hereafter cited as *The Infinity Puzzle*.

non-zero."[60] In the *QED* field we see how Nothingness and infinite bulwark a QMO and its symmetry to its system.[61]

The alpha number is derived by the mathematical correspondence of the magnitude of Planck's quantum ($\hbar$), the magnitude of the electric charge (e-), and the speed of light (c). The interactions of these three elements show Dirac how electromagnetic charges relate to the findings of the *Special Theory of Relativity* (via c) and QMT (via $\hbar$). Of particular interest in this regard, Close points out, is the fact that Dirac's equation theorizes that aspects of a QMO have both dimensionless and infinite properties. Dirac's work gives me an analogy for how the eternal presence of the Son exists in the substance of bread and wine after its consecration.

The relativistic nature of space-time encouraged Dirac to reconsider the observer's conceptualization of the QMO's role within its indigenous system, whereas the electron's operation within the electromagnetic field fueled his re-evaluation of the proportionate scales of the electron's dimensional identity. The electron's spin generates a centrifugal force as part of its individual self-reflexive identity. Furthermore, in this schema, the electrons' gyromagnetic ratio reveals the unending motion of a localized system. Together, the electrons create a magnetic field—allowing the parts to analogously "holo-somatically" image the whole. The mathematical equation of this ratio can be seen as a non-subsistent displacement point that drives the symmetry between its formal definition and the existing physical being. The hylomorphic principle of the manifestation of the form-matter correlation is analogous to this collection of electrons and their magnetic field.

The number of possible states for a quantum system is theoretically infinite as the quantum particle's spin state continually iterates and

[60] *Euclid's Window*, 72. Tomonaga-Schwinger's Relativistic Quantum Field Theory expands upon this correspondence in non-classical physics. It hypothesizes that a continuous three-dimensional surface corresponds to a four-dimensional space-time continuum. The points on each of their respective dimensional representations are separated—the "distance" between them cannot be nullified even at the speed of light and hence metaphorically it acts as "Nothingness" or a "null" space.

[61] See also Silvan S. Schweber, *QED and the Men Who Made it: Dyson, Feynman, Schwinger, and Tomonaga* (Princeton, NJ: Princeton University Press, 1994), 57–58 and Helge Kraugh, *Dirac: A Scientific Biography* (Cambridge, UK: Cambridge University Press, 1990), 54–56.

coalesces into fields in three-dimensional and relativistic spaces.[62] The particle's spin relates to its velocity, and it is therefore possible to understand its place within a larger field, indeed, metaphorically, the whole of the universe. In practice, the QMO's potential states statistically norm with view to its strict interior geometric ordering. Conversely, the accidental structures of the interior structure of the bread and wine are holographically spirited at transubstantiation into the divine essence and thus image the eternal in a contained spatial body.

Here, we see a hallmark of Dirac's genius—the ability to move between material being and its formalization by mathematical laws. These macro and micro instances of the QMO require Dirac to exceed the theoretical limitations of first-order mathematics and their categorization in vector analysis to explore material reality in higher mathematical dimensions. Wilczek remarks of Dirac's *QED* equations, that "[t]o construct such an equation, Dirac had to expand the terms of the discussion. He found that he could not get by with a single first-order equation—he needed a system of four intricately related ones, and it is actually this system that we refer to as 'the' Dirac equation."[63] Just as the material being exists within a network of relations in the phenomenal world, Dirac's work also shows that the QMO's non-subsistent correspondences have a broad range of spatiotemporal and dimensional contexts. Dirac's epistemological innovations turn on the fact that "[t]he most powerful method of advance that can be suggested at present is to employ all the resources of pure mathematics in attempts to perfect and generalize the mathematical formalism that forms the existing basis of theoretical physics, and after each success in this direction, to try to interpret the new mathematical features in terms of mathematical entities."[64] [15] An electron and its path within a larger electromagnetic field have their own unique but interrelated mathematical explanation of existing physical laws and probability's ever-present influence. There are electromagnetic waves or "immaterial" waves and matter waves, and both are needed to give an integrated picture of this phenomenon.

One can explain the physical quanta with reference to its mathematical representation and re-representation. Physical reality "holographically" or

[62] See Hartnoll, Sean A., Andrew Lucas, and Subir Sachdev, *Holographic Quantum Matter* (Cambridge, MA: The MIT Press, 2018), 286.

[63] See *It Must Be Beautiful: Great Equations in Science*, Edited by Graham Farmelo (London, UK; Granta Books, 2003), 137. Hereafter cited as *A Piece of Magic*.

[64] P. A. M. Dirac, *Proceedings of the Royal Society*, A, 133, 60 (1931).

virtually exhibits in its systematic meta-leveling or hierarchic dimensional contexts. The emergence of formal laws reflects the material object's evolution, and I argue that such duplex orders find their metaphysical basis in the eternal dimensions of transcendental truth. The Spirit's Ur-Kenotic hypostatic relations generate an analogous field that holds finite reality in its bosom.

The mutual distribution of quantum waves does not eradicate the identity of each wave because they are assimilated into a single field, even when they are differentiated by entropy. Dirac's equations explain how the QMO and by extension the recursive nature of identity can be understood as information or as a probable causal pattern.[65] Spatiotemporal identity or statistical pattern of a probability cloud identifies the symmetry of wave and pattern in the QMO. As Leonid Butov explains, "at the foundation of modern quantum physics, waves in nature were divided into electromagnetic waves, such as the photon, and matter waves, such as the electron. Both can form a coherent state in which individual waves synchronize and combine. A coherent state of electromagnetic waves is known as a laser; a coherent state of matter waves is termed a Bose–Einstein condensate."[66] [[16] cited in [17]] The egalitarianism between material and energetic waves confirms the feasibility of my own holo-somatic metaphysics as the holographic principle communes a physical being as an image, metric, field, phase state, configuration space, etc.[67] As Sutton comments, "these 'Field particles' are more than a convenient mathematical means of describing the field; in certain circumstances, they emerge from the field as detectable entities, as real as electrons or protons. In electromagnetic theory, the field particles are photons, which can emerge from the electromagnetic field as the light we see."[68] [18]

The relative degrees of the quantum particle's freedom find parallels in an electron's embodiment in the electromagnetic field, divergent

[65] See Brian Greene, *The Hidden Reality: Parallel Universes and the Deep Laws of the Cosmos*. EPub. eBooked (New York, NY: Alfred A. Knopf, 2011), also see hard copy, 123–124. Hereafter cited as *Hidden Reality*.

[66] Leonid Butov, "Solid state Physics: A Polariton laser," *Nature* 447 (7144) (31 May 2007): 540–541. Cited in John Stachel and Mihaela Iftime, presentation 06-25-07.

[67] See Gordon Belot, *Primitive Ontologies*, 72. See also David Wallace, "Everett and structure," *Studies in the History and Philosophy of Science Part B: Studies in History and Philosophy of Modern Physics* 34 (2003): 87–105. In Chap. 3, we saw in angelic nature a theological analogate for this principle.

[68] See Christine Sutton, "Hidden Symmetry: *The Yang-Mills Equation*", in *It Must Be Beautiful: Great Equations in Science*, Edited by Graham Farmelo (London, UK; Granta Books, 2003), 243. Hereafter cited as *The Yang-Mills Equation*.

mathematical nodes, or a being's kenotically ordering in my speculative metaphysics.[69] The possible infinite "uncertainty" of the quantum world can through observation and reason be "demoted" or "determined" as an empirical measure. Consequently, a QMO or a being's empirical facticity can be established by means of the negation or subtraction avoided in the act of measuring.[70] This is possible because symmetry holds between the photon's freedom and human vision. Human perception gives momentary custody to uncertainty for "[t]he objective world simply is; it does not happen. Only to the gaze of my consciousness...does a section of this world come to life as a fleeting image in space which constantly changes in time."[71] [19] When the spatiotemporal situates a single entity, it parlays probable freedom in a causal explanation of a particular electron's path from point *A* and its unhindered appearance at some point *B*. Consequently, Richard Feynman argues that the electron's path implies infinite routes or possible trajectories prior to its "subtracting" or "choosing" one.[72] The path from *A* to *B*, Feynman reasons, requires one map out, and thereby, it theoretically observes at the conceptual level all potential routes that the QMO can take.

When "kenotic" differing or its imaging as non-subsistence is taken as an active principle of physical reality, we have an example of "relational differing" that has been termed a "purification postulate."[73] Feynman developed earlier insights into the QMO by outlining a mathematically precise vindication of nonlocal instantaneous action not in terms of its immediate temporal properties. The quantum system is not limited to its spatial mea-

[69] In 1971 Gerard 't Hooft and Martinus J.G. Veltman "renormalized" or showed how quantum related infinities can be canceled out.

[70] See Stephen Hawking and Leonard Mlodinow, *The Grand Design* (New York, NY: Bantam Books, 2010), 106–110, see especially Feynman on subtracting all probable histories from their end point. Hereafter cited as *The Grand Design*. Following the success of QED, Abdus Salam and Steven Weinberg normalized the infinities of electromagnetism and the weak force, whereas the strong force was renormalized by the theory of quantum chromodynamics (QCD), see ibid., 109.

[71] Herman Weyl, *Philosophy of Mathematics and Natural Sciences* (New York, NY: Prager, 1984), 116.

[72] See *Hidden Reality*, 133. On the retro-causal properties of the quantum system see also Robert W. Spekkens, "Evidence for the epistemic view of quantum states: A toy theory," *Physical Review* A 75, no. 3 (2005):1–32.

[73] Chiribella, D'Ariano and Perinotti coined the term "purification postulate". Àlaslav Brukner, who explain its function as follows: "[e]very mixed state PA of system A can always be seen as belonging to a part of a complete system *AB* that itself is in a pure state ψ *AB*. This pure state is called purification and is assumed to be unique up to a reversible transformation on *B*," See Àlaslav Brukner, "Questioning the Rules of the Game," *Physics* 4, (55) (2011), https://physics.aps.org/articles/v4/55.

sure, but as Feynman showed, the quantum system can be represented historically.[74] Putting aside the question of its experimental plausibility, Feynman's defense of retroactive causality makes an equilibrist leap by suggesting that the randomness of the quantum system is spatiotemporally non-commutative. Feynman's expansive interpretation of space-time in a quantum system elucidates how causal events relate to human subjectivity. Weatherall notes that "in order to get the theory to work… [Feynman] needed to allow for particles to act on one another both forward and *backward* in time, so that the motion of a particle right here and now depends not only on the distributions of other particles in the past, but also on the distrubution of other particles in the future."[75] [20] Hawking and Mlodinow explain the indefinite nature of the future means every observation remains partial and incomplete.[76] Putting aside the question of its experimental plausibility, Feynman's defense of retroactive causality makes an equilibrist leap by suggesting that the randomness of the quantum system is spatiotemporally non-commutative. Hawking and Mlodinow explain "that no matter how thorough our observation of the present, the (unobserved) past, like the future, is indefinite and exists only as a spectrum of possibilities."[77] Our perceptual and rational "nets" partially and momentarily fix the tentative life of the quanta's spatial and temporal identity. A vertiginous instability exists for the observer in a universe where space curves and time are relative to one's position, making the complete tracking of the moving quanta nearly impossible. A quantum particle briefly appears at the behest of outside incitements before "re-cloaking" itself within reality's simultaneously moving spatiotemporal "folds."

Wilczek sees in Dirac's work on *QED* various foundational intuitions about how the void or its spatiotemporal equivalent is the key to understanding the historical identity of a subatomic particle. It is difficult for

[74] See *The Bride of the Lamb*, 57, for a theological vision of time.

[75] *Void*, 109.

[76] See *The Grand Design*, 60–72 and 90–95. See also Michael Redhead, *Incompleteness, Non-locality and Realism: A Prolegomenon to the Philosophy of Quantum Mechanics* (Oxford, UK: Clarendon Press, 1987), 5 and Wesley Van Camp, "Principal theories, constructive theories, and explanation in modern physics," *Studies in History and Philosophy of Science Part B: Studies in the History and Philosophy of Modern Physics* 42, (1) (Feb., 2011): 23.

[77] *The Grand Design*, 118. See also Michael Redhead, *Incompleteness, Non-locality and Realism: A Prolegomenon to the Philosophy of Quantum Mechanics* (Oxford, UK: Clarendon Press, 1987), 5 and Wesley Van Camp, "Principle theories, constructive theories, and explanation in modern physics," *Studies in History and Philosophy of Modern Physics*, vol. 42 (2011): 23.

created beings to comprehend negative spaces. The introduction of mass and momentum helps us begin to perceive the theoretical difference between negative and positive spaces.[78] Even when an electron is physically absent from a particular location, it "holographically" presumes the positron—its opposite charged partner.[79] Even though virtual particles have an incredibly short "life span," they have a profound influence on reality. The obliteration of a negatively charged virtual particle by its positive opposite returns a system back to its "normal" state. The balancing of negative and positive forces gives us a way to imagine an empirical actualization of nothingness as well as a theological exposé of how causality's own provocations of opposites are part of reality's own "kenotic" ordering of finite differing.

4.3 Nothingness and Quantum Structure

The act of seeing has long served philosophers and theologians with a metaphor for the human person's grasp of truth. Holo-cryptic metaphysics argues that a similar qualification is recognized in subjective consciousness's ability to equate an immaterial system of measure, for example, and apply this to the quantification of a material being. Space and time's causal derivation from the pure qualification of Nothingness (*creation ex nihilo*) stems from our analogous participation of the Creator Spirit's own intellectual quantification of the abyss. It is impossible to completely understand or detail how perception's quantification proceeds from the deeper mystery of consciousness' essential qualification. It is impossible to overcome this enigma as the human person's material existence finds its compass of the Transcendent within the interior mind.

With the Son's Ur-Kenotic incarnation (*μεταστοιχειώσασα*), Jesus Christ convenes the perfect holo-somatic refraction of the Transcendent in human form. Insofar as Jesus Christ is fully human, his place within his own Creation diverts against material quantification through non-subsisting correspondences to an eternal standard.[80] Human identity, Fern

[78] See Frank Wilczek, "A Piece of Magic: The Dirac Equation," in *It Must Be Beautiful: Great Equations of Modern Science*, Edited by Graham Farmelo, (London, UK: Granta Books, 2003), 142.

[79] Polykarp Kusch (1911–1993) and his collaborators verified the existence of the anti-electron or positron experimentally in the early 1940s. See *A Piece of Magic*, 140.

[80] For a Christian perspective on this issue see: Augustine, *The Enchiridion*, *NPNF3*, chapters 105–107.

observes, is a meontic and telontic drama that finds its final and perfect realization in the person of Jesus Christ. In Jesus Christ's seeing the created order as fully human and fully divine, a new, irreducible, and final observational reference point instantiates space-time. Prior to his redemptive Passion, Jesus Christ's perception of the created order as Creator begins to restore reality. As Christians, we must take seriously a Creator who deems to see reality as his creatures. Jesus Christ's perception of the created order makes possible the future Eucharistic person—the ecclesial man or woman—to share a similar vision of the universe. Bulgakov explains that "causality can be included in creativity but only as subordinate to its goals and plans. For itself, mechanical causality is blind and empty, in the sense that creative novelty is absent from it. Creativity is guided by a task: a goal; it is exemplary (*causa exemplaris* according to the ill-chosen expression of scholastic theology), entelechic."[81] [21] By holo-somatically viewing the formal metaphysical change of the Eucharistic species, one is able to participate in Jesus Christ's vision of the created order through the faithful's eyes. Transubstantiation affords us a materially embodied and virtual exemplification of the truth of hypostatic divine thought.

The physical, intellectual, and spiritual depiction of being depends upon our sensual and conscious perceptions. An icon is not a realistic picture, as the material order may occasion idolatry when the created order's link to the supernatural is denied. For "[t]he icon recognizes no other measure than its own and infinite excessiveness [*démesure*], whereas the idol measures the divine to the scope of the gaze of he who then sculpts it, the icon accords in the visible only a face whose invisibility is given all the more to be envisaged that its revelation offers an abyss that the eyes of men never finish probing."[82] The icon is not meant primarily to be a means to engender affective responses but rather a means to look into truth's infinite manifestations beneath the physical and aesthetic properties.[83] The writing of a religious icon "translates" and imaginatively represents the supernatural in two dimensions. Hence, the icon's artist, like the priest at the altar, "posits a mystery present in the work of art as the encounter with a metaphysical order beyond or hidden with the ordinary, sensuous world

[81] *The Bride of the Lamb*, 38.

[82] See *God Without Being*, 21.

[83] Mark Patrick Hederman, "Cinema and the Icon," in *The Crane Bag, Media and Popular Culture, Dublin*, Vol. 8, no. 2 (1984): 95.

is not right."[84] [22] A religious image or symbol ideally corroborates physical reality to give us access to the Transcendent—as is the case in the supernatural or "four-dimensional" liturgical space of the Mass. Religious symbols are never devoid of meaning, as their use is never merely deterministic but invites one to creatively perceive the transcendental dimensions of daily life. The spiritual and artistic appropriations of the hyper-dimensional frame become a means toward personal sanctification, "[f]or all concepts and abstract ideas can become icons or 'sacred images' when one considers them not as the end, but rather the beginning of the way of knowledge of spiritual reality."[85] [23]

Personal observations of the Transcendent's trace in created reality are part of the ecclesial community's own participation in physical reality's transfiguration by Christ's redemptive act of the Eucharist. When one truly sees the other in light of universal truth and beauty, a refutation against all titanic acts of egotistical invention is possible (2 Corinthians 10:12). To glimpse the Transcendent's image in reality, humans must engage in a form of "artistic" excavation as "[t]he artist does not himself invent the image, but only removes the covering from an image that already exists; he does not put paint on canvas, but, as it were, clears always the alien patina, the 'overpainting' of spiritual reality."[86] [24]

All created religious and scientific signs, symbols, and representations hinge on non-subsisting relationships, and hence, the subject needs to interpret their evolving meaning. The wooden panels upon which the icon is written or drawn or the computer screen that reveals photorealistic graphics are two material conduits for the symbolic. The artist, philosopher, and theologian understand that truth's union with being means that all that is fragmentary or fleeting exists as outlier notes of the universe's symphonic whole as "[t]here is no way the sum total of the represented object's geometric attributes can be available in the representation…The representation is always more unlike the original than like."[87] Subjects take an artistic object's geometrical and visual attributes and make their beauty personally meaningful.

[84] David Morgan, "Secret Wisdom and Self-Effacement: The Spiritual in Art in the Modern Age," in Richard Francis, *Negotiating Rapture: The Power of Art to Transform Lives* (Chicago, IL: Museum of Contemporary Art, 1996), 41.

[85] *Meditations on the Tarot*, 175.

[86] *Iconostasis*, 64–65 and 73.

[87] *Reverse Perspective*, 259.

The interior properties of a piece of spiritual art are not meant to be eclipsed by a subjective observer but rather discerned. On the spiritual essence's geometric, material, and graphic portrayal, Florensky rhetorically asks: "does the latter have enough points to correspond to the points of the former, or, in mathematical terms, can the power of a three-dimensional image and that of two-dimensional image be comparable? The answer that immediately comes to mind is 'of course not.'"[88] [25] The religious artist's work entails "writing" or symbolically representing even those negative spaces, points of absence, or silence that convey the Transcendent's passing over. Renato Poggioli's attribution of art gives us an example of a similar process at work in great art. Absent sound, the silence and motionlessness "transcends the limits not only of reality but those of art itself, to the point of annihilating art in an attempt to realize its deepest essence."[89] The fullest possible meaning—its holo-somatic presentation—reveals itself in proportion to the subject's openness and the strength of one's vision of the material and non-existing alike. The quantum ecosphere is analogous to the space created by a fugue by Bach or a Mass setting by Mozart. Religious art gives us some access to the supernatural space of the Transcendent just as the concordance of natural and complex numbers allows us to holographically represent a being or the contraction and inflation of cosmic space.[90]

The Transcendent makes Creation its repository, and therefore, embodied consciousness is able to perceive the image of the Transcendent through the Pure Divine Spirit's hypostatic continuing communication of the resurrected person of Jesus Christ. The material order radiates the Christological in conjunction with its own finite hylomorphic identity. By being open to physical and supernatural reality, holo-somatic consciousness emerges as a lived possibility for the human person. This mutual assent first takes place at the sensual level for "without the subject's sensory space, [the concept or object] would not be what it is; it would be incapable of fulfilling the *raison d'être*, the idea that it is supposed to embody."[91] [26] Conscious identity is not simply a theological excursion into the purely speculative but fully thrives when the Transcendent's rev-

[88] *Reverse Perspective*, 255.

[89] Renato Poggioli, *The Theory of the Avant-Garde* (New York, NY: Harper and Row, 1971), 201–202.

[90] See *Park*, 321–322.

[91] *TL* 1, 63.

elation in the Eucharist is made part of the believer's interior identity. Our spatiotemporal existence betrays an arrangement to the Transcendent given our conscious and free nature's superficial grasp of physical reality. We see in the QMO's identity an analogous proportioning of external influences as part of the measure of its complex interior spaces. The quantum particle's geometrical structure, for example, is isomorphic to the summation of the vectors in its subspaces.[92] The quanta's internal architecture is compatible with its geometric subspaces conceptualized as ortho-lattices (L (H)) within a closed system and yet is infinitely inflatable in abstract Hilbert space (H).[93]

A set of limitations is applicable to all formal systems and therefore must be intuitively or I would say "kenotically" weighed. The act of measurement is an eliminative or "kenotic" gesture, as it involves abstracting a property from an object. Thus, "in mathematics, as in scientific research, we find two tendencies present. On the one hand, the tendency toward abstraction seeks to crystallize the logical relations inherent in the maze of material that is being studied and to correlate the material in a systematic and orderly manner. On the other hand, the tendency toward intuitive understanding fosters a more immediate grasp of the objects one studies, a live rapport with them, so to speak, which stresses the concrete meaning of their relations."[94]

Science's attempt to provide a unified vision of reality is obscured when due diligence is not given to how a being's non-subsistent relations center a being's identity. Concurrently, the kenotic measure of the artistic object allows for the subject's self-transcendence. As David Morgan explains "the spiritual in human experience is the perception of humanity's limits and the proposal of a path for overcoming these limits. The spiritual in art, we might say, consists of the artistic configuration of human boundaries and

[92] These states are: |↑Z ⟩I |↑Z⟩ 2 = (1, 0,0,0,), |↑Z⟩I |↓Z⟩2 = (0, 1, 0,0), |↓Z⟩I |↑Z⟩ 2 = (0, 0, 1,0) and |↓Z⟩I |↓Z⟩2 = (0,0,0,1).

[93] See Michael Dickson, "Quantum Logic Is Alive (∧ It is True ∨ It Is False)," *Philosophy of Science* 68, no. S3 (Sep., 2001): S274–S287. https://doi.org/10.1086/392915. Accessed March 18, 2017 at http://mdickson.net/pubs/Quantum Logic.pdf: 2. See also Eddy Keming Chen, "Time's Arrow in a Quantum Universe: On the Simplicity and Uniqueness of the Initial Quantum State" (Dec. 12, 2017): 7–9.

[94] See David Hilbert and S. Cohn-Vossen, *Geometry and the Imagination,* Translated by P. Nemenyi (Providence, Rhode Island: AMS Chelsea Publishing, 1991), Preface.

their negotiation in the experience of a work of art."[95] [22] An individual's being's authentication of Nothingness and Being always owes something to chance.[96] An interminable symmetry between Nothingness and Being conceptually grounds any number of probable possibilities. Empty space does not exist in the physical universe. However, without the concept of Nothingness, the sense of space would be nonsensical. As Heisenberg's equations show, the uncertainty of a particle in a given space-time is not absolute insofar as space is never completely empty.[97] To understand the tension between Nothingness and Being requires a further Transcendent point of reference: "thus, here, for the first time, a certain transcendence is mixed into what till now has been immanence. It is not simply—as it has been to this point that—the principle of non-contradiction signifies (both noetically and ontically) a rhythmic middle. It is rather here that there is a stress upon 'end-directedness,' which becomes more clearly an 'end towards which [a being desires to achieve].'"[98]

Classical physics assumes an observer, can "cognitively detach" or "objectively abstract" themselves from the system they investigate. Current scientific theories argue that a subject's presence during an experiment is intractable. A sharp dividing line runs between these two approaches to the nature of reality as "[t]o entertain the idea that nature might express itself through the paradoxical unity of irreducible aspects was for classical physics almost impossible to believe."[99] [27] With the advent of Einstein's GTR, STR, and the encroachments of QMT, the claims that a being or system is completely isolated or objective was rejected.[100] Truth does not superlatively reveal when one removes oneself from what is observed; rather, the truth only manifests when apprehended by a free and rational

[95] David Morgan, "Secret Wisdom and Self-Effacement: The Spiritual in Art in the Modern Age," in Richard Francis, *Negotiating Rapture: The Power of Art to Transform Lives* (Chicago, IL: Museum of Contemporary Art, 1996), 36.

[96] On problems with reductionist versions of probability, see Gordon Belot, "Undermined," *Australasian Journal of Philosophy* 94, (4) (2016) 781–791.

[97] See *The Grand Design*, 179–180.

[98] *AE*, 210.

[99] Stanley L. Jaki, *The Relevance of Physics* (Chicago, IL and London, UK: University of Chicago Press, 1966), 92.

[100] See Albert Einstein, *Sidelights on Relativity*, Translated by. G.B. Jerry and W. Perrett, (London: Methuen Co., 1922), 3–24, Thomas S. Kuhn, *Black-Body Theory and Quantum Discontinuity*, 1894–1912 (Chicago, IL: Chicago University Press, 1978), and Olivier Darrigol, "The Quantum Enigma," in M. Janssen and C. Lehner, *The Cambridge Companion to Einstein* (Cambridge, UK: Cambridge University Press, 2014).

subject. Conscious subjectivity makes greater claims on individuals because they are part of an emerging physical reality.[101]

The subject must parcel out syntactical and semantic meaning without naively assuming that they have taken full possession of an ideal or complete knowledge of a being or concept. Consequently, I argue that the ideal analogously reveals itself in those meaningful non-subsistent and logical correspondences that image the Ur-Kenotic community. George David Birkhoff (1884–1944) and John von Neumann (1903–1957) outline how a QMO can represent logical propositions. Von Neumann explains that the projection of a quantum object's physical attributes makes them an excellent placeholder for logical propositions.[102] The connection between a material being (e.g., a computer) and the reentrancy or recursive use of quantum logical statements provides us with mathematical analogy to what past theologians explored in natural theology. Michael Dickson explains that "in quantum logic, the logical operations of 'and,' 'or,' and 'not' are modeled, respectively by the operations of meet ($\wedge$, *infimum*), [and] join ($\vee$, *supremum*) and [they] represent an orthnocomplement ($\perp$) on L(H). Given these operators (and some symbols, such as parentheses), one defines the formal sentences of quantum logic in the obvious way."[103] [28]

The QMO's unwavering geometric structure and its holographic spatiotemporal geography introduce the question of logic's sentential and predicative propositions. A quantum system can exist only as a localized closed system by "holographically" imposing an overall unity on the system. This state of unity-in-difference distinguishes itself from physical non-subsistence, as it qualifies difference in a unique way. Two particles, for example, U and V, are quantum complements if U commutes onto a

[101] See Argyris Nicolaidis, "Relational Nature," in *The Trinity and an Entangled World: Relationality in Physical Science and Theology*, edited by John. Polkinghorne (Grand Rapids, MI: Wm. B. Eerdmans Publishing Company, 2010), 100–101. See also *AE*, 77.

[102] See John von Neumann, *Mathematische Grundlagen der Quantenmechanik*, (Berlin, DE: Springer–Verlag, 1932) and English translation: *Mathematical Foundations of Quantum Mechanics* (Princeton, NJ: Princeton University Press, 1955), 253. See also George David Birkhoff and John von Neumann, "The logic of Quantum Mechanics," *Annals of Mathematics* 37 (1936): 823, reprinted and edited in C. A. Hooker, *The Logico-Algebraic Approach to Quantum Mechanics* Vol. 1 (Dordrecht, NL: D. Reidel Publishing Company), 1–26.

[103] Michael Dickson, "Quantum Logic Is Alive ($\wedge$ It is True $\vee$ It Is False)," *Philosophy of Science* 68, no. S3 (Sep., 2001): 2. S274–S287. https://doi.org/10.1086/392915. Accessed March 18, 2017 at http://mdickson.net/pubs/Quantum Logic.pdf: 2.

linear subspace of Hilbert space and *V* exists in orthogonal relationship to *U*. If *U* and *V* are not orthogonally arranged, their state has not been decided—they are in the anomalous probable state of being jointly "in" and "not in" relationship and this paradox requests we qualify some quantum unions as weaker than others.[104] This is not an academic postulate, as "quantum difference" must account for the unique type of uncertainty of the QMO that is not relevant to classical logic. The state of quantum superposition must be formally and empirically explained. Baltag and Smets explain that "quantum disjunction $P \sqcup Q$ is weaker than classical disjunction $P \vee Q$. In fact, quantum disjunction captures the physical notion of 'superposition': a state satisfies $P \sqcup Q$ if it is *a superposition (linear combination)* of states satisfying *P* or *Q* (or both)."[105]

In quantum superposition, a state exists that simultaneously satisfies the logical state of two QMO objects. [index quantum logic state] The logic of quantum superposition such as transubstantiation questions our daily experience of physical reality as a single frame of reference, for "the superposition that occurs in quantum mechanics is of an essentially different nature from any occurring in the classical theory as is shown by the fact that the quantum superposition principle demands indeterminacy in the results of observations in order to be capable of a sensible physical interpretation."[106] The logical discrimination of superimposed quantum properties is possible because the spatial layout of independent particles does not "sacrifice" but substantiates the QMO's physical properties as represented by its mathematical substratum. Consequently, fuzzy and paraconsistent logics seem to be a viable analogy for quantum superposition's material and geometric coordination within a closed system. Baltag and Smets explain that "[t]he physical significance of superposition is that, if a physical experiment cannot distinguish between (the system being in two, or more) possible states… then it cannot distinguish between them and any superposed states. Hence, the properties that can

[104] See C. Piron, *Foundations of Quantum Mechanics* (Reading, PA: W.A. Benjamin Inc., 1976).

[105] Alexandru Baltag and Sonja Smets, *Dynamic Logic*, 289. https://doi.org/10.1007s/1129-010-9783-6. The logic of quantum superposition will be explored in terms of the Eternal Logos-Eucharist-Jesus Christ relationship in Chaps. 6 and 7.

[106] Paul Dirac, *The Principles of Quantum Mechanics* (Oxford, UK: Oxford University Press, 1958), 14. Cited in *Smith*, 67.

be directly "tested" by experiments must be given only by sets of states that are closed under superposition: in Hilbert space H."[107]

The logic of quantum superposition, its physical state, and its virtual representation coexist within a quantum system. This collation affords us an analogy for the interior state of the human person's soul finding its natural superposition in Ur-Kenotic hypostatic life. Material reality acts as a formal "operating system" that abhors the stationary and the void. Everything that is, is in some way in motion that represents itself—"*Naturam esse est per se datum inquantum naturalla sent manifest sensu.*"[108] Aristotle argues that the "soul" is the nexus point for natural and supernatural motion, as "the soul is either the terminus or the origin of movement: in perception, movement starts from particular objects and ends in the soul; in recollection, the impulse begins in the soul and extends to movement in the sense organ… In both cases, the soul does not retreat from activity, but rather perfects it, since transitional movement (*kinesis*) is only on its way from potency to activity, whereas the soul as *entelechia* is perfectly realized, whose effectiveness is self-realizing and therefore beyond alteration."[109] [2] The human soul's created non-subsistent relations virtually image the Eternal hypostatic union of the Divine Persons as realized in the Spirit's allotment of the grace generated by the divine-human person, Jesus Christ. Beautiful art, eloquent mathematical equations, and scientific formulas give us a way to proportionally assimilate Nature's motion as moments of eternal transfiguration. Close explains: "The time variation of light is too fast for our substance to follow. Its frequency is too high. And so, to make the best of a difficult situation, our sensory system processes the information and encodes a small part of it perceived color. That code, by the end of the day, bears little trace of its origin!... when we see a symbol of change, not anything that changes."[110] [29]

[107] Alexandru Baltag and Sonja Smets, "Quantum Logic as a Dynamic Logic," *Synthese* 179 (2011): 289.

[108] See Aristotle, *Physics,* Books I and II, Translated with Introduction, Commentary and Notes by William Charlton, General Editors J.L. Ackrill and Lindsay Judson (Oxford, UK: Clarendon Press, 2006), Lesson 1, *Phys.* V, 1, 224, a 20 and *Metaph.* XI, 10, 1067b, 15–25.

[109] John Milbank, Graham Ward, *God and Beauty,* 18. See also Aristotle, *De Anima,* II, VII, 408c 11–18, in *De Anima with Commentary of St. Thomas Aquinas,* Translated by Kenelm Foster and Father Sylvester O.P. (London, UK: Routledge and Kegan Paul, LTD, 1941), 265.

[110] *A Beautiful Question*, 204.

4.4 Quantum and Holo-cryptic Spaces and Dimensions

Our post-Newtonian world requires that we expand how we think about a direct one-to-one correspondence between theory and a physical object. The multidimensional nature of reality allows for a range of formal descriptions that include multiple cross-disciplinary theories, each replete with its own unique subclasses and adjacent references. Theology and science realize truth and express it in their own way and across a wide dimensional spectrum. Moving away from a strict Aristotelian position that reduces every proposition to subject-predicate form, Leibniz's mathematical philosophy argues that phenomenal beings participate in a supra-sensible form he calls monads.[111] These monads act as a unifying force (*vires*) in physical reality without distracting from the unity of the whole: "Leibniz's more radical logical position, that the predicate of every proposition is 'contained in' the subject, is paralleled, on his side, by his famous metaphysical doctrine that the world consists of self-contained subjects—substances or monads which do not interact."[112] [30] With the concept of monads, past theorists were better able to construct a picture of the QMO. Gauge invariance, for example, conditions the system in concert with the mathematical descriptions of the QMO.[113] The mathematical concept of symmetry is at the heart of modern gauge theories, for example, say: "we have two different atoms, and no matter how we rotate them there is no measurable effect upon the individual atoms themselves... A similar situation may be imagined for two points on a piece of rubber; no matter how we rotate the piece, the measured scale between the two points is unaffected. Stretching the rubber changes the scale or gauge. Gauge symmetry is an abstract transformation theory involving gauge fields, which allows us to restore the invariance of the measurement."[114] [31]

[111] See G.W. Leibniz (2008). *Die Philosophische Schriften*. Hildesheim: Georg Olms (Reprint of edition C. I. Gerhardt, 1875–1890. Berlin, DE: Weidmann). Cited by volume and page e.g., *Leibniz*, 2008, V. 196 VII. 564).

[112] *The Philosophy of Mathematics*, 21.

[113] See "*Yang-Mills Equation*," 239. See also J. Bardeen, M. Carter and Stephen W. Hawking, "The four laws of Black hole mechanics," *Commun.Math.Phys.* 31 (1973): 171.

[114] Anthony M. Alioto, *A History of Western Science*, 2nd Edition, (Englewood Cliffs, NJ: Prentice Hall, 1993), 410.

Gauge theory can be seen as a specialized way to measure or scale a physical object and by extension a conceptual entity.[115] Similar to QED, bulk gauge schemes describe fields and their relationship to particles within a defined system.[116] Frank Close draws out the importance of Weyl's discoveries concerning the differences between gauge invariance for furthering our understanding of the physical and mathematical laws of the universe.[117] The concept of gauge invariance is important, for example, to explain massless vector bosons. The massless vector bosons give us an excellent way to explore how an ontological object can populate the void or exist in a complex system. Consequently, I take the correspondences between gauge fields and their topological manifold as a holographic proscription of the QMO's vector fields. In this way, the subunits of the quantum structure can be distinguished. These mathematical "flanges" are consistent with the QMO's identity as a wave and with their place within complex space. The numerical representation of an object's topology Weyl takes to be a direct analog for its relevant energy.[118] A QMO's probable manifestation as matter or energy is defined along the boundaries set by space-time.[119]

Again, art gives us an analogous way to understand how energy (light) can be thought of as an embodied being in space-time. The artist James Turrell explains that the most interesting aspect of the nature of light in his sculptures recalls the observer's role in the famed double-slit experiment. He writes that it seems light is "aware that we are looking at it, so that it behaves differently when we are watching it and when we're not, which it imbues it with consciousness."[120] [32] Radiating light from an unseen interior source reveals the essential properties of an object of art. Light is the most receptive of material mediums as it adapts to material objects, traces its surfaces, reflects, enfolds, and covers any physical sub-

[115] Sean A. Hartnoll, Andrew Lucas, and Subir Sachdev, *Holographic Quantum Matter*, (Cambridge, MA: The MIT Press, 2018), 19.

[116] See *The Infinity Puzzle*, 85–89.

[117] *The Infinity Puzzle*, 1–13.

[118] See Herman Weyl, *Space, Time and Matter*, Translated by Henry L. Brose (New York, NY: Dover Books, 1952), 64. See also Herman Weyl, "Field und Materie," [Field and Matter] *Annals of Physics* 65 (1921): 541–563 and Herman Weyl, *Raum—Zeit—Materie* [Space-time-matter] 4th edition, (Berlin: Springer, 1921).

[119] *See Holography and Emergence*, 306.

[120] EGG: The Arts Show, interview with James Turrell. Cited in Mark C. Taylor, *Refiguring the Spiritual: Religion, Culture and Public Life* (New York, NY: Columbia University Press, 2012), 142.

stances it comes across. Not all of the light or interior truth is contained within its material home as it pulsates on the outer surfaces or skin of our embodied existence. The light issues from the object and thus begins its journey to the most extreme possible distance from its point of release. Through the light's movement from within the sculpture to its external presentation in the lived spaces of human sensation and consciousness, the viewer's own presuppositions about the rigidity of reality are thrown into relief. Light pulsates in the sphere that its own illumination creates in Turrell's sculptures and analogously in the vector boson's manipulation of the spatiotemporal "surfaces" our reality or the electron's influence of the electromagnetic field it establishes.

A being's holographic or representational density must account for the manner of information's manifestation within a particular space or dimension. Abstract and hyper-dimensional geometries are not unlike viewing images of various definitional qualities on a computer or television screen. Those individuals who grew up without high-definition television know that the picture was less vibrant, as images lacked the definition common to today's highly pixelated screens. In the past, watching television was often comparable to walking in a snowstorm. Adding more pixels onto a television or computer screen allows for an increase in points of light and a higher flow of data and consequently a better picture result. The addition of information refines the definition of the picture and allows for a fuller palette of colors to be displayed. The intensification of color follows when its luminosity is increased—greater amounts of light allow for images to be projected in more depth. The latest virtual reality technologies achieve a similar effect given the mysterious nature of light, and no doubt the sophistication of this technology will continue over time. Barbara Maria Stafford notes that there are some drawbacks to our current technologically advanced images. The extreme amount of information that is transmitted by postmodern machines influences our senses and cognitive capacities. The exact detail and clarity of individual images taxes our eyes and diffuses our ability to perceive the totality of the whole that houses the individual picture. She states that "in our polymodal information age, one lesson of the episodic—with its crystalline detail, its distilling gesture, its precise coordination—is that the total phenomenal melt, by contrast, reduces the range and specificity of chromatic experience. Caving into an all-embracing blur prevents us from understanding how 'the well fenced out real estate of the mind' can, in fact, be one with the events observed and how observing, in turn, does not necessarily entail being one with the events."[121] [13] Informational and visual overload can decenter a human person's understanding of their own spatiotemporal and physical existence.

While it is impossible for the conscious mind to rationally comprehend what the eyes take in with every glance, the subconscious does "register" this information. This information remains unfiltered and in a potential state until the consciousness draws upon it. The symmetry existing between the conscious and unconscious minds rationally models what is found in reality's most primitive correspondences of particles popping in and out of existence in quantum foam. The holographic image of the QMO builds upon the networking of these fundamental elements and fields. The two most fundamental schemes that mutually define reality at the subatomic level are termed the Anti-de-Sitter (*AdS*) and Conformal Field Theory (*CFT*). *AdS/CFT* symmetry provides a useful tool for my conceptualization of a being's non-subsistent and symmetrical nature as long as one interprets their relationship from an analogous transcendental observational point. According to string theory, the "holographic" ordering of matter and energy argues for the existence of a correlating duality between four-dimensional space-time and five-dimensional *AdS* space. These symmetries are theoretical constructs that I see as an analogy for two divine exemplars emerging from the Nothingness. The *Ads* and *CFT* symmetries quantify matter within space-time from Nothingness and thus provide a metaphor for the divine idea's mysterious work in Creation or the essence's informing of a physical being. Springing from Nothingness like the beings born from the head of Zeus, *AdS* and *CFT* symmetries are believed to precede the birth of space-time. Teh notes that "it is fairly common in physics to use 'emergence' to indicate that the bottom theory describes the fundamental degrees of freedom, whereas the top theory describes the phenomenological degrees of freedom."[122] [33]

Werner Heisenberg (1901–1976) figured, that quantum mechanical matrices could be taken to be an analogy for a Fourier series.[123] From quantum research's use of the concepts of matrices and Fourier series, physicists developed the idea of branes. Fourier series are defined as a periodic function $f(x)$ of a wave that can be seen to infinitely expand or undulate. A parallel between the quantum wave and the Fourier series expands

[121] *Echo Objects*, 173.

[122] *Holography and Emergence*, 303–304.

[123] See Eric W. Weinstein, "Fourier Series." From Math World: A Wolfram Web Resource: A Wolfram Web Resource. http://mathworld.wolfram.com/FourierSeries.html.

our understanding of reality's fundamental properties. The Fourier series can be understood mathematically as the infinite sum of sines and cosines. This series like the quantum wave has associative infinite properties, but as it makes use of the orthogonal properties of sine and cosine functions, it also identifies with finite and unchanging properties.

Some scientists theorize that the universe is born at the quantum scale when relational symmetries emerge from Nothingness and null dimensions.[124] The "strings" in string theory propose an elementary quantification of Nothingness. Strings, like divine exemplars, are emergent structures "riding" the boundary edge between Nothingness and our phenomenal universe. The fact that these strings are seen to operate at hyper-dimensions further supports my theory of reality in that they are holo-somatically defining themselves along the edge of our perceptual range.[125] Radical "trans-elemental" change occurs in transubstantiation as supernatural and natural dimensions are seen to operate in tandem. Theorists commonly consider five different versions of string theory. However, in March 1995, Edward Witten's genius was once again made evident when he showed how the five previous predominant string theories could be understood as mathematical approximations of a single String Theory he termed M-theory.[126] M-theory mathematically expounds on these earlier insights in hopes of arriving at a grand unified theory (GUT) that explains everything "beyond the grand unified theory—'the Theory' that explains everything."[127] M-theory is simultaneously a retelling of our continuing tale of how Nothingness and symmetry realize new ways to pass on reality's fundamental principle of non-subsistence.

M-theory attempts to explain gravity's operations starting at the quantum scale—an unlikely pairing made possible by the persistent presence and explanatory value that the concepts of null dimensions, nothingness, or silence have for the premising of multidimensional universes from the harmony of its component parts.[128] Among other important discoveries made by M-theory relevant to the present work is the idea that a material

[124] See Sean A. Hartnoll, Andrew Lucas, and Subir Sachdev, *Holographic Quantum Matter*, (Cambridge, MA: The MIT Press, 2018).

[125] See Henning Genz, *Nothingness: The Science of Empty Space*, Translated by Karin Heusch, (Reading: Helix Books/Perseus Books, 1999), 257–259.

[126] *Euclid's Window*, 244–248.

[127] See Sheldon Lee Glashow, "Toward a Unified Theory of Physics," *Michigan Quarterly Review* 23 (Spring 1984), 220.

[128] See *The Grand Design*, 41, 87, 107–109, and 113–116.

being's quantified interior dimensional informational structure is what makes it unique.[129] At a more fundamental level, the spatiotemporal locations in *STR*'s four-dimensional readings of macro-particles correspond to higher dimensional branes in String Theory (M-theory).These membranes can be thought of as holographic representations of underlining matrices.[130] Mlodinow explains that M-theory's description of physical reality requires a radical new interpretation of the nature of space-time.[131] The interior vibration of a respective string echoes in the external universe.[132] If the event of transubstantiation is to be made sensible to this generation, our attention must move beyond past deterministic and accidental interpretations of being. M-theory provides us with a scientific analogy for past hylomorphic explications of how interior geometries give us a metaphor "substantial fingerprint" for a material object. With this quantifiable model, we can better enter into a theological discussion of the *creatio ex nihilo*, transubstantiation and supernatural dimensions.

The spatiotemporal "quantifies" and "qualifies" the QMO in its diverse configurations and movements on infinitesimal and cosmic scales. The mathematical prolixity behind this state need not concern us here. What is of great importance for our purposes is the fact that this status depends upon reality's non-subsistent or "kenotic" nature. One model analogously represents this state by arguing for virtual or quasi-particle interactions as part of local fields that can also be interpreted to appear as energy in space-time. At the quantum level, this event seems to extend the logic of Einstein's theory of photoelectric effect within the arena of abstract geometries of flat and curved spaces. In each of the "flat" and "curved" spaces, energy and matter behave differently.[133] The contraction of space-time can create gravitational waves by focusing mass or curved space at a particular point, thereby creating a spot of greater density than the surrounding area. I take these gravity waves to be a holographic image of the mass from which they emerge. Furthermore, as strange as it seems, curved space can influence other regions of space even if these spaces are empty of matter.

[129] According to M-theory, space-time has ten spatial dimensions and one temporal dimension.

[130] *Holography and Emergence*, 300.

[131] *Euclid's Window*, 258.

[132] See *The Hidden Reality*, 347.

[133] See Karen Crowther, "Emergent space-time according to effective field theory: From top-down and bottom up," *Studies in History and Philosophy of Science Part B: Studies in the History and Philosophy of Modern Physics* 44, (3) (2013): 321–328.

This is possible because both regions contain energy in the form of fields that function in a manner similar to matter. The same dynamic holds between a particle and QED fields. In the subatomic world of QMT, our experiential understanding of the relationship between matter and energy is tested. The classical view of matter and energy keeps the two concepts largely separate. While it is true that every object can be given an attributed value of potential energy, it is common that this designation determines its quantifiable nature within the scope of our physical world.[134] The potential energy of a QMO can be defined in terms of its local configuration of space-time. Forces in any system, the laws of conservation tell us, work toward a state of equilibrium. From the perspective of the position of classical statistical mechanics, for example, the thermal equilibrium of a dominant micro-state is reflected in a measure of the system's greatest volume. The establishment of the macro-state's equilibrium at its greatest volume mirrors the holographic projection of the intensity of its electrical charge as a field. Super-symmetry also gives us a way to understand how symmetries and the laws of conservation of energy and matter are logically commensurate to hierarchical holo-somatic representations of identity. At its core, super-symmetry conceptualizes and communicates something about the unification of physical reality.[135]

All spatial and dimensional structures lend themselves to similar mathematical and theological descriptions. The theological claim that the Transcendent interacts within the spatiotemporal dimensional confines of bread and wine thus finds nominal support in the nature of a fundamental particle's own structural ordering. Present-day scientific theories help us better understand how transubstantiation intertwines the natural and supernatural realms. The realization of a material being conveys the idea of motion and relational difference, and the universe is structured to receive these properties or, better said, symmetry in the four dimensions.[136] A QMO's instrumental and observational validation is commensurate with a believer's assent to the veiled reality of the consecrated bread and wine (the *res et sacramentum*, the *ex opere operato* of the sacramental sign).

[134] See *The Yang-Mills Equation*, 237.

[135] *The Grand Design*, 95–97. See also N. Arkani-Hamed, A.G. Cohen and H. Georgi, *Physical Review Letters* 86 (2001): 4757, arXiv:hep-th/0104005. See also Quantum Shift, 172–175. Quotation from Lee Smolin, *The Trouble with Physics: The Rise of String Theory, the Fall of a Science and What Comes Next* (Boston, MA: Houghton Mifflin, 2006), 248, Fig. 15.

[136] This formulation is in concert with the early philosophical claim that being is something that exists in the act, see *ST.* I, 46, 1.

The fact that fundamental reality can be relayed holographically again suggests something of how the infinite plays out as a constant to the ongoing division of dimensional framing. The super-symmetrical ordering of matter and anti-matter or the various links of the material and essential identities of subatomic particles dole out previous hylomorphic attempts in the conceptual garb of post-Newtonian physics.

The potentially infinite ways that a being could appear is made known in the concrete. However, the material and accidental are what deposit a being's essential identity. The essential is further described by its super-essential properties. As every being has its own essential and super-essential identities, the latter opens up a way for the comparison of unlike or even apparently self-negating relationships, such as matter's accord with anti-matter. In such cases, the super-essential functions as a holo-somatic image of the Transcendent.[137] Thus, QMT need not necessarily negatively impress our theological investigation of transubstantiation from the viewpoint of a secularized materialism. Rather, it can allow for new speculative approaches and vistas. Applying John Conway and Simon Kochen's interpretation of QMT as outlined in their "The Strong Will Theorem," it is possible to argue for a modified realist account of the world insofar as one takes as almost interchangeable "properties," "events," and "information," manifest as subatomic particles in one of the four elementary relational interactions of Strong Nuclear Force, Electromagnetic Force, Weak Nuclear Force, or Gravitational Force found in Nature. It would be an uncritical exaggeration to claim that the drama of fundamental particles explains human freedom, and yet if theologians do not offer a critical account of physical reality, it remains little more than a mythic structure. Consequently, the probability laws of QMT and materiality come together in free and conscious subjectivity.[138] The human observer freely settles on the actual parameters of quantum ordering in the Conway-Kochen theory. All creatures, St. Paul maintains, live in the world as finite, limited, and free, that is, as "super-essential," "holo-somatic," or "spiritual" beings. At the highest level, the rational being's holo-somatic freedom makes possible their embodied identity.

[137] See *Homo Abyssus*, 398–399.

[138] See *The Strong Free Will Theorem*, 8–9.

References

1. Nicolas Gisin, *Quantum Chance: Non-locality, Teleportation and Other Quantum Marvels*, forward Alain Aspect (Cham, Heidelberg: Springer, 2014), 90–91. Hereafter cited as Quantum Chance.
2. John Milbank, Graham Ward, and Edith Wyschogrod, *Theological Perspectives on God and Beauty* (Harrisburg, PA: Trinity Press International, 2003), 18.
3. David Park, *The How and the Why: An Essay on the Origins and Development of Physical Theory*, (Princeton, NJ: Princeton University Press, 1988).
4. Wolfgang Smith, *The Quantum Enigma: Finding the Hidden Key*, Forward by Seyyed Hossein Nasar (San Rafael, CA: Angelico Press, Sophia Perennis, 2005), 146.
5. Arthur I. Miller, *Imagery in Scientific Thought*, (Cambridge, MA: MIT Press, 1986), 146.
6. Valia Allori, "On the Metaphysics of Quantum Mechanics." In *La philosophie de la physique: d'aujourd'hui a demain*, Edited by Soazig LeBihan. (Devon: Editions Vuibert, 2013), 4.
7. Max Born, *Natural Philosophy of Cause and Chance*, (Oxford, UK: Oxford University Press, 1949), 109.
8. Max Tegmark, "The Mathematical Universe." *Found. Phys.* (Sept., 2007): 2, arXiv:0704.0646v2 [gr-qc]. Accessed 01-01-18.
9. Homi K. Bhabha, "Aura and Agora: On Negotiating Rapture and Speaking Between," in Richard Francis, *Negotiating Rapture: The Power of Art to Transform Lives* (Chicago, IL: Museum of Contemporary Art Chicago, 1996), 12.
10. Gordon Belot, "Quantum States for Primitive Ontologies," *European Journal for the Philosophy of Science* 2 (2012).
11. Aleksandra I. Zecevic, *Truth, Beauty, and the Limits of Knowledge: A Path from Science to Religion* (San Diego, CA: University Readers, 2012), 52.
12. Frank Wilczek, "A Piece of Magic: The Dirac Equation," in *It Must Be Beautiful: Great Equations of Modern Science*, Edited by Graham Farmelo (London, UK: Granta Books, 2003).
13. Barbara Maria Stafford, *Echo Objects: The Cognitive Work of Images*, (Chicago, IL: The University of Chicago, 2007).
14. P.A. Ligomenides, "Scientific Knowledge as a Bridge to the Mind of God," Edited by John Polkinghorne in *The Trinity and an Entangled World: Relationality in Physical Science and Theology* (Grand Rapids, MI: Wm. B. Eerdmans Publishing Company, 2010), 89.
15. Paul A. M. Dirac, *Proc. R. Soc.* A., 133, 60 (1931) *Reality*, 123–24.
16. Leonid Butov, "Solid-state Physics: A polariton laser," *Nature* Vol. 447, no. 7144 (31 May 2007): 540–41.
17. John Stachel and Mihaela Iftime, presentation 06-25-07.

18. Christine Sutton, "Hidden Symmetry: The Yang-Mills Equation, in *It Must Be Beautiful: Great Equations of Modern Science*," Edited by Graham Farmelo (London, UK: Granta Books, 2003), 243.
19. Herman Weyl, *Philosophy of Mathematics and Natural Sciences* (New York, NY: Prager, 1984), 116.
20. James Owen Weatherall, *Void: The Strange Physics of Nothing*, (New Haven, CT: Yale University Press / Templeton Press/Templeton Press, 2016), 109.
21. Sergei Bulgakov, *The Bride of the Lamb*, Translated by Boris Jakim, (Grand Rapids, MI: Wm. B. Eerdmans Publishing Company, 2002).
22. David Morgan, "Secret Wisdom and Self-Effacement: The Spiritual in Art in the Modern Age," in Richard Francis, *Negotiating Rapture: The Power of Art to Transform Lives* (Chicago, IL: Museum of Contemporary Art of Chicago, 1996).
23. Anonymous, *Meditations on the Tarot: A Journey into Christian Hermeticism*. Afterword by Hans Urs von Balthasar, (New York, NY: Penguin Putnam, Inc., 2002).
24. Pavel Florensky, *Iconostasis*. Translated by Donald Sheehan and Olga Andrejev. Introduction by Donald Sheehan, (Crestwood, NY: St. Vladimir's Seminary Press, 1996).
25. Pavel Florensky, "Reverse Perspectives," in *Beyond Vision: Essays on the Perception of Art*, Edited by Nicoletta Misler and Translated by Wendy Salmond (London, UK: Reaktion Books LTD., 2002).
26. Hans Urs von Balthasar, *Truth of the World* Vol.1, Translated by Adrian J. Walker (San Francisco, CA: Ignatius Press, 2000), 192.
27. Stanley L. Jaki, *The Relevance of Physics* (Chicago, IL: University of Chicago Press, 1966), 92.
28. See Michael Dickson, "Quantum Logic Is Alive ($\wedge$ It is True $\vee$ It Is False)," *Philosophy of Science* 68, no. S3 (Sep., 2001): S274-S287. https://doi.org/10.1086/392915. Accessed March 18, 2017 at http://mdickson.net/pubs/Quantum Logic.pdf: 2.
29. Frank Wilczek, *A Beautiful Question: Finding Nature's Deep Design*, (New York, NY: Penguin Books, 2015).
30. Stephan Körner, *The Philosophy of Mathematics: An Introduction*, (New York, NY: Harper Torchbooks, 1960).
31. Anthony M. Alioto, *A History of Western Science* 2nd ed. (Englewood Cliffs: Prentice Hall, 1993), 410.
32. James Turrell, "EGG: The Arts Show," interview cited in Mark C. Taylor, *Refiguring the Spiritual: Religion, Culture and Public Life* (New York, NY: Columbia University Press, 2012), 142.

33. Nicholas J. Teh, "Holography and Emergence." *Studies in History and Philosophy of Science B: Studies in History and Philosophy of Modern Physics* 44, no. 33 (2013): 300–311.
34. Frank Close, *The Infinity Puzzle: Quantum Field Theory and the Hunt for an Orderly Universe: With a New Epilogue Covering the Discovery of the Higgs Boson*, (New York, New York: Basic Books: A Member of the Perseus Books Group, 2011), 29.

CHAPTER 5

Black Hole Entropy and the Holographic Universe

5.1 The Holographic Universe

The discovery of a "black hole" is one of the most stunning discoveries of modern physics. I turn now to explore how theological and metaphysical holo-cryptic principles reveal in one of Nature's most extreme of physical entities, a black hole. This cosmological beast gives us a metaphor for Nothingness. It is difficult to conceive of nothingness, and even more problematic is seeing it as important for sacramental reality. Each sacrament, Thomas reason, is "derivative" of the Eucharist, as this involves immediate contact with the Risen Lord, Jesus Christ, an event that presupposes his defeat of the 'Nothingness' of death and Hell.[1] The Eucharist is comparable to a geometric point upon which all other sacramental "lines" converge. One can imagine lines radiating from the Eucharist serving as the universe's sacramental connective "tissue." The universe emerges from the "Nothingness" of zero dimensions at the Big Bang.

The Neo-Platonic tradition allows for immutable forms of nothingness, the void, silence, and absence. Henri Poincaré (1854–1912) critiques this Platonic spirit as found in Kant's comparing thought to its mathematical structure. Poincaré writes that "geometry is not an experiential science; experience forms merely the occasion of our reflecting upon the geometrical ideas which preexists in us. However, the occasion is necessary; if it did

[1] See *ST.* III, 52, 6.

M. P. Fusco, *The Physics and Metaphysics of Transubstantiation*,
https://doi.org/10.1007/978-3-031-34640-8_5

not exist we should not reflect; and if our experiences were different, doubtless our reflections would also be different. Space is not a form of our sensibility; it is an instrument which serves us not to represent things to ourselves, but reason upon things."[2] [1] Poincaré sees consciousness as a mutual and ongoing rational exercise of experimental "geometric" displacement. Distancing himself from materialist notions of consciousness, he sees in the geometric form a definitive example for physical being and reality, one that establishes that "all that is not thought is pure nothingness."[3] [2]

The universe's fundamental symmetries (*AdS/CFT*) influence the universe's geometric shape and, by extension, mass and energy. Concurrently, the presence of mass or energy either positively (by inflation) or negatively (by contradiction) changes the spatiotemporal topology. The relationship between abstract curved geometric space and the spatiotemporal fabric of the universe under the force of gravity and energy speaks to this creative tension and is central to Einstein's General Theory of Relativity (GTR). There is much that can be said about Einstein's insight into how geometry provides us with a propaedeutic to explain gravity's function as a core cosmological principle. Although a full explanation of gravity still escapes theorists, they look to this preliminary work as the first stage in the explication of quantum gravity.[4]

Karl Schwarzschild (1873–1916) set out to mathematically delineate the geometry created by mass and energy as set out by GTR (e.g., an *AdS*-Schwarzschild Black Hole). To imagine a purely geometric space, Schwarzschild had to remove the physical entities of mass and energy from space-time geometries from his model and abstract out their influences.[5] By translating the physical entities of mass and energy into mathematical

[2] Henri Poincaré, "On the Foundations of Geometry," Translated by T.J. McCormack, *Monist*, 9 (1898): 41.

[3] Henri Poincaré, *The Value of Science* (1904) in *The Foundations of Science: Science and Hypothesis, the Value of Science, Science and Method*, Translated by G. B. Halsted (New York, NY: Science Press, 1913), 355. See also Linda Dalrymple Henderson, "Mysticism, Romanticism, and the Fourth Dimension," in Maurice Tuchman and Judi Freeman, Edited et al., in *The Spiritual in Art: Abstract Painting: 1890—1985* (New York, NY: Abbeville Press, 1986), 224.

[4] Marc Holman, "Foundations of quantum gravity: The role of principles grounded in empirical reality," *Studies in History and Philosophy of Science Part B: Studies in History and Philosophy of Modern Physics* 46 (May 2014):147.

[5] See Sean A. Hartnoll, Andrew Lucas, and Subir Sachdev, *Holographic Quantum Matter*, (Cambridge, MA: The MIT Press, 2018), 128–131.

entities, a model of greater predicative power is possible than available by simple astronomical observation of physical objects. By placing a "perfect" sphere in "absolute" Nothingness, Schwarzschild's thought experiment uncovered a new "type" of space-time. He mathematically designated a hypothetical sphere of space-time free of all matter and energy and thus anticipated the discovery of a "Black Hole." I see in Schwarzschild's entity the holographic image of an existing black hole.

Schwarzschild's work proves that some dimensional topologies can be profitable understood as negative and hence retroactively suggests the physical universe's preexisting relationship to nothingness as the infinite point of the Big Bang argues. When a star collapses under the centripetal pull of its own gravitational forces, it becomes a black hole. The extreme force of the star's gravity eradicates the geometric structures of its component elements and thus pulls the existing star toward the "null-dimensional" point. As the horizon line of the astronomer's view recesses, distortions increase in their viewfinder. Consequently, a black hole's horizon fades from view due to its curvature—an enfolding that perhaps extends infinitely. However, mathematical explanations of the negative and positive curvature of space-time provide us with a way to theoretically model the effects of an object falling into a black hole.

Adapting Hermann Minkowski's theories, Einstein was able to advance his own intuitions about space-time. Shlain explains that "in 1908 Herman Minkowski, a German mathematician and former teacher of Einstein, expressed in equations this reciprocal relationship, and recognizing that it comprised the fourth dimension, named it the *spacetime continuum*. The new phrase coined for this revolutionary mind-expanding concept joined two old words, space and time, fusing them in order to emphasize the fact that each, which for millennia had been held separate, was in truth a magnificent unity."[6] Minkowski mathematically expressed the reciprocal nature of space and time as a single space-time continuum. Physical objects can be mathematically represented not only in Euclidean geometry but also in non-Euclidean and hyper-dimensional spaces. Minkowski's theory of space-time gives us a "simpler" model of how three-dimensional Euclidean space and time act as a four-dimensional manifold by expanding the information of a three-dimensional space onto a higher dimension. This geometry is ideal because it assumes that an independent inertial frame of

[6] *Shlain*, 132.

reference exists between any two events.[7] Einstein's insights concerning the replacement of time as positive (infinite) and space as negative (curved) relate to my theological proposals concerning the sacred nature of liturgical practice and ritualistic space. In GTR and sacramental theology, researchers and the faithful are asked to imagine how space-time is less a fixed physical boundary than an ever-changing mental image. Given GTR findings, any temporal designations in the universe are only possible when one realizes that paradoxically no such measures preexist in the universe.[8]

It is only when the spatiotemporal is "translated" or mapped into a higher dimension or an ideal surface that Einstein is able to explain the immediate and simultaneous from the viewpoint of a moving observer and universe. For example, the GTR gives us a way to account simultaneously for a quarterback's perspective, that of the receiver running a pattern, and the fans in the stadium. More precisely stated, Einstein projects three-dimensional events onto an infinitely differentiated smooth manifold (e.g., C^{∞}). This move conserves the topological properties of both the system and its observation. Historical accounts and infinite space-time are posited and diagrammed as potential unlimited differentiation on a four-dimensional manifold. The unique properties of four-dimensional space open up a new way to envision a material being's identification with infinite time and space as part of the physical topology of the Eucharist, as "R^4 is the only Euclidean space that admits 'non standard differentiable structures' and moreover it does so infinitely many times."[9] [3] Presumably, a believer experiences something similar when participating in the Eucharist's holo-somatic actualization. The smooth manifold gives us another view of the artist's "blank canvas" or the null dimension at the heart of nothingness's preference for the void in a black hole.

The geometrical representation of non-classical space-time is not restricted to any particular manner of division or segmentation (e.g.,

[7] See *Surfaces and Essence*, 443–444. We assume that the Minkowski space-time is discrete and topologically homogeneous and can therefore account for the invariance of light speed and the principles of relativity.

[8] See *Void*, 59. Given the stability and flexibility of Minkowski's space-time, it supports such theoretic framing as those outlined in The General Theory of Relativity, as Einstein explicates the covariance of motion under those diffeomorphisms that define gravitational and inertial mass as part of the same equivalence principle. Minkowski's space-time also gives us a way to understand how cultic sacrifice is reenacted in subsequent liturgies.

[9] Marc Holman, "Foundations of quantum gravity: The role of principles grounded in empirical reality," *Studies in History and Philosophy of Modern Physics* 46 (2014):152. See also S.K. Donaldson and P.B. Kronheimer, *The Geometry of Four-Manifolds* (Oxford, UK: Oxford University Press, 1990).

hours, seconds, meters, volume, etc.), and thus taken as malleable to our interpretation of nothingness. Furthermore, the birth of space-time at the hands of subjective endeavor is seen as an archetype for material being's future revisions as part of the ubiquity of the principle of non-subsistence. Schwarzschild's work gives us a mathematical equivalent for such differing's influence within an idealized system. The absence of mass and energy at lower dimensions within a black hole communicates as malleable relationships in three dimensions.[10] Matter and energy can be taken as non-subsistent equivalents that coexist as two binary informational modes ($E = mc^2$) in proportion to matter's communication as light in a holo-somatic representation (e.g., c2). The informational identity of matter and energy mutually changes. A midway point theoretically exists between mass and energy, as they can be differentiated. The symmetrical exchange between the two states is possible because the actual state of either mass or energy is *a priori* presumed. Matter's motion in three-dimensional space-time (e.g., mc^2) finds it perfect balance in energy's (E) "momentum" as a holographic projection in multidimensions.

When matter enters a black hole's event horizon, it is imprisoned in "a point of no return: once you pass the event horizon, you can never pass back out…You cannot even send a signal out, because light cannot escape either…[you] will just keep falling until eventually [one] simply pops out of existence."[11] [4] However, matter takes on a new dimensional configuration within the black hole. This new being holographically appears at the event horizon as light that radiates into the universe's three-dimensional space. The black hole's debility of a physical being's integrity does not eradicate its essential or informational identity because it virtually conserves it as energy—as an "image" on the event horizon.[12] Matter's transformation into energy and images from the black hole is termed "Hawking radiation" by Physicists. As noted previously, this essential disclosure is a holo-somatic property of all beings. To this end, Balthasar notes that the essence realizes an "inexplicable active irradiation of the center of being into the expressive surface of the image, an irradiation that reflects itself in the image and confers upon it a unity, fullness and depth surpassing what

[10] See C. V. Vishveshwara, Scattering of Gravitational Radiation by an Schwarzschild Black hole, *Nature* 227 (1970): 936–938.

[11] *Void*, 74.

[12] The so-called Hawking radiation follows the second law of thermodynamics and verifies the conservation of information under the unique space-time structure of a black hole. Ideally, this information communicates as light on a gravitational boundary. See Gerard 't Hooft, "Dimensional Reduction in Quantum Gravity," (1993), arXiv:gr-qc/9310026.

the image as such contains."[13] [5] A being's informational identity analogously shares in this event. A system's elementary disorder or randomness is symmetrical to its relative entropy in both classical and non-classical physical systems, Euclidean or complex spaces.[14] Identity is never truly lost, as a physical being's mass and energy can be translated into its relative holo-somatic principle. Thomas speaks of the conservation of mass and energy holomorphically in terms of act and potency's unified composite, for "in whatever thing we find two, one of which is the complement of the other, the proportion of one of them to the other is as the proportion of potentiality to act; for nothing is completed except by its proper act."[15] [6] The transformation of matter into energy accounts for the relative identity of a physical particle and its spatiotemporal existence, as well as its metaphysical identity directed by Ur-Kenosis. The mutual relationship between mass and energy within a single physical entity is congruent to the meaning of a subject's act-potency determinations. The subject perceives and consciously records another being and in so doing "actualizes" its existence as part of one's cognitive structuring.[16] These operations stage a more developed comprehension of spatiotemporal differing's dependencies on Ur-Kenosis in the created order because they involve the conscious creation of mental images (e.g., cognitive holographs).

Space serves as matter's receptacle or perhaps better stated as its two-dimensional blanket, whereas time indicates a determination or measure of potency or non-subsistence in physical reality. The entropy of the black hole is calculated by taking the square of roughly one-quarter of the total area of the event horizon in Planck lengths.[17] Consequently, assuming

[13] *TL* 1, 142.

[14] See Max Planck, *Eight Lectures on Theoretical Physics* (New York, NY: Columbia University, 1915), 50 and on the role of the theory of thermodynamics, for instance, on the formulation of special relativity, see M.J. Klein, "Thermodynamics in Einstein's Thought," *Science*, 157 (3788) (1967): 509–516.

[15] *SCG*, II, 53, 2.

[16] Thomas Aquinas, *Commentary on De Anima*, III, Lect. 2 , §583, Translated by Kenelm Foster, O.P., and Silvester Humphries, O.P., and Introduction by Ralph McInerny (Notre Dame, IN: Dumb Ox Books, 1994).

[17] See Jacob D. Bekenstein, "Universal upper bound on the entropy-to-energy ratio for bounded systems," *Physical Review* D 23 (215) (1981): 287–298. This geometric expression of the entropic at the event horizon distinguishes it from previous hierarchical versions that imply that the effect participates as a lesser power or efficacy than its cause. See *SCG* 1, 2, 2, 15. In both approaches, the effects are entropic to their cause. However, disciplines interpret this equation differently.

generic energy we find that the total amount of information possible in a given area of a black hole is calculated by dividing its area by Planck's constant raised to the number of degrees of freedom for "the covariant entropy bound".[18] It is the area of this region and not its interior volume that fixes the limits of information in a black hole.[19] The so-called Bekenstein-Hawking boundary of the event horizon gives us a specialized view of the fundamental symmetry within the spatiotemporal dimensional context of the black hole.[20] The difference or rift between the black hole and the universe is considered to be a zero boundary. The amount of matter within the black hole is proportional to the amount of holographic information that can be stored.[21] Because the total amount of information that can be stored in a black hole is determined by computing a two-dimensional area, its entropy can be imagined to exist in a subtractive or "kenotic" relationship to three-dimensional space-time.[22] Understood in relationship to the principles of holographic information, the universe can be accorded emergent properties and freedom given the positive and negative values of time and space and thereby a holo-somatic depiction of matter; thus, "the holographic principle in a sense questions locality (whereas it preserves quantum-mechanical unitarity). This has led to the claim that the locality of our macroscopic world is either an emergent

[18] See Raphael Bousso, Oliver DeWolfe, and Robert C. Meyers, "Unbounded Entropy in Space-time with Positive Cosmological Constant," *Foundations of Physics* Vol. 33, (2) (2003): 299.

[19] See Jacob D. Bekenstein, "Information in the Holographic Universe—Theoretical Results About Black Holes Suggests that the Universe could be like a Giant Hologram," *Scientific American* (August 2003), 59, *Holography and Emergence*, 302 and Robert M. Wald, *Space, Time, and Gravity: The Theory of the Big Bang and Black Holes*, 2nd edition, (Chicago, IL: The University of Chicago Press, 1992), 136–137.

[20] See J.M. Maldacena, "Eternal black holes in Anti-de-Sitter," *Journal of High Energy Physics* 0304, 021 (2003) [arXiv:hep-th/0106112].

[21] *Hidden Reality*, 248–249. See also Jacob D. Bekenstein, "Universal upper bound on the entropy- to-energy ratio for bounded systems," *Physical Review* D Vol. 23, no. 215 (1981): 287–298.

[22] The informational limit is evident with a "unitary" description of black hole evaporation as suggested by AdS/symmetries at the event horizon. On the Anti-de Sitter (*AdS*) boundary conditions for radiation in four dimensions in a black hole, see Edward Witten, "Anti-de Sitter Space and holography," *Advances in Theoretical and Mathematical Physics* Vol. 2, 253–291.

phenomenon, or may be retained by implementing some sort of gauging which leaves just the right amount of degrees of freedom."[23] [7]

Information is indestructible, and as Bekenstein remarks, "the physical world [is] made of information, with energy and matter as incidentals."[24] [8] Two important facts follow from this, namely, (1) an image of a being falling into a black hole serves as information in the form of energy, and (2) the essential or holographic attenuation of a being as a mathematical construct is possible given the uniformity of the universe and the universality of its laws. Lacking the necessary surface area to encode, the external communication of information would be impossible between beings; silence would ensue. Luciano Floridi asserts that the erasure of all information leads to a meaningless uniformity or an empty entropic state without difference. Information such as the Ur-Kenotic act is "a difference that makes a difference."[25] Without reality's "finitization of Ur-Kenosis" as non-subsistent differences, the distinction between form and content is impossible. Information that serves as my stand-in for essential truth can be perceived without difference in a system.[26] I see these theological and philosophical considerations reflected in scientific theory and experimental results. Space at the infinitesimal level, Weyl points out, is *a priori* homogeneous and becomes meaningful with the introduction of some distortion.[27] The unremarkable state of the event horizon boundary becomes meaningful when it and Eucharistic interiority are informed by a physical being's representation as light-like geodesics.

The stability of the event horizon changes with the introduction of new geodesics. A metaphoric density can be attributed to the collection of the two-dimensional geometry of the geodesics given the observed distortions at the event horizon. Furthermore, changes in an emergent system imply a specialized motion of probability; for "quantum uncertainty across the horizon of the black hole generates fluctuations of the horizon geometry. However, 'fluctuations' imply 'probability,' and 'probability' implies

[23] Norman Sieroka and Eckehard W. Mielke, "Holography as a principle in quantum gravity?—Some historical and systematic observations," *Studies in History and Philosophy of Modern Physics* vol. 46 (2014):172. See also Raphael Bousso, "The Holographic Principle," *Review of Modern Physics* vol. 74 (2002): 838.

[24] Jacob Bekenstein, "Information in the Holographic Universe," *Scientific America*, Vol. 289 (2003): 59.

[25] *Information*, 23.

[26] *Information*, 23–24.

[27] See Herman Weyl, *Mathematische Analyze des Raumproblems: Vorlesungen gehalten in Barcelona und Madrid* (Berlin: J. Springer,1924), 24.

motion or thermodynamics.[28] Concealing from us a part of the universe, a black hole makes its quantum fluctuations detectable in the form of heat."[29] [9] Here, we have a suitable scientific analogy for the beauty of a being's holographic or virtual presentation. Our ability to observe this radiation even light years away suggests that the two-dimensional image at the event horizon is projected to the "ends" of the universe. Theoretically, this emitted information or holographic entities will project toward the very edge of the universe.[30]

5.2 Nothingness, Infinite Differing, and Eucharistic Interiority

In 1998, Dr. Juan Maldacena published a scientific article on complex spaces in the universe. Maldacena shows the correspondence of near infinitivally small strings mathematically (holographically) forecast onto a higher-dimensional "world sheet."[31] A being's representation as an informational holograph seems consistent with the tenets of String Theory, QMT and black hole entropy.[32] The Nobel laureate physicist Gerard 't Hooft furnishes us with a mathematical description of matter and energy's holographic rendition. Holographic information influences other beings, and proof of this conjecture is seen in matter creating a gravitational field that distorts space-time.[33] Leonard Susskind's research confirms and devel-

[28] See Gerard 't Hooft, The Holographic principle: Opening Lecture in 37th course: Basic and highlights of fundamental physics 29 August–7 September 1999. Erice, Italy: *International School of Subnuclear Physics*. Hep-th/000304.

[29] *Hidden Reality*, 226.

[30] See *The Holographic Principle and M-Theory* at http://www.dampt.cam.ac.uk/research/gr/public/holo/ accessed 2/15/17.

[31] Maldacena hypothesized about the operation of strings on a 3 + 1 dimensional type (IIB) membrane. See Juan Martin Maldacena, "The Large-N Limit of superconformal field theories and supergravity," International Journal ofTheoretical Physics 38, (1999):1113–1133 . https://doi.org/10.1023/A:1026654312961.

[32] See Ron Cowen, "Simulations back up theory the Universe is a hologram," *Nature News* (10 December 2013). https://doi.org/10.1038/ptep/ptu028 and Masanori, Hanada, Yoshifumi, Hyakutake, Goro Ishiki, and Jun, Nishimura, "Holographic description of a quantum black hole on a Computer," *Science* Vol. 344, no. 6186 (April 2014): 882–885.

[33] In String Theory, a world sheet is a two-dimensional manifold that describes how Strings are the smallest of subatomic elements in space-time. See Gerard 't Hooft, "Dimensional Reduction in Quantum Gravity" (1993). arXiv:gr-qc/9310026 and J.M. Maldacena, *The Large N limit of superconformal field theories and supergravity, International Journal. of Theoritical. Physics.* 38 (1999) 1113, [hep-th/9711200].

ops 't Hooft's insights into the holographic principle and gravity's "catabolic" constrictions in a black hole.[34] The holographic image is latent in the physical identity of a being in three dimensions and only fully conveys at higher dimensions. Consequently, we can retroactively see space-time as a divisive or "kenotic" measure of informational fields and geometric metrics. A being's informational or formal intensity (e.g., in-forming), spatial structure, and movement mutually interact because "[s]pace tells matter how to move. Matter tells space how to curve."[35] As lower dimensional geodesics have been shown, mathematics makes certain that the interior "essential" truth of being precipitates in its holographic and topological calibrations of a being regardless of its dimensional status or placement.[36] The holographic principle does not translate exactly to other doctrines of transubstantiation. However, in accounting for the role of Nothingness' predication in the "subsisting, concrete *res*," a way opens to expand the doctrine of transubstantiation and the liturgy to include how matter-energy holographically configures to communes in a black hole.[37]

The topology of a black hole's interior gives theologians a new way to interpret Nothingness' continued presence in Creation, the Cross, and in ritualistic remembrance at the Mass. The concept of the division of matter or space-time is central to liturgy and is an inescapable part of human conscious experience. Florensky speaks of three divisions that correspond to the perception of higher truths. Broadly understood human perception grasps physical space, interior sensual dimensions and theoretical abstract spaces. One can divide each of these three areas into further subdivisions.[38] Theoretical physics states that the first of these dimensions is more expansive. M-theory argues, for example, that matter exists in a universe that can be divided into ten dimensions. Developing this insight, the theorists Andrew Strominger and Cumrun Vafa mathematically derive the dimen-

[34] See Leonard Susskind, "The World as a Hologram," *Journal of Mathematical Physics* Vol. 36, no. 11 (1995): 6377–6396. Leonard Susskind, Tom Banks, Willy Fischler, and Stephen Shenker reformulated M-theory using a holographic description of black holes. Ideally, this activity communicates information as light on a gravitational boundary. See Raphael, "The Holographic Principle," *Review of Modern Physics* Vol. 74, no. 3 (2002): 825–874.

[35] See Charles W. Misner, Kip Thorne, and John Wheeler, *Gravitation* (New York, NY: W. H Freeman, 1973), 5.

[36] See *Theoretical Writings*, 175.

[37] See *Homo Abyssus*, 200–202.

[38] *Reverse Perspective*, 265.

sional "branes" of physical reality.[39] The authors develop how information more precisely relates to the dimensional space. The concentration of information is limited to the dimensional capacity of a given region of space-time. So too, the authors point out, there is a limit to the amount of information that can be removed, "subtracted or abstracted from a given region of four-dimensional space."

Through conscious action, individual beings are able to interconnect across all levels in reality. The act of knowing oneself and the other is metaphorically a physiological, sensual, and semantic discovery. The receptive individual shares their truth as a moment of virtual resurrection—the gifting and re-gifting of holo-somatic identity, to others.[40] [28] Given this universality, rational subjects are able to apprehend the "other" and ultimately the finally Non-Other (*Non-aliud*) of the crucified Lord, Jesus Christ. As a closed system, reality avails itself of perceptual and cognitive growth when created difference has a future rendezvous with an Ur-Kenotic identity. Material, formal, and efficient differing are means to reenact in analogous fashion that Transcendent simple unity which is also a "perfect subtraction." This state of affairs exists as part of created reality, because matter's most essential form, the mathematician-philosopher tells us, is "number." Badiou states that "the traditional notion of initial conditions entails that our universe 'started out' in some particular state…mathematical structures do not exist in an external space or time, are not created or destroyed, and in many cases also lack any internal structure resembling time."[41] [11] Given a Neo-Platonic interpretation of number, it is only possible to perceive and comprehend a physical being's identity when perceived within spatiotemporal frames of reference.

When one partner of an inseparably paired or an entangled particle is sucked into a black hole, its twined partner outside of gravity's direct reach continues as a virtual image of its decimated partner. The destruction of the material form of one particle immediately presents in the informational structure of its entangled other at another location. Various geometric and dimensional spaces can be used to depict an image of two perfectly aligned beings. Holo-somatic identity argues that the human person analogously images the Divine in this manner. Nyssa's depiction of

[39] Andrew Strominger and C. Vafa, "Microscopic Origin of the Bekenstein—Hawking Entropy," *Physics Letters B* Vol. 379, Issue 1–4 (June 1996): 99–104.

[40] See *After Writing*, 150 and 266, on how ecclesial space images in the world.

[41] Tegmark, *The Mathematical Universe*, 10.

Moses' contemplative experience of the divine is instructive. He envisions Moses within a cloud of radiating light that informs his prayer and inspires how he will direct his people like the sun's rays directs the hidden seed to grow from the dark earth. Spiritual leadership is multidimensional because it presumes the existence of other free and conscious persons. The contemplative life of prayer exemplifies the natural law and a communal purification, for if it were otherwise, a necessary separation would exist between the sacramental life and the created order or matter from space-time. The etymology of *latreia* (worship) captures this meaning, as the term refers to one's duty to political and social communities. The ancient practice of the liturgical act requires that individuals renounce something of their isolation and status by donating time or money for the public's welfare.

The dark night of the soul and human suffering is transformed by those graces won by Jesus Christ and thus introduces a temporal dimension to the human person's holo-cryptic identity as imago Dei. There is a way in which imagination and memory, for example, work in tandem at the Mass to interpret all that is suffering and fallen nature in the hyper-dimensional language of grace, for "[t]he celebration of the Eucharist is itself *anamnesis*, which means that it is contemplation in love and the communion of love with love; and it is only from such celebration that a Christian mission goes out into the world: *ite missa missio est*."[42] [12] Human love supremely entrusts this primitive form when two people consciously and somatically embrace their differences in unity purpose: "it is not by means of one isolated faculty that man is open, in knowledge and in love, to the Thou, to things and to God: it is as a whole (through all his faculties) that man is attuned to total reality."[43] [13] Love hermeneutically mirrors one's participation in truth as part of the incarnational reality of the Eucharist.[44] [5] The mystic penetrates into this mystery by probing into the question of existence and non-existence. With contemplation, one experiences darkness and abyss as a means to better judge how one's desires are congruent with Creation's restoration and humanity's common goal to transcend all limitations.[45] Contemplation leads to that truth that defines one's personal

[42] *Love Alone*, 109–110.

[43] *GL* 1, 243.

[44] *TL* 1, 116, 118 and 123.

[45] See Nyssa, The Life of Moses, Translation and Introduction by Abraham J. Malherbe and Everett Ferguson and Preface by John Meyendorff (New York, New York/Mahwah: Paulist Press, 1978), 91–94 [nos. 152–161].

identity by embracing the manifestation and absence of the Father, Son, and Spirit in one's daily and prayer life.[46]

The wholeness of self that a person seeks is answered in the Eucharistic species unification of the immanent and the transcendent Jesus Christ. The Eucharist introduces into one's epistemological framing the possibility of a transcendent supra-observational point. Jesus Christ does not subvert the Divine interior life but brings about in his person and actions the union of Creation's infinitely differing immanence and the Ur-Kenotic inner life of the Trinity, for "multiplicity is not something that supervenes upon being from the outside. Instead, it springs from being's properly enacted movement into subsistence, insofar as it is precisely the infinite positivity of the 'One' God that allows being to be 'itself' in the multiplicity of its finite subsistence."[47] [14] The Incarnation creates a space whereby transcendent and finite beings exist in a consonant and free or "kenotic" relationship.[48] The dialectic between immanent being and virtual representations of the transcendent are seen as part of our conscious structuring as imago Dei (John 17:20–3). There is a self-reflective component of the Eucharistic celebration that reveals the deeper mystery of created existence.[49]

The mystery of created identity finds its supernatural conclusion in the post-apocalyptic landscape where the Eternal Lamb reveals Godself in the "hidden manna" of the heart (Revelation 2:17).[50] Once fully restored with the Son of Man's return, Creation will be as sanctified as the inner sanctuary of the "Holy of Holies" (Isaiah 66:20).[51] The physical world is sympathetic to information's entropic unfolding and ritualistic spaces to transubstantiation's folding of the space-time within the Eucharistic species. A sacramental view of reality correlates to a holo-somatic understanding of a being's identity.[52] The substances of the bread and wine are converted into those of Jesus Christ. The bare material informational content of bread and wine

[46] See *The Pillar and Ground of Truth*, 12–13.

[47] *Homo Abyssus*, 75.

[48] Hans Urs von Balthasar, *Epilogue* (San Francisco, CA: Ignatius Press, 1987), 89. Hereafter cited as *E*. See also Augustine, *On the Trinity, NPNF3*, XV, 14, 23–24.

[49] Augustine, *Sermons on the New Testament, NPNF1*, vol. 6, Sermon XLV, Translated and Introduced by Philip Schaff, D.D., LL.D., (Grand Rapids, MI: Wm. B. Eerdmans Publishing Company, 1980).

[50] See *God Without Being*, 175 and A.N. Whitehead, *Science and the Modern World* (New York, NY: The New American Library/A Mentor Book, 1925), 22.

[51] Nyssa, *The Life of Moses*, Translation and Introduction by Abraham J. Malherbe and Everett Ferguson and Preface by John Meyendorff (New York, New York/Mahwah: Paulist Press, 1978), 97–101 [nos.17–183] .

[52] See *Emblems of Mind*, 220 on ritualistic dimensions of mathematics.

actualizes the Divine's stratagem for all creation. Grace does not annihilate being's essential truth or informational structure at the moment of consecration but transforms and perfects the natural order, including the subjective consciousness (*Gratia perficit naturam, non supplet*) of those aware of this transformation event. The Mass holographically implements Christ's sacrifice on the Cross but does not override or undermine or diminish its original historical incidence. Furthermore, the Son's eternal position within the Transcendent Trinitarian community, the Eternal Word's role at Creation or its apocalyptic raveling confirms but does not displace the truth of the Eucharistic sacrifice, for "[t]he end of the world is not physical but metaphysical. In reality, the world does not end but is transfigured into a new being, into a new heaven and a new earth."[53] [15]

The Spirit distributes the divine grace that virtually unifies Creation and makes relational differing commensurate with human consciousness and freedom and therein, historical reality.[54] This process finds theological justification in the Spirit's gifting of Pure Groundlessness, that is, the "Simple Nothingness" that can appear as an antinomic ripening of space-time when introduced into human experience lacking a spiritual commitment. It is in the finite's meeting of the Eternal in the guise of universal Nothingness that material being co-inheres in its natural processes and higher symbolic renditions. In ritualistic acts, these two dimensions come together in the faithful's participation in the creative, sacramental, and spiritual act. Open to the divine's presence, the person is able to invest their existence and its myriad of symbolic representations with meaning. In this way, the rational consciousness appropriates meaning based on sensual experiences, for "[w]hen they are put in the past, these symbols, are, in the plane of the empirical world, called remembrances. When they are put in the present, they are called imagination."[55] [16] The imagination uses symbols to recognize truth's ability to manifest across all spatiotemporal moments. Human memory is a foundation for the operations of human consciousness. The imagination has trans-historical capabilities, as it not only recalls but also serves to make the future present. Subjective imagination announces the eschatological for "the past in its form and of the present in its matter."[56] [17] Ultimately, these mental orchestrations

[53] *The Bride of the Lamb*, 401. See also *DS* 1754, Canon 4 at Council of Trent: Twenty-Second Session (1562).

[54] See, *TD* 4, 237–242, *TD* 2, 206–210.

[55] *The Pillar and Ground of the Truth*, 149.

[56] Henri Bergson, *Mind-Energy*. Translated by H. Wilson Carr (New York, NY: Palgrave Macmillan, 2007), 133.

are only possible by the Pure Spirit's self-imposed Ur-Kenotic concession. The Groundless Transcendent is recognized in the created order's holo-somatic truth in human consciousness.

5.3 Jesus Christ's Holo-somatic Identity in the Eucharist

Like the silence following a musical note, nothingness gives being the space to project beyond their immediate boundaries to the theoretical ends of the universe.[57] Mathematics in the concept of the set gives us a logical equivalent to exceed immediacy with a view to the infinite. The mathematical set is open to defining any members, even infinite members. In this apparatus of set a, we can "actively unfold the structure of nothing" through the ideas of "pure spontaneity," "inconsistency," "multiplicity," and "infinity."[58] The empty set $Z = \emptyset$ mathematically conceptually situates the ideas of nothingness and the void. The logic of set theory promotes its own understanding of how truth and being are inseparable. It is adapted to "domicile" an infinite succession of terms for in the "null-set axiom in set theory, the assertion that $\emptyset$ exists, is the ontological decision for the void. In set theoretic terms, the void is the fundamental set that contains no elements, on which all other consistent sets are built."[59] The null or empty set serves as a mathematical envoy for Nothingness' protraction in the infinite modes of a being's evolution. A number's intensive reflexive and extensive relationships are communicated by the empty set just as the ontological being is first defined by nothingness. The abyss makes possible future emergence, as Badiou states, "every thinkable being is drawn from operations first applied to the void alone. A multiple will be all the more complex the longer the operational chain which, on the basis of the void, leads to its determination."[60] [18] The mathematical set gives us a way to understand that every being includes an implicit albeit recondite reference to Nothingness for this acts as a backdrop or principle of the non-being that inheres in individual identity in Creation. Theological claims about the metaphysical equilibrium that exists between finite being and non-subsistent being are possible because of a prior arrangement with

[57] *Being and Event*, 58.

[58] The scission Badiou is advocating here is abstract and not the material differentiation of physical beings. See *Theoretical Writings*, 53, 124, and 196.

[59] *Hidden Void*, 226.

[60] *Logics of Worlds*, 112. See also *Being and Event*, 281–285 and on the possible moral implications of determination and negativity see *De pot.* 7, 5, resp. 1–6.

Nothingness.[61] Being serves as a metaphoric empty set or a dimensional-less space upon which factual meaning unfolds in multiple modes of exigency and causal interactions.[62]

Relational differing is closely aligned with the idea of the infinite, as both share a conceptual relationship to the dimensionless and Nothingness. The partial nature of a being's disclosure means that every event is incomplete; hence, "that there is a being of nothing, as form of the unpresentable."[63] [19] Reason's interrogation is born from the free and conscious person's creative power. The aesthetic sensibility of the conscious person's imagination creates images by distinguishing between sensual apprehensions and analytic and synthetic intellectual judgments—an operation that presupposes analogous "space" between images and ideas within the mind. Formulating images from sensual apprehension is only possible with the implicit application of absence—a kenotic wayward point. A similar dynamic is evident in Badiou's mathematical philosophy, as Sean Bowden states, the "consistency of presentation thus requires that all structure be doubled by a meta-structure which secures the former against any fixation of the void that is against any inconsistency."[64] [20] By virtue of hidden Nothingness, identity emerges only to be reformulated, presented, and infinitely re-represented ("redoublement") within human consciousness. Analogous to the QMO's depiction in Hilbert space, the primordial representation of a physical being requires conscious reflectivity and a spatiotemporal platform based on some reference to the void. The created being is fundamentally a gift and good according to theologian. Consequently, even a created being's extreme distance from the divine in Hell, a being is never completely annihilated—energy and information are never destroyed. Analogously, the number one represents a being's irreducible identity. Badiou explains that it is "impossible to subtract oneself from the proper name if this subtraction's uniqueness provides the basis for the propriety of a name. As a result, there would seem to be no proper of the proper, which is to say no singularity of that which subtracts itself from all self-doubling through the name of its singularity."[65] Through

[61] *Homo Abyssus*, 206–207.

[62] See *Homo Abyssus*, 125.

[63] *Being and Event*, 54. See also ibid., 55 and 68–69.

[64] Sean Bowden, "Badiou and Lautman," in *Badiou and Philosophy,* Edited by, Sean Bowden and Simon Duffy (Edinburgh, UK: Edinburgh University Press, 2012), 50. See also *Being and Event*, 93–94 and *Post-Continental Philosophy*, 86.

[65] *Theoretical Writings*, 109. See also *Logics of Worlds*, 324.

self-reflective consciousness, a person holo-somatically creates a self-image of their soul and this remains even after death—the soul can never be equated with zero or nothingness.

Boundaries afford sensual perception with the means to define interim spaces and objects within a larger sense of the categorical whole. Objective existence and subjective perception are governed by difference.[66] The subject cannot determine truth's discharge prior to its historical revelations, its ontological ingress in the world. Some remnant of the void remains, as every conscious image or symbol is incomplete or "kenotically" ordered. The defined line and blackness give the artist the means to positively represent nothingness in its various modalities. The darkened edge of a picture or blank space in a work of art mirrors the interior work of the imagination. The holo-somatic image virtually grounds the correspondence between imagination and object of art, for the "metaphysical and epistemological view of darkness as mere lack or emptiness informs certain 'virtual reality' theories of mind based on the model of a hallucinatory inner landscape or penumbral theatre. Specifically, the metaphor of a dim architectural interior—a shifting perceptual arena housing no one, yet cinematically registering the transits of neural dynamics on an occasionally spot lit stage."[67] [21] The method of conceptualizing transcendent identity and nothingness functions as a metaphor for that receptivity that opens the conscious subject to reality. Subjective experience and knowledge flourish to the degree that one can remove or "subtract" one's perceived egotistical needs and disordered desires from every perception and act. When one consciously opens up to the foreign and strange, one begins to become spiritually attuned to the Nothingness of Christ's death on the Cross and the Eucharistic sacrifice.[68]

A being's holo-somatic identity is interpreted to be an eschatological mode of Jesus Christ's own crucified and resurrected state. The inevitability of death becomes meaningful for the Christian when read as part of the mystery of the immanence and transcendence of the divine. Przywara offers a way forward for "the differences between *a priori* and *a posteriori* metaphysics, 'God in the creature' (as the final form common to both) is differentiated into two versions. In the first, the 'in' within the formula 'God in the creature' (understood as the positivum of *a priori* metaphysics) tends

[66] See *Logics of Worlds*, 210.
[67] *Echo Objects*, 213.
[68] See *Homo Abyssus*, 489–490.

to mean that the creature is simply a (passive) 'manifestation of God.' *A priori* metaphysical thought goes from the ground, end, and definition to the grounded, directed, and determined. In this respect, it therefore involves in its ultimate depth a reproduction of God's own standpoint (of his knowledge *per modum quo artifex cognoscit artificiata*—that is, of his knowledge from the vantage of the ideative and the creative, ...in the second version of 'God in the creature,'...the 'in' means a kind of (active) 'potency' of the creature 'towards God,' insofar as the creature is able to measure out the expanse of being (from below to above), even to its ultimate foundations."[69] [22] The life and death of the finite man Jesus Christ makes possible the historical revelation of the grace of the risen Christ.

The Spirit shares the historical person and mission of Jesus Christ with the entire created order—a movement received and celebrated in the fullness of ecclesial life. The philosophy of nature, especially its quantum and relativistic interpretations, can aid in our grasp of grace's operations. The postmodern fondness of defining reality by means of repetitious re-representing is here critiqued by the theological point of view that argues that grace recreates (evolves) and redeems (reorientates) the natural order. The incarnation and resurrection of Jesus Christ are intrinsic parts of a single history that comes together in the liturgy and directly argues for how grace-nature corresponds at the quantum level.

A theologian may argue that it is only as created as imago Dei, that is, as rational and free beings, that reality holographically witnesses to a "Christological naturalism" that inspires the human species. Richard Fern explains that this ordering is possible because "God creates the world from nothing means the power and presence of God, the divine *energeia*, lies at the core of Creation, in every place and every time. In this sense, creation resembles that Holodeck; shut off the program and you are left with nothing, an empty room."[70] The yoke between matter and energy summons to our debate the questions of how mathematical dimensions mimic some of the characteristics evident with our own working theological hypothesis on the nature of the potential, void, or nothingness. The indefatigable motion of being (matter-energy) echoes in the indwelling oscillations of a being's own interior essential-existential "in-divisions." Przywara labels *negativium* and *positivum* as the two termini of this interior non-subsisting

[69] *AE*, 158.

[70] See Richard L. Fern, *Nature, God and Humanity: Envisioning an Ethics of Nature* (Cambridge, UK: Cambridge University Press, 2002), 147.

"field" or "In-Sein" of being. Actual existence resides in the neighborhoods of pure potentiality and Nothingness' metaphysical modulations.

Being and the mathematical concept of the 1 cannot be reduced to Nothingness or zero for "becoming does not tend unilaterally away from nothing toward being, but remains constantly turned toward nothing and inwardly unassured; it is saved from falling into nothing at each instant precisely by the creative hand of God."[71] A being essentially or holographically embodies Nothingness by consciously incorporating nonbeing as pain, suffering, sickness, and ultimately death. The believer also freely acts when she recognizes that the noumenon's reflexive orchestration of her interior structuring is proportionate to their acceptance of the Spirit's enlivening or divinization of the faithful and creation. Reason's grasp on reality depends on its understanding of controlled differing in reality. Reason and faith participate in non-subsistence's analogous mirroring of Divine Wisdom, because "the derivation of perfection from God is therefore possible only because being created does not belong in an absolute sense to being as being. Otherwise the finite would not be able to participate in God' uncreated infinite perfection, and, because in this case all being would be created, God would have to perfect Himself in the created entities."[72] The Spirit positively conducts a being away from various threats of nonbeing by introducing grace, that is, new supernatural information to a willing subject (Ezekiel 28:14-15).[73] The believer's consciousness spiritually informs their perceptions of the world by means of "pneumatic" reason.

Conscious dehiscence defines and limits the Adamic sarx from the omniscient God only because Jesus Christ bridges this impossible distance. [23] The Eucharist allows one to foreground one's personal resurrection or "theonification," as this event is experienced in those redemptive moments of love achieved from the graces won at the Passion by the Spirit. With the New Covenant and the plenitude of the Marian fiat, Adam's solitude is transformed into a Eucharistic solidarity, where one's communal existence opens to the Logos sarx of Jesus Christ (John. 1:14). Our material identity necessarily puts a barrier between, or distances us from, every other creature. The divine-human person of Jesus Christ actualizes how impossible "nonlocal" positions are made whole in love. The Church is home (*communio sanctorum*) where differing correspondences between the relative

[71] *TL* 1, 251. See also *AE*, 168 and 231–232.
[72] *Homo Abyssus*, 142.
[73] See *ST*. I, 38, 1.

and infinite are given to the faithful as an opportunity to tailor one's life around the principle of *metanoia* (sacramental efficacy) and the Eucharist's power to transmogrify matter toward its resurrected dignity. Jesus Christ's and return home to His Trinitarian home anticipates the Divine's final judgment of Creation and the promise of resurrection (1 Peter 3:18).

Jesus' eschatological return announces the ensuing eternal banishment of the "dragon" to the abyss (Revelation 20:2) and the arrival of the New Jerusalem. In finalizing humanity's response to the Passion of Jesus Christ, the last vestiges of the created order outside its current spatiotemporal framing will be endorsed. Complex space gives us a way to understand how human life in the Spirit transpires at the trans-spatiotemporal dimension, as space and time are creatures formed to serve divine purposes. Whenever the spatiotemporal ceases to inform history, human individuality is lost, for "[t]his misinterpretation gets concretely expressed in the fact that finite reason attains to the unfolding of an ontological spatiotemporality out of the hypostasis of being, a spatiotemporality that therefore, insofar as it is sustained by the negation of being's exinanitio [kenosis], comes to light as anti-time and anti-space."[74] [14] In the Parousia the meaning behind the universe's foundations is unanimously armed in natural order's restoration—a conclusion achieved by the Spirit's transformation of all limitations. Absolute Groundless Being acts universally "in the transfiguration of the world on 'the day of the Lord' (2 Peter 3:7–13), as well as, in the judgment of the Lord, to which *all* peoples will come."[75] [15] The sanctification of the infinitely divisible, whether temporally or spatially rendered, is part of the eternal choice to holo-somatically house the Spirit until the Parousia and Jesus Christ's return.[76]

The liturgy introduces its own agency as it accelerates a sacramental logic that orders spatiotemporal reality and the eternal. The proclamation of the Last Supper evinces the Eternal Word's love—a soteriological act, for the Eucharist is a "holo-cryptic" mystery that "transubstantiates everything" (John 17:5).[77] The human person is called into relationship with the Divine as an invitation to see the past as a "holo-cryptic" anamnesis, that is, encompassing an eternal future, but one that exists in virtual expe-

[74] *Homo Abyssus*, 175.

[75] *The Bride of the Lamb*, 267.

[76] See *Homo Abyssus*, 111–112. See also Dionysius, "Divine Names" in *The Works of Dionysius The Areopagite: Part 1*, Translated by Rev. John Parker, M.A., (London, UK: James Park and Co., 1897), Chapter IV, section X.

[77] See *The Pillar and Ground of the Truth*, 117.

riences of finitization that is fully revealed in the resurrected state. Immanent reality continues to imagine the Transcendent until the death of Jesus Christ is fulfilled in history with his Parousia. The Christian commits to the future by living a memory, a state of "ever-present anamnesis ('Do this in memory of me,' 1 Corinthians 11:25) of the self-sacrifice of God's love (*unde et memores*), the living and resurrected Christ becomes present (Matthew 18:20)-but present 'until he comes again,' (1 Corinthians. 11:26), and therefore [is] not looking backward, but with eyes set forward, into the future and full of hope."[78] [12] Because the Spirit does not unduly influence the natural teleology of physical objects, the Spirit's action can be likened to an absolute positive "holographic" projection of Nothingness—the Pure Groundlessness of Divine Freedom that acts as a creative source for making Being infinitely generative. The aporic (ἄπορίᾱ) dimensions separating the immanent order from the Divine are infused with the Spirit's grace. [24]

The Spirit makes the abyss separating the created from the uncreated into a concept made available to limited consciousness. The Spirit heralds the person to contemplative the dimensional recesses and supernatural dimensions of reality, as well as, their interior depths. The Spirit's prognostication of Eternal trans-dimensional spaces creates places for the human soul and consciousness to flourish. The Groundless Spirit creates the space for immaterial consciousness to find the means to experience the nature of unity. For "in God there are things that exist only to provide love with every opportunity for development, to give it the room which it would lack if everything were stale foreknowledge—room which it needs, for it cannot exist without self-surrender, movement and flight."[79] [25]

If a traditional Aristotelian philosophy of Nature drives one's doctrine of the Eucharist, one nominally attains the untenable position that accidents inform a material body without a substance (subject) of inherence. Disturbingly "free," "holistic," and "relativistic," the quantum world gives us an exit strategy from the alarming conclusions that hylomorphism may have for a sacramental theology. For the Catholic theologian, the Eucharist is not simply a material object whose meaning is conventionally and *de facto* established. Christ's kenotic culmination in the event of the Paschal sacrifice and its ritualized "mechanism" of transubstantiation results in the real presence of Jesus Christ. The Eucharist virtually displays

[78] *Love Alone*, 89.
[79] *TD* 5, 96.

the ultimate living being; Balthasar states that "the divine and human person in whose… interior design is carried through by the formative power of the unique, distinctive, personal nature of Jesus Christ: he is the valid 'expression-image' (*Ausdrucks—Bild*) of the invisible God (Colossians. 1:15)."[80]

Christ takes on human nature (*και ο λογος σαρχ εγενετο*, John 1:14), in an all-important way, namely, by being "made sin."[81] [10] The fall of human nature is taken up by the Son in the person of Jesus Christ as an economic virtualization of obedience and thereby of the Son's imaging of the Father within the Trinity. So dedicated, one can say that Jesus Christ, the Savior, existed eternally as part of the Divine's Providential designs. At the moment of transubstantiation, the somatic presence of Christ is realized in the Eucharist. The material and substantial identity is not deterministically replaced at the moment of consecration as if material identity is problematic. Rather, the Eucharist is also a real symbol of Christ that makes possible the soul's invisible sacrificial self-gifting at the Eucharistic table. The created order continues holographically to exude its original source as a divine exemplar. The act of being "made sin" does not effect a change in Trinitarian life.

Each individual is only virtually isolated within their sin, as evil has no transcendental reach beyond its immediate pragmatic interests. Nature and physical beings are ordered to evolve beyond the given. Stapp notes that "everything that we know about Nature is in accord with the idea that the fundamental process of Nature lies outside space-time…but generates events that can be located in space-time."[82] [[26] cited in [27]] The sacramental, sacrificial, or donative actions are metaphysically dependent on the spatiotemporal—be it the liturgical actualization of Israel's memory of the Lord's saving historical work (1 Corinthians 5:7), the reality of the Holy Land, the sacred space of the heavenly Jerusalem, or the tabernacle's housing of the Divine (Revelation 21:10, 23). Space and time must be capable of dramatizing the event of the Eucharist not because they conform to an abstract calculus but rather because their formulation allows for historicity and personal growth to coexist as transcendental possibilities.

[80] Hans Urs von Balthasar, "Christliche Kunst und Verkündigung," *Mysterium Salutis* vol. 1 Eds., J. Feiner and M. Löhrer (Einsiedeln and Cologne, 1965), 710–711.

[81] See *Covenantal Theology*, 318–320.

[82] Henry P. Stapp, "Are Superluminal Connections Necessary?" *Nuovo Cimento*, vol. 40B (1977): 191. Cited in *Smith*, 79.

Jesus Christ in his entirety, Thomas teaches, is "embodied" within the Eucharist. This proposition, I have argued, is near impossible to justify from the viewpoint of classical physics: a causally deterministic operating system and the formless infinity of the ancients will have, at best, an uncomfortable marriage. I am not claiming that the act of transubstantiation is capable of scientific description, verification, or falsification: such is impossible for any miracle or act of God. However, the spatiotemporal is more adequately explained as an efficacious sacramental space when we make quantification the bride of the infinite. Put another way, the invisible hyper-dimensional realms of QMT and M-theory better symbolize how the Son is in the Eucharistic substance (*per modum substantiae*). The Eternal Logos exists as Jesus Christ in the Eucharist as a matter of nonlocal supernatural and efficacious causal relations. Entropic principles show that even when matter is destroyed in a black hole, its essential or informational identity radiates as a holographic object. Indeed, at its most primitive level, the material order projects its geometric, topological, and formal structures as a holographic principle that supports traditional hylomorphic and quantum theories. The interior constrictive forces of the black hole model the "pressing out" of the substantial identity of bread and wine at consecration to make room for the "holographic image" or essence of Jesus Christ.

References

1. Henri Poincaré, "On the Foundations of Geometry," Translated by T. J. McCormack, *Monist*, 9 (1898): 41.
2. Henri Poincaré, The Value of Science (1904) in The Foundations of Science: Science and Hypothesis, the Value of Science, Science and Method, Translated by G. B. Halsted (New York, NY: Science Press, 1913), 355.
3. Marc Holman. "Foundations of quantum gravity: The role of principles grounded in empirical reality." Studies in History and Philosophy of Science Part B: Studies in History and Philosophy of Modern Physics 46 (May 2014): 142–153, https://doi.org/10.1016/j.shpsb.2013.10.005.
4. James Owen Weatherall, *Void: The Strange Physics of Nothing*, (New Haven, CT: Yale University Press/Templeton Press, 2016), 74.
5. Hans Urs von Balthasar, *Truth of the World* Vol.1, Translated by Adrian J. Walker (San Francisco, CA: Ignatius Press, 2000).
6. Thomas Aquinas, *Summa Contra Gentiles*: Book Two: Creation, II, 53. Translated and Introduced by James F. Anderson (Notre Dame, IN: University of Notre Dame Press, 1956).

7. Norman Sieroka and Eckehard W. Mielke, “Holography as a principle in quantum gravity?—Some historical and systematic observations,” *Studies in History and Philosophy of Science Part B Studies in History and Philosophy of Modern Physics* 46 (May 2014):172.
8. Jacob Bekenstein, “Information in the Holographic Universe,” Scientific America, Vol. 289 (2003): 59.
9. Brian Greene, *The Hidden Reality: Parallel Universes and the Deep Laws of the Cosmos* loc. 118–19 of 226. EPUB eBook ed. (New York, NY: Alfred A. Knopf, 2011).
10. Donald Keefe S.J. *Covenantal Theology: The Eucharistic Order of History* (Novato, ON: Presidio Press, 1996).
11. Max Tegmark, “The Mathematical Universe.” *Found. Phys.* (Sept., 2007): 2, arXiv:0704.0646v2 [gr-qc]. Accessed 01-01-18.
12. Hans Urs von Balthasar, *Love Alone is Credible*, Translated by David C. Schindler, (San Francisco, CA: Ignatius Press, 2004).
13. Hans Urs von Balthasar, *Glory of the Lord: A Theological Aesthetic*: Vol.1, Edited by Joseph Fessio, S.J and John Riches. Translated by Erasmo Leiva-Merikakis and John Riches (Edinburgh: T&T Clarke, 1982).
14. Ferdinand Ulrich. *Homo Abyssus: The Drama of the Question of Being*, Translated by D.C. Schindler, (Washington, DC: Humanum Academic Press, 2018).
15. Sergei Bulgakov, *The Bride of the Lamb*, Translated by Boris Jakim, (Grand Rapids, MI: Wm. B. Eerdmans Publishing Company, 2002).
16. Pavel Florensky, *The Pillar and Ground of the Truth: An Essay in Orthodox Theodicy in Twelve Letters*, Translated by Boris Jakim. Introduced by Richard F. Gustafson. (Princeton, NJ: Princeton University Press, 1997).
17. Henri Bergson, *Mind-Energy*, Translated by H. Wildon Carr and Introduced by Keith Ansell Pearson (New York, NY: Palgrave Macmillan, 2007), 133.
18. Alain Badiou, *Logics of Worlds: Being and Event*, 2. Translated by Alberto Toscano (London, UK: Continuum, 2009).
19. Alain Badiou, *Being and Event.* Translated by Oliver Feltham (New York, NY: Continuum, 2010).
20. Sean Bowden, “Badiou and Lautman,” in *Badiou and Philosophy*, Edited by Sean Bowden and Simon Duy (Edinburgh, UK: Edinburgh University Press, 2012), 50. Footnote 728—inserted as a citation partially in *Theoretical Writings*, 109.
21. Barbara Maria Stafford, *Echo Objects: The Cognitive Work of Images*, (Chicago, IL: The University of Chicago, 2007).
22. Eric Przywara, *Analogia Entis: Metaphysics: Original Structure and Universal Rhythm*, Introduced and Translated by John R. Betz and David Bentley Hart (Grand Rapids, MI: Wm. B. Eerdmans Publishing Company, 2014), 158.

23. Hans Urs von Balthasar, *Mysterium Paschale: The Mystery of Easter*, Translated and introduced by Aidan Nichols, O.P. (Grand Rapids, MI: Wm. B. Eerdmans Publishing Company, 1990).
24. Hans Urs von Balthasar, *Presence and Thought: An Essay on the Religious Philosophy of Gregory of Nyssa*, Translated by Mark Sebanc (San Francisco, CA; A Communio Book, Ignatius Press, 1995), 27–35.
25. Hans Urs von Balthasar, *Theo-drama: Last Act*: Vol.5. Translated by Graham Harrison (San Francisco, CA: Ignatius, 1998).
26. Henry P. Stapp, "Are Superluminal Connections Necessary?" *Nuovo Cimento*, Vol. 40B (1977): 191.
27. Wolfgang Smith, *The Quantum Enigma: Finding the Hidden Key*, Forward by Seyyed Hossein Nasar, (San Rafael, CA: Angelico Press: Sophia Perennis, 2005).
28. Catherine Pickstock, *After Writing: On the Liturgical Consummation of Philosophy* (Oxford, UK: Blackwell Publishers Inc., 2003).

CHAPTER 6

Transubstantiation and Quantum Mechanical Theory

6.1 Quantum Entanglement, Non-locality, and Teleportation

In the Ancient Near East, by entering into covenants, rights and duties could be accepted by unrelated persons. These covenants were often ratified with a sacrifice. The blood of the ritualistic sacrifice symbolized the curse of death that would befall any partner who violated the stipulations set out in the covenant. God reveals Godself to Abram and establishes a sacred covenant with him. Thus, the relationship between God and Abraham was sacred and bestowed blessings or curses upon the people to the degree they were true to the covenantal terms. Jesus Christ universalizes this marriage of the created order and the absolutely other realm of Trinitarian existence by gifting the totality of his person in the Eucharistic sacrifice (*ex opere operato*) of his Passion. The covenant and Christian liturgy confirm a peoples' identity and renew and restore an individual's relationship with the Transcendent (Exodus 29:45). The hylomorphic theory can and should be developed in light of revised quantum models of reality. In their own ways, this project advances when theological and scientific accounts recognize their own respective "faith" tenets.[1] According to Catholic dogma, consecration takes place when a recognized priest repeats Christ's own words at his Last Supper. Jesus Christ is believed to

[1] *The Relevance of Physics*, 136.

M. P. Fusco, *The Physics and Metaphysics of Transubstantiation*,
https://doi.org/10.1007/978-3-031-34640-8_6

speak across the ages, lending his spiritual ecstasy to the priest. The trans-historical character of the words of consecration finds their physical analogate in quantum superposition, where the immediate renovation of two entangled partners occurs at superluminal speeds across nearly unlimited distances.[2] Quantum superposition states that if one knows the probable energy and location of one entangled quantum particle, one can deduce the likely momentum and physical properties of its entangled partner. With quantum non-locality, we have another way to imagine how prayers transmit from the faithful to the Transcendent. Divine grace works to bring about eternal purposes through the free actions of individual beings, just as quantum particle nonlocal relations fundamentally order the entire universe.

Einstein balked at what quantum non-locality meant for his beliefs about a physical being's axiomatic or lawful behavior in the universe.[3] In the 1935 paper "*Can Quantum Mechanical Description of Physical Reality Be Considered Complete*?" (hereafter cited as *EPR*) coauthors, Albert Einstein, Boris Podolsky, and Nathan Rosen provided the stated explanation and justification for nonlocal quantum entanglement.[4] Brian Greene explains that they "wanted to show that every particle does possess a definite position and a definite velocity at any given instant of time, and thus they wanted to conclude that the uncertainty principle reveals a fundamental limitation of the quantum mechanical approach."[5] [1] The EPR authors wanted to defend the independence of every fundamental particle in the universe by offering a thought experiment that critiques the nonlocal influence of the QMO at a distance. At heart, the EPR argument takes exception to any defense of apparently non-causal correlations that create new states of affairs.[6] Quantum non-locality seems to dispute the

[2] See Erwin Schrödinger, "Discussions of probability relations between operated systems," *Proceeding of the Cambridge Philosophical Society* Vol. 31, 55 (1935).

[3] See *Mind, Matter and Quantum Mechanics*, 88.

[4] See A. Einstein, B. Podolsky, and N. Rosen, "Can Quantum-Mechanical Description of Physical Reality be Considered Complete?" *Physical Review* 47 (1935): 777. Paper reprinted in *Physical Reality: Philosophical Essays in Twentieth Century Physics*, Edited by Stephan Toulmin (New York: Harper and Row/Torch Books, 1970). See also Niels Bohr, "Discussion with Einstein on Epistemological Problems in Atomic Physics," in P.A. Schilpp, *Albert Einstein: Philosopher-Scientist* (La Salle, IL: Open Court Press, 1949), 199–242.

[5] Brian Greene, *The Fabric of the Cosmos:Space, Time and the Texture of Reality*, (New York, NY: Alfred A. Knopf, 2004), 99.

[6] On Thomas' view of Aristotelian causality see *ST.* I, 2, 3. On instantaneous connection in quantum system, see John S. Bell, "On the Einstein-Podolsky-Rosen paradox," *Physics* I.3 (1964): 195–200.

GTR's hypothesis that a physical entity cannot move faster than light. However, a distinction can be made, says John Polkinghorne, between instantaneous quantum action and the violation of the GTR, for "[t]he latter forbids the communication of information faster than the velocity of light, but analysis shows that the EPR process cannot be used for the immediate transmission of the details of the state of affairs from one point to another point. Quantum entanglement is a subtle form of internationality."[7] [2] The mystery of the Absolute Groundless Spirit presents for theologian the example of how true intentional harmony is made possible.

The EPR authors searched for a "hidden variable" to explain how two entangled quantum particles only "appear" to be spatiotemporally independent. This "yet-to-be-discovered" hidden variable or force, they argued, would show how two QMOs at a distance were actually causally and physically linked under universal laws in a single material tapestry.[8] The EPR authors are commonly categorized as realists given their aims and presuppositions. The question of nonlocal influence questions the axiomatic status of uncertainty in the universe. On the essential realism of the EPR authors, Menas Kafatos and Robert Nadeau write that "[t]his strategy allows one to assume that although quantum indeterminacy may be a property of a quantum system in practice, it need not be so in principle. It also allows one to view physical attributes of quantum systems, such as spin and polarization, as objective or 'real' even in the absence of measurement and to assume, as Einstein did, a one-to-one correspondence between every element of the physical theory and physical reality."[9] [3] Paradoxically, even though quantum non-locality suggests that "coupled" particles simultaneously influence each other even though they are causally isolated and remote, this determined result does not override the interior freedom of the QMO. According to QMT, the measurement of a particle's momentum and that of its position are exclusive. If one knows the

[7] John J. Polkinghorne, "The Demise of Democritus," in *The Trinity and an Entangled World: Relationality in Physical Science and Theology*, Edited by John J. Polkinghorne (Grand Rapids, MI and Cambridge, UK: William B. Eerdmans Publishing Company, 2010), 7. See also David Jennings and Matthew Leifer, "No Return to Classical Reality," arXiv:1501.03202.v2 (2015) and Henry P. Stapp, "S Matrix Interpretation of Quantum Theory," *Physical Review*, (1971): 1307–1308.

[8] See David Bohm and B. J. Hiley, "On the Intuitive Understanding of Non-Locality as Implied in Quantum Theory," *Foundations of Physics* Vol. 5 (1957): 96.

[9] Menas Kafatos and Robert Nadeau, *The Conscious Universe: Parts and Wholes in Physical Reality* (New York, NY: Springer–Verlag, 2000), 66. Hereafter cited as *Conscious Universe*.

value of a subatomic particle's position, for example, the calculation of its momentum is impossible.

Bohr criticizes the EPR paper for its "ambiguity" and its failure to account for the subject's role in any scientific experiment.[10] The very act of measuring, he notes, introduces a tacit influence on the system investigated. Bohr reminds the EPR authors that the measurement of the first entangled electron influences "the very conditions which define the possible types of predictions regarding the future behavior of the system." [4] He notes that scientists always risk mistakenly conflating the QMO's material existence and its theoretical rationalizations by means of quantification. However, it was not until 1964 with John Stewart Bell's mathematically rigorous defense of quantum entanglement and nonlocal interactions that a compelling mathematical response to EPR was put forward.[11] Bell's work shows that the interior identity of two entangled particles perseveres regardless of the distance separating them. Bell's focus on the mathematical and logical structure of the QMO illustrates how holocryptic metaphysics interprets more traditional notions of the essential as an informational construct.

Slavoj Žižek (1949–) sees the difficulties of the "open" ontology of QMO addressed by the EPR authors. More specifically, he takes exception at any implicit claims that perfect knowledge or measurement is possible. The quantification of reality demands its conscious conceptualization and re-representation by a historically bound individual. Žižek writes that "Heisenberg, Bohr, and others,… insisted that this incompleteness of our knowledge of quantum reality indicates a strange incompleteness of quantum reality itself, a claim that leads to a breathtakingly weird ontology. When we want to simulate reality within an artificial (virtual, digital) medium, we do not have to go to the end: we just have to reproduce features which make the image realistic from the spectator's point of view."[12] [5] To understand a quantum system, one must consider how conscious behaviors influence how one determines and symbolically represents the external accidental and interior mathematical properties of a QMO. To

[10] Niels Bohr, "Can Quantum-Mechanical Description of Physical Reality be Considered Complete?" *Physical Review* (48) (October 15, 1935): 700.

[11] See John S. Bell, "On the Einstein-Podolsky-Rosen paradox," *Physics* vol. 1, no. 3 (1964): 195–200 and Cynthia Sue Larson, "Primacy of Quantum Logic in the Natural World," Cosmos and History, *The Journal of Natural and Social Philosophy* vol. 11, no. 2 (2015): 332.

[12] Slavoj Žižek and John Milbank, *The Monstrosity of Christ: Paradox or Dialectic?*, Edited by Creston Davis, (Cambridge, MA: MIT Press, 2009), 89.

observe a QMO, one must further understand how mathematical probability operates within the system as a whole. Bell's explication of the mathematical arrangement of the QMO achieves such an aim. In this way, a comparison between the two philosophical approaches of Einstein and Bohr is suggested. This is achieved if the mathematical entities representing the QMO are implicitly understood as possessing a virtual or holographic existence. As Kafatos and Nadeau argue, Bell shows "certain kinds of measurement could distinguish between the positions of Einstein and Bohr. One set of experimental results would prove quantum theory incomplete and Einstein correct."[13] [6] Bell's work shows how quantum entanglement's superluminal speeds can be understood as part of the ontological properties of the QMO, the nature of probability, and the theoretic hermeneutical suppositions an observer brings when attempting to make sense of experimental data. Consequently, these so-called Bell inequalities promote a conception of non-locality without imposing a naive limitation on a mathematically rigorous conveyance of QMT. Bell's approach earmarks space for the independent viewer and quantifier of the quantum system and I would say, the holo-somatic truth of the number. Bell's theorem demonstrates "how no physical theory of local or reclusive variables can ever reproduce all the predictions of quantum mechanics" but mathematically proves how the EPR particles connect.[14] [7] Bell's theory advances a convincing case for quantum non-locality, one that has been experimentally verified.[15]

Bell's scaling of non-locality does not necessarily preclude our interpretation of Nothingness' paradoxical causal transcriptions within Creation. The subject's ability to perceive an object presumes the existence of temporal, spatial, and geometric continuities—which are only possible by assuming the "existence" of difference or non-subsistence. The Eucharist empirically and spiritually establishes a prototype for quantum non-locality in its historical reenactment of an event that simultaneously mirrors the transcendent truth of the Eternal Logos. Furthermore, through sacrificial

[13] *Conscious Universe*, 60.

[14] See Cynthia Sue Larson, "Primacy of Quantum Logic in the Natural World," *Cosmos and History: The Journal of Natural and Social Philosophy*, vol. 11, no. 2, (2015): 332. See also: A. Aspect, J. Dalibard, and Gérard Roger: Experimental test of Bell's inequalities using time-varying analyzers, *Phys. Rev. Lett.* 49, (1982): 1804–1807.

[15] See Marissa Giustina, A. Mech, S. Ramelow, B. Wittmann, J. Kofler, J. Beyer, A. Lita, B. Calkins, T. Gerrits, S.W. Nam, R. Ursin, A. Zeilinger, "Bell violation using entangled photons without the fair-sampling assumptions," *Nature* 497 (09 May 2013): 227–230. See *Mind, Matter, and Quantum Mechanics*, 94–95.

dimensions of Jesus Christ's death and its remembrance at the Eucharist, a way opens for an individual to experience Creation's primordial Nothingness as part of an eternal redemptive plan. The symbolic codification of the mystery of Transcendent non-locality is not flatfoot, as its representational gestures remain partially apophatic in nature.

The "mathematical formula bears necessarily a visual and syntactical aspect"[16] [8] which gives us a way to see a QMO's formal description as nonlocal. In our holo-cryptic metaphysics, the intellect's ability to imagine the non-subsistence of bread and wine parallels the attempt to conceive the form or kernel of an object's interior mathematical properties and physical identity by means of specialized processes and symbolic representations. The difference between a material object, its conscious recognition, and its symbolic representation must be kept distinct but harmonized. When the metaphysical "edge" between a physical object and its conscious appropriation is conflated, both are put at risk. Confusing the living presence of Jesus Christ in the Eucharist and the bread and wine's material properties, for example, risks the idolatrous.

The interior-exterior duality of a physical identity builds upon the observer's fundamental intuition of the *a priori* differences between nothingness and symmetry. Tegmark gives us an allegory for how an observational point of view helps qualify the difference between an object's mathematical structure and its placement in reality. On the one hand, he explains how an observer's perception can be from above in a "bird's eye" perspective. This viewpoint is comparable to the work of QMT as a whole. On the other hand, an observer can also investigate reality below from a "frog's point of view." He explains that this perspective follows a classical scientific approach to reality, as the frog's perceptions are temporally and spatially limited by three-dimensional Euclidean space and the laws of gravity set out by Newton. Tegmark abstracts from these two observational frames to present an analogy for the movement of quantum particles. He explains that "in four-dimensional space-time of the bird perspective, these particle trajectories resemble a tangle of spaghetti. If the frog sees a particle moving with constant velocity, the bird sees a straight strand of uncooked spaghetti. If the frog sees a pair of orbiting particles, the bird sees two strands intertwined like a double helix."[17] [9] The more immediate viewpoint of the frog takes into account how space curves in

[16] *Smith*, 46.
[17] Max Tegmark, *The Mathematical Universe*, 3.

relation to a given mass. The universe of looping pasta that situates the frog and bird's viewpoints symbolically represents how the clustering of quantum particles reveals the informational state of subatomic gravity.

The frog's spatiotemporal location is important when considering the relationship between gravity and the GTR. Tegmark notes that in a closed system under the directives of GTR, the frog would perceive a plate of pasta as a sort of flying spaghetti monster. GTR allots a different observational perspective and, therefore, a new cognitive state. From a bird's viewpoint flying around a ceiling, the atoms of Newton's theoretic universe appear as a metaphoric plate of pasta on a table. In a quantum and relativistic universe, a frog would not perceive the pasta as plated and stationary but rather as flying rigatoni in space-time.[18] Tegmark models how a "pseudo-transcendental" view of the quantum world is possible. This is possible because his approach references both Schrödinger's deterministic quantum-wave perspective of the quantum world and Heisenberg's insights into the nature of quantum uncertainty. These two theories come together in Tegmark's artistic depiction of the "transcendental measure" of the frog looking at the bird and vice versa, for "if the bird sees such deterministic frog branching, the frog perceives apparent randomness from the bird's view is by definition banished."[19] [9]

The rarefied apprehension of the supernatural is analogous to the indirect assessment of quantum waves, subatomic particles, or atoms by means of mechanical measurement. The interior ratio of matter and form and external accessibility of a being from a "transcendent" or "historical" examination is critical to our interpretation of transubstantiation's harmonious ordering of the substantial and the essential.[20] The correlation of physical and formal "spaces" that bread and wine inhabit is congruent with its holographic actualization in hyper-dimensions. On the meeting point of the Transcendent's supernatural dimensions and other complex dimensions, Florensky's position is instructive. He argues that "space itself is not merely a uniform structureless place, not a simple graph, but is in itself a distinctive reality, organized throughout, everywhere differentiated, possessing an inner sense of order and structure."[21] As outlined in the previous chapter, the extreme conditions found in a black hole where mat-

[18] See ibid., 4.

[19] Ibid., 4.

[20] See *ST.* I, 75, 6, *ST.* III, 75, 3.

[21] *Reverse Perspective*, 218.

ter's physical dimensions are transformed into its informational substantial identity or the moment of transubstantiation give two examples of this process. The Spirit presents Christ's essential identity within the metaphorical "event horizon" of the phenomenal accidents of the bread and wine fractured for the ecclesial community.

The Eucharist's multifarious warp and weave of space and time serves to define the essential "threads" that the spiritual consciousness forms into meaningful conceptual patterns. The bread's "theonification" confirms the divine's physical, "anti-gravitational" and "non-curvature" holographic fractal presence of the Mystical Body of Christ within bread and wine. Being's multi-variant meanings in the created order and their relationship to the Creator make these material elements a suitable recipient in the Eucharist of the holo-somatic Jesus Christ as is the Eternal Son and a human person.[22] [10] Spatiotemporal local presence and non-local presence are closely related to the concept of human freedom, as here "the idea of events that are intrinsically unpredictable" reveals itself in physical reality.[23] [11] This arrangement is made even more complicated when one considers the fact that non-locality and chance are themselves predicated by the global uncertainty of reality as a whole. Thus, non-locality and chance theoretically "subtract" from the observer's position, guaranteeing one's incomplete but free positioning in a universe driven by chance. The probable and telic are wed in the quantum cosmos, making subjective consciousness and metaphysical non-subsistence revelatory.[24] [11]

Uncertainty, non-subsisting relationships, and multipart dimensional spaces together formulate the descriptive parameters of an existing being's identity or "base code."[25] Robert Oppenheimer narrates a being's strange disclosure at the subatomic level, when professing that "if we ask, for instance, whether the position of the electron remains the same, we must say 'no'; if we ask whether the electron is at rest, we must say 'no'; if we

[22] See *GL* 7, 425. The presence of Jesus Christ is in every "bit" of the Eucharist, and demands reverence. See also *DS* 885 and *DS* 1653.

[23] See Nicolas Gisin, *Quantum Chance: Nonlocality, Teleportation and Other Quantum Marvels*, Forward by Alain Aspect (Cham Heidelberg, New York, Dordrecht, London: Springer, 2012), 7. Hereafter cited as *Quantum Chance*.

[24] Ibid., 50. The fact that each quark can have three "colors" associated with it gives us an analogy for sub-atomic non-subsistence.

[25] See Niels Bohr, Jan Faye, Henry J. Folse, *The Philosophical Writings of Niels Bohr, Causality and Complementarity, Philosophy of Science* IV (Woodbridge, CT: Ox Bow Press, 1999): 293–294.

ask whether it is in motion, we must say 'no.'"[26] [12] The irreducible identity of a QMO can be seen as a particular disclosure of that randomness that defines Nature.[27] [13] The pliable margin between the observer and a QMO is local and momentary to the subjective observer. However, Stapp comments that "the abrupt change associated with the process of observation or measurement is not a real process, but merely an artificial [one] of man's theorizing about nature, dependent upon where he places an imaginary cut."[28] One's spatiotemporal perception of the QMO in the universe confirms the workings of the subjective imagination and the importance of the role point of view in unimpeded knowledge.

The transfer of information at speeds exceeding that of light propagation is possible if the entirety of known space acts as an implicate order.[29] [14] The idea of an implicate order says that all space-time acts as the elusive hidden variable that the EPR paper sought. Two putative points are reimagined to be part of a single material and relational continuum. If reality is an ensemble of connected space-time points, Bohm conjectures, then "a new notion of unbroken wholeness which denies the classical idea of analyzability of the world into separately and independently existing parts" results.[30] [15] Physicists argue that Bohm's position raises questions about how multiple individual events can take place within a single frame of reference if only one universe exists—where does the final towering turtle look downward? An unlimited number of potential universes seems necessary in the implicate order's explanation of superluminal exchanges at work in a causally based universe. Stapp explains that no "metaphysics not involving faster-than-light propagation of influences has been proposed that can account for all of the predictions of quantum mechanics, except for the so-called many-worlds interpretation, which is

[26] Robert Oppenheimer quoted in Werner Heisenberg, and further quoted in James B. Conant, *Modern Science and Modern Man* (New York, NY: Columbia University Press, 1953), 271.

[27] Peter J. Lewis, *Quantum Ontology: A Guide to the Metaphysics of Quantum Mechanics* (Oxford, UK and London, UK: Oxford University Press, 2016), 50 and 151.

[28] *Mind, Matter and Quantum Mechanics*, 113.

[29] See David Bohm, *Quantum Theory* (Englewood Cliff: Prentice-Hall, 1951), David Albert, *Quantum Mechanics and Experience*, (Cambridge, MA: Harvard University Press, 1992) and Valerio Scarani, and Nicolas Gisin, "Superluminal hidden communication as the underlying mechanism for quantum correlations. Constructing Models," *Brazilian Journal of Physics* 35, (2005): 328–332. Hereafter cited as *Superluminal Hidden Communication*.

[30] David Bohm and B. Hiley, "On the Intuitive Understanding of Non-Locality as Implied in Quantum Theory," *Foundations of Physics*, vol. 5 (1957): 96.

objectionable on other grounds."[31] [16] In other words, the implicate order is theoretically commensurate with my own holo-cryptic metaphysics' argument for the virtual identity of a being or the universe as a whole insofar as the Transcendent stands outside all as Creator and Eternal Judge.

To this day, scholars remain dubious of Bohm's explanation of non-locality, as it insinuates the need to reject the constant speed of light and suggests the existence of many worlds or a multiverse.[32] The existence of a multiverse may not account fully for experimental evidence concerning the entwined quantum.[33] Multiverse theories imply that every quantum particle exists in an entangled state, and if this is the case, a scenario results that may not take into proper account the random quantum action. Gisin explains that the many-worlds interpretation implies a totalitarian form of determinism. Indeed, according to this interpretation, "entanglement is never broken but continues to spread further and further. Therefore, everything is entangled with everything and it leaves no room for anything like free will."[34] [11] This of course revises the question of how the Transcendent's will can control and direct Creation without eradicating secondary causality and human freedom. Further, the complete explanation of quantum non-locality must include a way to explain the relative independence of fundamental particles, just as a valid liturgical theory implicates the relative autonomy of the natural order. John Polkinghorne confirms that physics such as theology depends on knowledge claims and thus a metaphysical decision. He notes that even indeterminacy and deter-

[31] Henry P. Stapp, "Quantum Theory and the Physicist's Conception of Nature: Philosophical Implications of Bell's Theorem," in *The World View of Contemporary Physics*, Edited by Richard E. Kitchener (Albany, NY: SUNY Press, 1988), 40. Hereafter cited as *Quantum Theory and Nature*. Cited in *Universe Mind*, 134.

[32] See Henry P. Stapp, *Quantum Theory and Nature*, 40, David Bohm and B.J. Hiley, *The Undivided Universe* (London, UK: Routledge, 1993), 347, David Bohm, *Quantum Theory* (Englewood Cliffs: Prentice-Hall, 1951), David Albert, *Quantum Mechanics and Experience*, (Cambridge: Harvard University Press, 1992) and *Superluminal Hidden Communication*, 328–332.

[33] See David Wallace, *The Emergent Multiverse: Quantum Theory According to the Everett Interpretation* (Oxford, UK: Oxford University Press, 2012), David Deutsch, *The Fabric of Reality* (New York: Allen Lane, 1997), Michio Kaku, *Parallel Worlds: A Journey Through Creation, Higher Dimensions, and the Future of the Cosmos* (New York, NY: Anchor, 2006) and Max Tegmark, "The Interpretation of Quantum Mechanics: Many Worlds or Many Words?," *Fortschritte der Physik* 46, nos. 6–8 (November 1988): 855–862.

[34] *Quantum Chance*, 94.

ministic decisions in various quantum theories may share the same empirical conclusion, but the "choice between them has to be made on meta-scientific grounds, such as judgments of economy and naturalness of explanation."[35] [2]

Quantum teleportation assumes that an informational channel exists between two entangled quantum particles such that the extirpation or transformation of the informational state of one particle immediately affects the informational state of its entangled partner regardless of its placement in some noncontagious region of the universe. With quantum teleportation, it is possible to transfer the essence truth or information of a QMO regardless of the distances involved.[36] The theory of quantum teleportation reorders Aristotelian hylomorphic theory in harmony with my own holographic speculations.[37] The teleportation of an object's informational or essential marrow leaves its original substance as a formless mass. Quantum teleportation requires precise measurements, as only two objects of the same substance, size, and weight can exchange information in this way—the dimensional context of individual beings or universes remains an important factor. The shapes of the two objects to be teleported do not have to be the same, but as black hole entropy has shown—information's dimensional context is critical. At the local level and within the same dimensions, what is important for teleportation is that both objects are materially and informationally or "essentially" equivalent. The essential information's relationship to its "configuration space" is key.

In teleportation's holographic manifestation in three dimensions a normalization is possible along a common spatiotemporal frame of reference.[38] The interior reorientation of either entangled QMO is possible without the need for the addition or reduction of the emitter's mass.[39] Teleportation transfers the "quintessence" of one object into an awaiting

[35] John Polkinghorne, "The Demise of Democritus," in *The Trinity and an Entangled World: Relationality in Physical Science and Theology,* Edited by John Polkinghorne, (Grand Rapids, MI and Cambridge, UK: William B. Eerdmans Publishing Company, 2010), 9.

[36] See J. Bardeen, M. Carter, and Stephen W. Hawking, "The four laws of Black hole mechanics," *Commun.Math.Phys.* 31 (1973): 161–170.

[37] On the QMO's relationship to quantum teleportation, see *Quantum Chance*, 68.

[38] On the difference between configuration space and the spatiotemporal, see David Wallace, "Everett and structure," *Studies in History and Philosophy of Science, Part B: Studies in History and Philosophy of Modern Physics*, 34 (2003): 86–105.

[39] Ibid., 68–69.

formless mass that serves as a proper configuration space. Likewise, once an object's information is teleported, it loses its external and internal geometric structure and then "deconstructs" into a "formless" mass. Just as the symmetry between mass and energy defines a particular being's identity, the correspondence between inertia (mass) and configuration space is central to understanding the nature of quantum teleportation as the velocity and position of the QMO is impossible to observe simultaneously. Thus, quantum teleportation confirms that the essential "states" and material aspects of entangled beings are symmetrical but open to forceful spatiotemporal change and communication without rarefaction.[40] Once the internal dimensions of the receiving object are restructured by the arrival of the informational identity of the teleported object, its interior identity is altered. The formed exterior edge or accidental appearance of the teleported object mimics that of the past state of the transmitted object—the teleported object is now physically "identical" to the new object.[41] The individuated dimensional and configurational structure of the teleported object can transfer its past identity into the "future" because the space-time continuum acts a meta-structuring force that normalizes its "re-actualization." The past potential inertial identity of the teleported object is symmetrical to "reincarnation" as the mass of the object teleported. This is only possible if all space-time and its dimensional metastructure are fundamentally similar—the dimension and configuration of the two objects are symmetrical from the universe's "perspective."

The co-penetration and teleportation of the essential identity or information matrix as quantum superposition give us a way to build upon existing Aristotelian form-matter approaches and to develop the paradox of the question of transubstantiation. Where science speaks of the teleportation of information between objects, Catholic theologians argue for the sacramental transport of the essence of Jesus Christ's body, blood, soul, and divinity from heaven into the ordinary material elements of bread and wine. Thus, Christ becomes formally and substantially present in bread and wine (*forma substantialis and conversio substantialis*). This supernatural event is possible as it is willed by the Transcendent and by faith we have an analogous share in this manner of the eternal making the spatiotemporal meaningful.

[40] See *Quantum Chance*, 68.

[41] See *Quantum Chance*, 51–52, 68–70, and 98–100. Gisin Alain Aspect

6.2 Hell, Nothingness, and the Black Hole

According to the earliest biblical accounts, the unifying principle of the human person is closely aligned to the idea of "*nephesh*." The concept of "*nephesh*" is an early predecessor to the Christian idea of the soul—the embodied meeting point of the transcendent and immanent.[42] The human soul serves as the nodal point for the recording of the effects of one's actions and the intentional penumbra surrounding these decisions.[43] The existential and communal value of a person's life is revealed by the divine's judgment, which are "not only the judging gaze of justice but also the loving gaze of mercy."[44] [17] The human person has a natural inclination toward the resurrection. In comparison, the damned have a monomaniac fixation on that violence of sterile emptiness emulated in Satan's anti-kingdom. Hell's anti-creation tries to "negate" the abiding goodness and infinite holo-somatic generative capacities of the created order. The paradoxical non-spatial and a-temporal geography of Hell labors to distance being from its natural telic and "meontic" fulfillment in bodily resurrection and eternal life as announced with *creatio ex nihilo*. Our daily life incorporates this logic in our understanding of time as an infinitely divisive.[45] Those unable to intuit temporal sequence experience space as distorted. The saints in heaven presumably appropriate temporal succession as perfected and incorporated into the eternal Ur-Kenotic life of the Divine Persons, whereas damned souls experience the empty time as spatially constricted and meaningless repetition. Hell's anti-time continually reduces a created being's capacity for any genuine or creative thought. Berdyaev states that "Hell is the state of the soul powerless to come out of itself, absolute self-centeredness, dark and evil isolation, i.e., final inability to love."[46] [18] Hell is entropic in nature, whereas heaven is ordered to be eternal and unlimited. In heaven, the beatific vision acts as a negentropic force that spurs the resurrected to that ever-greater holiness characterized by infinite wholeness and eternal happiness.

The creative force of divine thought brings Creation into existence through the redeeming force of the Spirit's groundless presence. An event

[42] See *ST* I, 85, 3 and *ST* III, 175, 1.
[43] See *Homo Abyssus*, 274–284.
[44] *TL* 1, 78.
[45] See *The Bride of the Lamb*, 74.
[46] Nikolai Berdyaev, *The Destiny of Man* (London, UK, 1937), 351.

repeated in the Eucharist—two events completely foreign to Hell.[47] Jesus Christ's hallows Hell as he holds the "keys of death and hell" (Revelation 1:18). Jesus Christ experiences the infinite generative force of the Father's eternal love as the answer to evil's inestimable weight. At his death, even the love of Jesus Christ goes to the "depths," to the "uttermost" region of Hell's god-forsakenness (John 13:1). The universe participates, Maximus the Confessor tells us, in a cosmic liturgy where conversely, Hell is antithetic to the very temporal and spatial structure of the created order. In Hell, one cannot properly speak of space or time, as it lacks a sacramental and ecclesial dimension. The redemptive force of Christ's Passion etiolates Hell's futile stand against Creation by realigning it in full judgment to its ultimate heavenly terminus (Hebrews 9:24).

The unending punishment of Hell's diabolic negating forces cannot fully erase a being's memory of Jesus Christ (1 Corinthians 11:25). As a human being, Jesus Christ experiences the generative force of the Father's eternal love as the answer or proper informational construct to restore a fallen cosmos. Christ's love for the Father and the Godself's Creation inspires him to take upon himself the sins of the world and to bring them vicariously before God for condemnation (2 Corinthians 5:14). Conversion and the cathartic power of the human person's life are symbolically subsumed in the image of divine fire in scripture (1 Kings 18:22). Fire and light represent the sign of eternity and the mysterious nature of divine judgment that, Florensky states, separates the sinner from the dross of her sins: "The Spirit of Christ, coming to sinful creatures, will be that fire of trial which will purify everything, save everything, and fill everything with itself."[48] [19] This so-called Second Death Florensky continues is a spectacle of natural life that comes through ecclesial existence, a way to establish how one's sins are objectively "nothing." Armed with this grace that is also knowledge, an individual is free to repent, that is, no longer be subjectively self-enclosed around the disordered negations of sin and evil.[49] [19]

The "Eucharistic" person participates in evil's final overthrowing (Colossians 3:3–4), as the breaking and distribution of the Eucharist is not an act of eradication but a revelatory meeting of created internal and external forces that assert what divine love unifies "[f]or we being many are one bread, one body, all that partake of one bread" (1 Corinthians

[47] See *The Pillar and Ground*, 88, 132 and 155.
[48] Ibid., 182.
[49] Ibid., 91 and 161.

10:17). The Eucharist is a foretaste of Christ's resurrected life and a means for us to hope for mystical union with the Divine. The distribution of the Eucharist preambles the perfect unity of all Creation announced in the Mass' universal prayers as well as the faithful's call to evangelize the world (Mark 14:27).[50] The Spirit radiates Jesus' presence in the Eucharistic materially and thereby sacramentally feeds the faithful and transforms the created order (2 Corinthians 5:17–21). Sharing her faith in the world, the believer delivers her own person as Eucharistic presence and thereby partakes in Christ's single offering, for "he had achieved the eternal perfection of all whom he is sanctifying" (Hebrews 10:14). The priest is a "typos" or "*icon*" of Christ and an image of the Church (Hebrews 5.1, 12:24, 13:20).[51] The priest confects the Eucharistic sacrifice in the person (*persona*) of Jesus Christ as part of the universal call to holiness for "you are a chosen people, a royal priesthood, a holy nation, God's special possession, that you may declare the praises of him who called you out of darkness into his wonderful light"[52] (1 Peter 2:9). One participates in this reality at the Spirit's invitation. This call empowers one's interior spiritual growth and allows us to participate responsibly in Providence's patterning of Ur-Kenotic life in history, for "the accent that renders the Eucharist comprehensible falls on the real presence, in which the living Christ makes himself present to the Church by means of his deeds of power; but this deed of power by no means neglects the realization of the community which, as it realizes him by remembering him also realizes itself. *Anamnesis* is always a breakthrough to objective truth."[53] [20]

The presence of the proper material species must be used in the Mass as these admit (*epiousios*) the previous work of Jesus Christ at the Last Supper (*ex opera operato*).[54] The Eucharist is at once natural nourishment and "super-substantial" or "super-essential" nourishment, "for as the bread, which is produced from the earth, when it receives the invocation of God, is no longer common bread, but the Eucharist, consisting of two realities, earthly and heavenly; so also our bodies, when they receive the Eucharist, are no longer corruptible, having the [holo-somatic] hope of the resurrec-

[50] See Pope Pius XII, *Mediator Dei: On the Sacred Liturgy*, nos. 17 and 119, http://www.vatican.va/content/pius-xii/em/encyclicals/documents/hf.

[51] See Council of Trent, *DS* 1743 and Karl Rahner, "The Word and the Eucharist," 267.

[52] See *LG 10*.

[53] GL 1, 573. See also Peter Lombard (1160) in *Sent.* IV, 1, 4 and *SCG*, IV, 56, 4.

[54] The phrase "*ex opera operato*" is Latin for "the work of the worked" and refers to the belief that the efficacy of the sacraments derived from the actions of Jesus Christ and the proper material elements (e.g., bread, wine, water, oil, etc.), see *DS*, 783.

tion to eternity."[55] [21] The Eucharist makes the Eternal Word historically present and thus opens the faithful's limited view of reality to its completion in immortality (Romans 6:9f), and with this a rejection of all the devil's lies. The ecclesial assembly (*synaxis*) celebrates Jesus Christ, who is "the true vine" (John 15:1), the one who in the Eucharist pneumatically offers his body and blood (1 Corinthians 10:29).[56]

The supra-sensible dimensions of the sacrament are in harmony with its natural properties. The failure to see the Divine's hand in Creation leads to that untenable proposition of the short-sighted materialist. Jesus warns against this ideology and encourages that "unless you see signs and wonders you will not believe" (John 4:48). The supra-sensible properties of the Eucharist are made accessible through signs, bread, and wine, for its celebration is to be an experience lived among the members of a single ecclesial body. The sacraments realize grace in material reality but in a manner that demands the free response of faithful individuals—"*gratia quae sacramentorum virtus est.*"[57] [22] The aesthetic dimensions of the liturgy serve to ground one's personal commitments to charity and love (John 14:15–20).[58] The Eucharist disdains the vestige of idolatry (evil and sin) given material reality's original imprint by divine thought. In so doing, its potential infinite manipulations conform to the Son's eternal decision to incarnate as Jesus Christ and holo-somatically as Eucharist.

6.3 Ritualistic Sacrifices and the Virtualization of the Real

In 1439, the Council of Florence promulgated a hylomorphic explanation of the doctrine of transubstantiation. This theological tactic provides a great deal of interpretative flexibility on the part of the philosopher and theologian alike, as "in Aristotelian language one may say that essential definition expresses in matter while nominal definition expresses the matter presupposed by the form."[59] [23] In this spirit, the Nobel Laureate

[55] Ireneaus, *Ante-Nicene and Post-Nicene Fathers*, Series 1, vol. 1, "Against Heresies", Edited by Alexander Roberts D.D. and James Donaldson, LL.D, Revised with notes by A. Cleveland Coxe, D.D., Catholic Logos Edition (Buffalo, NY: The Christian Literature Company, 1885), Book 4, chap. 18, sec. 5. See also *Rom.* 6:23.

[56] *My Work*, 118.

[57] See Augustine, *Expositions of the Psalms, NPNF8*, lxxvii, 2.

[58] See *GL* 1, 413–14.

[59] Bernard Lonergan, *Collection: Papers by Bernard Lonergan*, S.J., Edited by F. E. Crowe, (New York, NY: Herder and Herder, 1967), 98.

physicist Frank Wilczek proffers that the traditional form-matter correlation can be taken as analogous terms for the scientist's quest to describe the active forces of Nature by means of measure. Hylomorphic theory is an earlier attempt at explaining the unending motion of form and matter. However, matter's correlation to form or essence has not been universally accepted, and among its detractors, one can list (1) materialists who deny the Divine altogether, (2) Christian believers who see the Eucharist symbolically, and (3) those whose Nominalist or Occamite leanings resist any attempts at explaining the meaning of consecration. Scientific advancements give us a way to add the concept of virtual information to the Church's traditional centering of a doctrine of transubstantiation around the triad of form, matter, and substance. Three elements are seen as necessary for the proper celebration of a sacrament: a specific material object (water, bread, wine, oil, etc.), an exact verbal formula (form), and a professed priest who performs the sacramental rite.[60] Within the liturgy, the spoken word is taken as an "immaterial" sign that directs the faithful to each other and thereby to the Divine, for "supernatural reality can display itself only through the medium of the human word, as long as it cannot present itself in its own proper reality… in the immediate vision of God."[61] [24] The sacramental form of each Eucharistic celebration depends on the efficacious power granted by the repeatable word (informational signs), whose meaning is reflected in salvation history's advance by free persons.

The domain of the symbolic and metaphoric (*metapherein*, "to carry over") resounds the interior properties of a physical being and its elusive supernatural heritage, for "If a symbol *as carrier* attains its end, then it is inseparable form the superreality it reveals: hence it is more than self-referential. If a symbol does not manifest a reality, then it attains no end; thus, we should not see it in any pattern or organization of 'carrying over' or transference; and in the absence of such, the thing is not a symbol-i.e., it is not a spiritual instrument—but it is merely empirical matter."[62] [25] Physical beings have a sacramental character given their ability to communicate or signify (*signatura rerum*) supernatural purposes or the existence of a Transcendent Creator. A sacrament bestows an "informational scaffolding" (grace) on a physical object by culminating its dual immaterial signification or virtual identity as material body and source in divine exemplar. A sacrament is a sacred sign—"*sacramentum, id est sacrum signum*"—that weds a particular physical object and a for-

[60] See *DS* 695.
[61] Karl Rahner, "The Word and the Eucharist," 267.
[62] *Iconostasis*, 65.

mal formula (*forma verborum*).[63] A sacrament visibly presents in an external physical object the unseen (*virtus sacrmenti*) grace of the Transcendent.[64] The material and virtual aspects of a sacrament are united theologians argue, as all created reality finds its actual and symbolic explanation in the Eternal Word's Incarnation.[65]

When orientated to the Divine, a physical being functions like a religious icon or symbol, as it bears the "energy" of the archetype it represents. In this holo-somatic imaging of the Divine, a physical being presumes its supra-mundane source as a divine exemplar.[66] The prototype and icon instantiate finite non-subsistent relations that act as apophatic markers of the hypostatic Trinitarian relations—the Divine Persons are the first iconoclasts.[67] Christian theurgy privileges the symbol's ability to orientate individuals to a deeper truth about reality and ultimately the Divine.[68] The Eucharist aligns the human person through the use of reserved signs and ritualistic actions that are symbolically prefigured for the "preparation of the holy food was thus prescribed by the Law in these symbols, and this prefiguration was prescribed in the view of its use to us."[69] [27] The promise of the Transcendent's apotheosis in bread and wine is metaphysically written into Creation.[70] Bonaventure proclaims, "that the entire world is like a single mirror full of lights presenting the divine wisdom, or like a charcoal emitting light."[71] [28] A sacrament refers to a holy or sacred physical object (*res sacrans* or *res sacra*), whose mystifying divine meaning is acclaimed by the faithful in lives that are the light of the world.

My interpretation of the Divine follows from polycentric perspectives drawn from disciplines such as theology, philosophy, science, and

[63] See Augustine, *The City of God*, *NPNF2*, Book X, chapter 4–5. See also, Augustine, *Reply to Faustus the Manichaen*, *NPNF4*, Book XIX, chapter 2, and Augustine, *On Christian Doctrine*, *NPNF2*, Book 2, chapter 2-3.

[64] See *ST*. III, 60, 4 and Augustine, *On Christian Doctrine*, *NPNF2*, Book 2, chapter 2–3.

[65] See Emmanuel Falque, *The Metamorphosis of Finitude: An Essay on Birth and Resurrection*, Translated by George and Edited by, John Caputo, *Perspectives in Continental Philosophy* (New York, NY: Fordham University Press, 2012), 82. See also *SCG*, 4, 61.

[66] Ibid., Emmanuel Falque, *The Metamorphosis of Finitude*, 65.

[67] See G. Ladner, "Origin and Significance of the Byzantine Iconoclastic Controversy," *Medieval Studies* 2 (1940): 144.

[68] See *The Bride of the Lamb*, 271.

[69] Cyril of Alexandria, *Catechetical Lectures On the Mysteries: "On the Body and Blood of Christ," NPNF2*, vol. 7, Lecture XII, 1–9, Translated and Introduced by Philip Schaff, D.D., LL.D., and Henry Wace, D.D. (Grand Rapids, MI: Wm. Eerdmans Publishing Company, 1983) and *Holy Pasch*., XIX (*PG* LXXVII)728 B.

[70] See *DS*, 1636 **Dz** 874.

[71] Bonaventura, *Hex*., 2.27.

mathematics.[72] An underlying analogous or anagogic ordering is found operative throughout cross-disciplinary accounts given essential truth's enduring grasp of material reality. Human reason depends on this symmetrical informing, as it makes possible our receptive nature. Balthasar states that the "locus of this first analysis and synthesis is prior to all discursive knowledge; it lies in it the still unrestricted openness of the horizon of being. It is thus the abiding foundation of every particular, limited intellectual activity that, reflecting the structure of man's cognitional powers, moves between the senses and the concept."[73] [17] The spiritual being is open to truth whatever disciplinary domain or specific language it chooses to reveal its presence.

Mathematical symmetry displays in analogies and this gives us a way to form theories about the physical and immaterial structure of beings and the universe. The quantitative and qualitative both serve to objectively explore individual beings and the order of being as a whole.[74] A presumed symmetry exists between theological and mathematical analogies. The concept of analogy (ἀνὰ λόγον λέγειν), like its mathematically based concept of symmetry provides a way to objectively structure the phenomenal universe; it is an "ordering that not only (intentionally) announces an objective 'order of being' (ἀνὰ λόγον τοῦ ὄντος λέγειν), but that, in itself (structurally as 'principle'), is that wherein this 'order of being' declares itself (the ontic law of the ἀνὰ λόγον τοῦ ὄντος εἶναι as the noetic law of the ἀνὰ λόγον λέγειν, and only thus as a fundamental law)."[75] Mathematics offers a mimesis of reality by means of a "Pythagorean Analogy" with physical reality through the formalization of structures through symbols.[76] Matter's potential "spiritualization" perverts into an ideological "religion" of dialectical materialism when it equates a physical being completely to its mathematical descriptions or the universe to physical laws.

Where apophasis and kenotic descriptions of the Transcendent and created order are hallmarks of the theologian's work, observation,

[72] A surface reading suggests that Western theologians tend to lean on such hermeneutical framings with greater confidence than their Orthodox counterparts, see *ST.* I–II, 112, 1, 1 in comparison to St. Gregory of Nyssa's take see: *The Great Catechism*, *NPNF2*, vol. 5, chapter 3, Translated and Introduced by Philp Schaff, D.D., LL.D., and Henry Wace, D.D. (Grand Rapids, MI: Wm. Eerdmans Publishing Company, 1983).

[73] *TL* 1, 166. See also ibid., 189–190.

[74] See *AE*, 192.

[75] *AE*, 187.

[76] See Mark Steiner, *The Applicability of Mathematics as a Philosophical Problem* (Cambridge, MA: Harvard University Press, 1998), 4, 7, and 116.

measurement, and mathematical descriptions of reality define the work of the scientist. Both approaches depend on some application of the concepts of subtraction and relational differing. By holding that grace never replaces nature but perfects it, the implicit wholeness and totality of reality chronicles in what often appear as unrelated and perhaps even dichotomous correspondences. Paradoxically, particular non-subsisting relationships and notions of Nothingness, image, the infinite and eternal. A finite being participates in divine life by means of the "spatiotemporal hypostasizing of its being," that is, Ulrich notes, as a finite creature. He elucidates that the "substantializing of being as being always implies the exaggeration of the 'nothing' in the sense of *non subsistens*, which hands being over to the essence insofar as it hypostasizes itself as the same time as absolute 'nothing.'"[77] [29] Furthermore, the divine essence does not necessarily imply nihilistic dread when understood as taken up in the divinized "transnihilation" of the Eternal Logos' incarnation.[78]

Jesus Christ's presence in the Eucharist, Thomas argues, is not to be understood in the same way that all other beings appear to exist spatiotemporally.[79] Prior to their natural process of rotting, the transient accidents of the bread and wine are sacramentally infused with substantial properties of the Eternal Logos and human person Jesus Christ. The Church encourages interpretative latitude in regard to the precise meaning of the terms "matter," "form," and "substance," as she has declined to give an exact definition to these terms in regard to Eucharistic theology.[80] Traditional metaphysical hylomorphism and understandings of substance in regard to the physical identity of the Eucharist can only benefit from the advances forwarded by post-Newtonian physics. As shown, theories of black hole entropy, QMT, information theory, and so forth give us a way to show how past philosophical concepts relating the nature of being and the spatiotemporal order are advanced in my holo-cryptic metaphysics.

The Mass is a sacramental sign of Jesus Christ's sacrificial death on the Cross and not a monotonous repetition or dramatic cloning of a past historical event (Hebrews 9:14). The bread and wine function as symbols or figures (*τυποι*) for the body and blood of Jesus Christ. Thus, sacramental signs relay the "holographic" identity of the eternal and finite natures of

[77] *Homo Abyssus*, 197.
[78] Ibid., 202–203.
[79] See *ST.* III, 75, 2 and ibid., 76, 4.
[80] See *DS* 1642 **Dz**.

Jesus Christ as an entitative property of the consecrated bread and wine. The holographic is not understood as a stand-in for some oblique philosophical gnostic notion. The Eucharistic act always admits the possibility of idolatry or the sterilization of the creative power of the symbol and sign's transfigurative power by participants. Each liturgy must reposition apophasis at the center of each of its communities' actual and historical celebrations. The faithful embrace a historical event as part of the lived moments of the liturgy.

The Eucharistic celebration is an effectual and repeatable sign of the sacral "renewal" (*renovatur*) of Christ's Passion. The hoped-for eternal life that the participant at the Mass anticipates is seemingly undermined by what inspires its reenactment—the death of the Divine. The efficacy of the sacrament (*signum rei scarae*) includes one of the most negative and unsettling outcomes, namely, "one will prefer to say, not that his death was a consequence of his birth, but that the birth was undertaken so that he could die."[81] The sacrificial propitiation of the Cross symbolically co-signifies the Trinitarian space where love "never ceases" (1 Corinthians 13:8) and is reflected in the emergent properties of the created being and the irreducibility of information, as evident in quantum superposition.[82] Personal conversion and acts of mercy further testify to love's trans-spatiotemporal influence intimated in Jesus Christ. He is the willing scapegoat, the "sin-offering" who redeems Creation from the moment of its first separation from its perfect holy origins.[83] At a deeper level, the "scapegoating" or substitution dramatized in cultic sacrifices witnesses the meta-ontological purposes of reality's own kenotic structuring in space-time. The liturgy reminds us that human sin's deprivation of the good is always an infinitely lesser force than the continued trace of divine kenosis in reality.

Christ's love redirects all to its eternal divinely ordained conclusion. As fully divine and human, Christ's death expiates all human self-disfigurement, as his sacrifice is made "once and for all" (Hebrews 7:27).[84] [30]

[81] Nyssa, *The Great Catechism, NPNF2* vol. 5, chapter 32, Translated and Introduced by Philp Schaff, D.D., LL.D., and Henry Wace, D.D. (Grand Rapids, MI: Wm. Eerdmans Publishing Company, 1983). and *Dominicae cenae*, no. 9. see: https://vatican.va/content/john-paul-ii/en/letters/1980/documents/hf-jpii_/let-19800224_dominicae-cenae.htmll.

[82] See *ST* III, 46, 1, TL 3, 176–178, *Mediator Dei: On the Sacred Liturgy*nos. 27, 32, 34 and 36, at vatican.va/content/pius-xii/en/encyclicals/documents/hf_p-xii_enc_20111947_mediator-dei.html

[83] *LG*, nos. 136.

[84] See also T. A. Buckley, *The Canons and Decrees of the Council of Trent* (London, UK; George Routledge and Co., 1851), 142–143. *DS* 1637 and 1638 **Dz** 875.

The liturgy chronicles and includes contemporary participation in all that is entailed in Jesus' death. The full *modus significandi* of Christ's Passion arms divine supereminence (1 Peter 3:18).[85] The Eucharist perfects and calibrates a physical being's deep-rooted vitality, its free and conscious response to the hope of future resurrection (1 Peter 1:3).[86] Focusing our subjective identity as a Christian means rewriting our non-subsistent natures against the eternal positivity of Christ's Passion. The grace offered with the fracturing of the Eucharistic bread perfects the congregation and gives us a counterexample of all evil or negative relationships in the mereological unity professed between individual believer and the Trinitarian communio. Where Eucharistic grace is already integrative and life-giving nourishment, evil alienates and isolates. Evil desires only the "here" and "now," as it is an intentional in-hospitality that refuses all relationships and contact with the "other." It is an "aspiring to self-deification, evil self-assertion does not remain identical even to itself but falls apart, decomposes, fragments in inner struggle. In essence, evil is a 'kingdom divided against itself.'"[87] [19] An individual who defines their life in terms of a "covenantal" relationship with the Divine recognizes in one's self and others, not only a fallen nature in need of redemption but also a stranger and reality who are holographic icons of the Divine.

By interpreting reality as an *analogia Christi*, an individual integrates their adoration of the Divine with a spiritual arrogation of Creation. In refusing to do evil or sin, an individual works toward the created order's transfiguration. The "Incarnation does not divide form from content or medium from message or the signifier from the signified," Graham Ward notes, "for beauty manifests truth and together call upon a mutual perfection for 'even knowing all the sins X may have done by commission or omission does not prevent the priest bowing to X in adoration of the image of God within him or her.'"[88] [31] The ritualistic meal has a symbolic value in Semitic cultures. It unifies a family or community. Individuals are invited to this meal, and their attendance is freely given or not. The

[85] See also Hebrews. 9:11, 9:25–26 and From the Council of Trent see *DS* 938.

[86] See *Constitution of the Sacred Liturgy*, 106 at vatican.va/archive/hist_council/iivatican.council/documents/vat-ii_const_19631204_sacrosanctum-consilium_en.html.

[87] *The Pillar and Ground of the Truth*, 127–128.

[88] Graham Ward, *The Beauty of God*, 44ff.

sacramental meal respects those who refuse to attend and thus set the boundaries for individuals and communities (Genesis 14:18).

Blessing the wine at his last Seder meal, Jesus prayed: "[t]his is my blood of the covenant, which is poured out for many for the forgiveness of sins" (Matthew 26:28). Jesus identifies his place within sacred history through ritualized sacrifice by becoming the eternal lamb of Passover (Exodus 12: 27). God's boundlessness manifests in Christ's Eucharistic body given "for the multitude" (Luke 22:19, 1 Corinthians 11:24, and Hebrews 13:12).[89] Originally, the term "Eucharist" was used to highlight mutually generative human relationships (Wisdom 18:2, Maccabees 2:27, Acts 24:3, Romans 16:4). One can only enter into an authentic relationship with another person by risking those boundaries—realized or imagined—that equally serve and impede our autonomy. In its essence, the Eucharist sees human flourishing as a communal event, for it gifts reality with the ONE without Boundaries, "the name of the Lord" (Genesis 4:26, 12:8, 21:33). By means of the power of the spoken word, a dramatic replaying of the past from its theological execution in the Jewish covenant through its perfection in the liturgy takes place, as "now all these things happened to them as a type and, they were written for our correction" (1 Corinthians 10:11).[90] The sacramental memorial on the altar anticipates, renews, and realizes the ecclesial nature of, that its grace-filled common ordering to (Romans 8:32, John 3:16).[91]

A logical dilemma follows if Jesus Christ as supra-Word and materially present in the consecrated hosts throughout the world is viewed from the rubrics of a classical understanding of physical reality. Thomas' resolution to this problem highlights the positive role ritual gives the sign. He notes "Christ's body is not in this sacrament in the same way as a body is in a place, which by its dimensions is commensurate with the place; but in a special manner which is proper to this sacrament. Hence, we say that Christ's body is upon many altars, not as in different places, but sacramentally and thereby we do not understand that Christ is there only as in a sign, although a sacrament is a kind of sign; but that Christ's body is

[89] See John Chrysostom, 880 (1194) *Homilies on First Corinthians*, 24: 2, *MG* 61, 200, NPNF XII, 140.

[90] See *LG*, nos. 5 and 6.

[91] See *Dominicae cenae* 11–12.

hereafter a fashion proper to this sacrament."[92] [32] The sacred symbol (*Signum sacrum*) and sign reveal physical reality's virtual truth. Supernatural signs, theories, and models exist for individuals to sensually and intellectually apprehend the Transcendent in created realities. These signs are "observed" (Luke 17:20) and "seen" (John 2:23) when one spiritually connects these conscious representations to one's personal life lived in the world. The faithful are embodied (spiritually embodied transducers) for evangelization (essential information). One's ability to see the Creator Spiritus' presence in created beings and their signatory and symbolic representations demands morally good judgments.

The mathematical normalization of immaterial multidimensional multifarious spaces gives us insights into the Transcendent-immanent existential correspondences. Neither formulation closes an autonomous and conscious beings from their supra-mundane source, as the truth reveals in relational differing, logical, or symbolic attribution. In this way, a symbolic parallel can be drawn between Neo-Scholastic *duplex ordo* and quantum entanglement, as both implicitly hold that locality "assumes that signals or energy transfers between space-like separated regions cannot occur at speeds greater than light. In addition, realism assumes that physical reality exists independently of the observer and that the state of this reality is not dependent on acts of observation or measurement."[93] [6] The *epiclesis*' prayer, set between the anaphora or Eucharistic Prayer and the Lord's Prayer in the Mass, introduces for some theologians, the Holy Spirit's descent at the consecration. The Lord's Prayer sums up the universal Church's prayers and the continuing work of the body of Christ in the world—to be part of the pattern of the Divine's providential plan.[94] Through the reciting of Scripture, the recorded memory of past spiritual journeys is "virtually" brought into the reality of the present day. The *anamnesis* or main Eucharistic prayer renews and announces a future in which divine grace once again becomes an unhindered source in which the entire created order's taxonomy participates in a trans-temporal fashion (Galatians 2:20).[95]

[92] *ST.* I, 74, 1, 3.

[93] *The Conscious Universe*, 60. See also Matthew F. Pusey, Jonathan Barrett, and Terry Rudolph, "On the reality of the quantum state," *Nature Physics* 8.6 (2012): 475–478.

[94] See *The Bible and The Liturgy*, 142.

[95] See Pope Pius XII, *Mediator Dei: On the Sacred Liturgy*, no. 16 and Louis Bouyer, *The Eternal Son: A Theology of Word and God*, (Indianapolis, IN: Our Sunday Visitor, 1979), 405.

At Mass, the priest and the congregation vocally and prayerfully recall the story of God's saving history on behalf of humanity, and thus, "spiritually" or "holo-somatically" annex the whole of the cosmos.[96] This confirms Thomas' view that a sacrament "is, a sacred sign, of an invisible sacrifice."[97] [33] Christians freely gather together to worship, and this liberty testifies to finite being's analogous involvement in the virtual negation of evil and sin realized by discerning the Spirit's ongoing work in reality. The Spirit's presence can be understood as groundless or non-subsistent relationships, as the Spirit exists across all dimensions—temporal, spatial, dimensional, and conceptual. Human reason in concert with created reality allows one to engage the question of this disclosure freely. Perhaps it is with this in mind that Leibniz relates the Spirit with complex numbers. He writes, that "the Divine Spirit found a sublime outlet in that wonder of analysis, that portent of the ideal world, that amphibian between being and not-being, which we call the imaginary root of negative unity." The concept of transcendental numbers offers many mathematicians and philosophers a way to understand the Transcendent and in cases some proof of God's existence.

Religious symbols and signs function because reality's conceptual differentiation is symmetrical to one's implicit and rational recognition of its manifestation in actual existing beings. For "esse as act points to an inexhaustible fullness and supra-essentiality at the heart of created reality, as non-subsistent it is 'affected' by the otherness of essence without for all that losing its 'simple completeness.' Indeed, this 'affection' is intrinsic to every 'nature as plenary actuality in the first place.'"[98] [34] Through sacramental participation, it is possible to tailor one's life consciously around a principle of conversion (*metanoia*) or non-subsisting differing (Colossians 5:7). One's "inner assimilation to the being of Jesus by faith… is sealed the sacraments… [and] infused into the believer's mind and being."[99] [26] The Church houses eternity holo-somatically, for she "carries in [herself] only the seed, the germ of the new and higher nature (in the Eucharist), 'the absolute form of transfigured matter,' with which we communicate in the organic existence of the Church, so that, starting

[96] *See* Pope Paul VI, *Decree on Ministry of Priests: Presbyterorum Ordinis* (Dec. 7, 1965) at vatican.va/archive/hist_councils/ii_vatican_council/document/va-ii_decree_19651207/document_presbyterorum-ordinis_en.htm#.

[97] Augustine, *On the City of God*, *NPNF3*, Book 10, chapter 5.

[98] Nicholas J. Healey, *The Eschatology of Hans Urs von Balthasar: Being as Communion* (Oxford, UK: Oxford University Press, 2005), 25.

[99] *TD* 4, 406–407.

from there, "freed in principle from the servitude of the material order, we can actively help the world to be transfigured."[100] [35] The believer acts as an eschatological agent for the Divine's inexplicable plan by aligning their life to the Gospel and the redemptive power such a union makes in their life (Philippians 2:17, 2 Timothy 4:6).

The Feast of the Tabernacles prefigures Jesus Christ's transfiguration on the Mount Tabor, as this celebrates the divine intent to convey history and creation to its completion in resurrection and the community of the heavenly court.[101] The congregation is both the Transcendent's living memory and acting hands working eschatologically to see the earth's fulfillment as the New Jerusalem (Romans 8:19–24).[102] The material substantiation of this truth in the Eucharist enjoins a conscious response, as does the law and prophetic word. The material order and sacred words predict the body's resurrection and the eternal holographic image of the Divine to be perfected in the beatific vision. As the prophet reveals, "The Lord says this: I am now going to open your graves; I mean to raise you from your graves, my people, and lead you back to the soil of Israel. In addition, you will know that I am the Lord, when I open your graves and raise you from your graves, my people. In addition, I shall put my spirit in you, and you will live, and I shall resettle you on your own soil; and you will know that I, the Lord, have said and done this—it is the Lord who speaks" (Ezekiel 37: 12–14). The Eucharist is consumed within the familiarity of a meal that simultaneously images the eternal heavenly feast. The Mass is at once a celebration of the resurrected spiritual body and a denunciation of its ultimate defilement when considered a ritualistic sanctioned act of idolatrous (cannibalistic) digestion.[103] The body and blood's digestion, by means of the acidity of gastric juices and peristaltic waves, reenact in physical processes what is ritualized in the remembrance of Jesus Christ's violent death. The physical act of food being mixed in the stomach signifies the metaphysical and spiritual connotations of intermingling and assimilation of the Divine and human beings. The body,

[100] *GL* 3, 331–332.

[101] *The Bible and Liturgy*, 338–339.

[102] See *ST.* III, 73, 3 and ibid., 79, 1.

[103] See Marie José Mondzain, "The Holy Shroud: How Invisible Hands Weave the Undecidable," in *Iconoclash*, Eds., Bruno Latour and Peter Weibel (Cambridge, MA and London, UK: MIT Press, 2002), 326.

soul, and divinity of Jesus Christ received at the table of the Lord perfects the natural order to its future completion in the Eternal Spirit—the material body and universe move together toward spiritual embodiment's final end.

Created in the Eternal Logos with the Spirit, freedom *in principio*, the Eucharist community commits all to its conversion and ultimate resurrected state. As High Priest, Jesus' self-sacrifice is the One Sacrifice of the ONE Flesh of divine and human natures—Creation's symbolic Tabernacle (Hebrews 9:24), that the Spirit orientates back to its original state virtually as a replica of the heavenly sanctuary.[104] Christ's Passion and the Eucharist celebrate God's providential plan to return Creation to the Father in praise and thanksgiving.[105] This Apocatastasis is universal insofar as it is constitutive of Jesus Christ's self-reflective identity. The pattern of "*exitus—reditus*" defines created reality as "in some sense the soul knows itself through itself, inasmuch as to know is to possess in itself knowledge of itself and in some sense it knows itself through a species of an intelligible object insofar as knowing implies thinking and distinguishing of self."[106] [36] Eucharistic grace moves beyond the immediate confines of the Church when the faithful act mercifully and justly in the world, for "[t]he salvific movement in the New Covenant is not that of in-gathering but of Eucharistic radiation. For this is the very movement of Jesus, who in the Eucharist holds nothing back in his act of distributing himself to the world, [and] which he is poured out to reconcile the world to God."[107] The Spirit opens an aperture on the nature of ecclesial mission by promoting the evangelization of the world by bridging all divisions that work against their proper unification and respective freedoms of the Transcendent and immanent orders: "for this is my blood of the covenant, which is poured out for many for forgiveness of sins" (Matthew 26:28).

[104] See *GL* 1, 438–439.
[105] See *ST.* II–II, 84, 1–2 and 106, 2.
[106] *De ver.*, 2, 2, 2.
[107] *ExTheo.*, 4, 277.

References

1. Brian Greene, *The Fabric of the Cosmos: Space, Time and the Texture of Reality* (New York, NY: First Vintage Book Edition, 2005), 99.
2. John Polkinghorne, "The Demise of Democritus," in *The Trinity and an Entangled World: Relationality in Physical Science and Theology*, Edited by John Polkinghorne (Grand Rapids, MI: Wm. B. Eerdmans Publishing Company, 2010).
3. Menas Kafatos and Robert Nadeau, *The Conscious Universe: Parts and Wholes in Physical Reality* (New York, NY, Springer–Verlag, 2000), 66.
4. Niels Bohr, "Can quantum-mechanical description of physical reality be considered complete?" *Physical Review* 48 (1935): 700.
5. Slavoj Žižek and John Milbank, *The Monstrosity of Christ: Paradox or Dialectic?* Edited by Creston Davis, (Cambridge, MA: MIT Press, 2009), 89.
6. Menas Kafatos and Robert Nadeau, *The Conscious Universe: Parts and Wholes in Physical Reality* (New York, NY: Springer–Verlag, 2000), 60.
7. Cynthia Sue Larson, "Primacy of Quantum Logic in the Natural World," *Cosmos and History: The Journal of Natural and Social Philosophy* Vol. 11, no. 2 (2015): 332.
8. Wolfgang Smith, *The Quantum Enigma: Finding the Hidden Key*, Forward by Seyyed Hossein Nasar (San Rafael, CA: Angelico Press: Sophia Perennis, 2005), 46.
9. "The Mathematical Universe," 4 *Found. Phys.* (Sept., 2007): 4, arXiv:0704.0646v2 [gr-qc]. Accessed 01-01-18.
10. Hans Urs von Balthasar, *Glory of the Lord: Vol. 7, Theology of the New Covenant.* Edited by Joseph Fessio, S.J and John Riches. Translated by Brian McNeil C.R.V and John Riches (San Francisco, CA: Ignatius Press, 1990).
11. Nicholas Gisin, *Quantum Chance: Non-locality, Teleportation and Other Quantum Marvels*, Forward by Alain Aspect (Cham Heidelberg: Springer, 2014).
12. Robert Oppenheimer quoted in Werner Heisenberg, and further quoted in James B. Conant, *Modern Science and Modern Man* (New York, NY: Columbia University Press, 1953), 271.
13. Peter J. Lewis, *Quantum Ontology: A Guide to the Metaphysics of Quantum Mechanics* (Oxford, UK: Oxford University Press, 2016), 25.
14. David Bohm and Basil J. Hiley, *The Undivided Universe* (London, UK: Routledge, 1993), 347.
15. David Bohm and B. Hiley, "On the Intuitive Understanding of Non-Locality as Implied in Quantum Theory," *Foundations of Physics*, 5 (1957): 96.
16. Henry P. Stapp, "Quantum Theory and the Physicist's Conception of Nature: Philosophical Implications of Bell's Theorem," in *The World View of Contemporary Physics*, Edited by Richard E. Kitchener, (Albany, NY: SUNY Press, 1988), 40.

17. Hans Urs von Balthasar, *Truth of the World* Vol.1, Translated by Adrian J. Walker (San Francisco, CA: Ignatius Press, 2000).
18. Nicholas Berdyaev, *The Destiny of Man* (London, UK: 1937), 351.
19. Pavel Florensky, *The Pillar and Ground of the Truth: An Essay in Orthodox Theodicy in Twelve Letters*, Translated by Boris Jakim. Introduced by Richard F. Gustafson. (Princeton, NJ: Princeton University Press, 1997).
20. Hans Urs von Balthasar, *Glory of the Lord: A Theological Aesthetic*: vol.1, Edited by Joseph Fessio, S.J and John Riches. Translated by Erasmo Leiva-Merikakis and John Riches (Edinburgh, UK: T&T Clark, 1982), 515.
21. Irenaeus, *The Ante-Nicene Fathers Translations of the Writing of the Fathers down to A.D. 325*, Series 1, vol. 1: *The Apostolic Fathers with Justin Martyr and Irenaeus*: "Against Heresies," Edited by Alexander Roberts, D.D. and James Donaldson, LL.D. Revised and chronologically arranged, with brief prefaces and occasional notes by A. Cleveland Coxe, D.D. *Catholic Logos Edition*. (Buffalo, NY: The Christian Literature Company, 1885), IV, 18, 5.
22. Augustine, "Exposition of Psalms," *Nicene and Post-Fathers*, vol. 8. Edited by Philip Schaff, D.D., LL.D. Logos Catholic Edition (New York, NY: Christian Literature Edition, 1888), lxxvii, 2.
23. Bernard Lonergan, *Collection: Papers by Bernard Lonergan, S.J.*, Edited by Frederick Crowe (New York, NY: Herder and Herder, 1967), 98.
24. Karl Rahner and Angelus Häussling, *The Celebration of the Eucharist* (New York, NY: Herder and Herder, 1961), 267.
25. Pavel Florensky, *Iconostasis*. Translated by Donald Sheehan and Olga Andrejev, Introduction by Donald Sheehan (Crestwood, NY: St. Vladimir's Seminary Press, 1996), 65.
26. Hans Urs von Balthasar, *Theo-drama: The Action*: vol. 4, Translated by Graham Harrison (San Francisco, CA: Ignatius Press, 1995), 406–407.
27. Cyril of Alexandria, *Holy Pasch* vol. 19: *Patrologiae cursus completes: Series Graeca* (Paris, France: Jacques Paul Migne, 1857–66), 728B.
28. Bonaventura, *The Works of Bonaventure: Cardinal Seraphic Doctor and Saint II: The Breviloquim*. Translated by José de Vinck, *Catholic Logos Edition* (Patterson, NJ: St. Anthony Guild Press, 1963), ii, 27.
29. Ferdinand Ulrich. *Homo Abyssus: The Drama of the Question of Being*, Translated by D.C. Schindler (Washington, DC: Humanum Academic Press, 2018), 197.
30. Pope Pius XII, *Mediator Dei: On the Sacred Liturgy*, nos. 76, 77, and 78.
31. Graham Ward, "The Beauty of God," in John Milbank, Graham Ward, and Edith Wyschogrod, *Theological Perspectives on God and Beauty* (Harrisburg, PA: Trinity Press International, 2003), 44.
32. *Summa Theologiae* I, 74, 1, 3 (Garden City, NY: Doubleday Company, 1964–1976).
33. Augustine, *The City of God, Nicene and Post-Nicene Fathers*, First Series, vol. 2. Edited by Philip Schaff D.D., LL.D., *Catholic Logos Edition*, (Buffalo, NY: Christian Literature Company, 1887), Book 10, chapter 5.

34. Nicholas J. Healey, *The Eschatology of Hans Urs von Balthasar: Being as Communion*, (Oxford, UK: Oxford University Press, 2005), 25.
35. Hans Urs von Balthasar, *Glory of the Lord vol.3: Studies in Lay Styles*, Translated by Andrew Louth, John Saward, Martin Simon, and Rowan Williams (London, UK: T&T Clarke, 1986), 310–332.
36. *Thomas Aquinas, Truth* Vol.1: Questions I-IX, Translated by Robert W. Mulligan, S.J. (Chicago, IL: Henry Regnery Company, 1952), 2, 2, 2.

CHAPTER 7

The Heart of Matter

7.1 Ur-Kenosis in a Holo-cryptic Universe

An ontological chiasmus exists between the QMO's interior architecture and its momentum (energy) and its position, that is, the difference between its immediate placement and the whole universe. In a quantum system, we can know with great accuracy an electron's magnitude or its location, but never both at the same time. A QMO discrete "jump" from wave to particle is reversible, as its previous state is "remembered" as a holographic inscription in the particle. The quantum wave's energy projects as a field (QED) that accounts for all possible future states. The QMO has a unique probability density that relates to its present and future repositioning and localities within its contextual field (QED).[1] Expanding upon how the QMO relates to its QED, a way forward to see how these define the universe. The QMO's individual properties pattern all of reality by the presence of its waveform. The nature of quantum superposition makes this state near impossible to sensually observe because "the wave function [is] a continuous superposition of all macroscopic possibilities, the result is an amorphous superposition of a continuum of different states. Because this translates mathematically into zero probability, the existence of a conscious observer registering specific measurements in quantum mechanical experiments is quite improbable."

[1] See *Mind, Matter, and Quantum Mechanics*, 54.

M. P. Fusco, *The Physics and Metaphysics of Transubstantiation*,
https://doi.org/10.1007/978-3-031-34640-8_7

Stapp further notes that "states characterized by local observable properties have the required characteristics of endurance and reproducibility, whereas superpositions of such states do not: the intention of these later states with their environment quickly destroys the phase state connections that define them."[2] [1] Quantum symmetries imply a certain vibrancy and temporal designation that dictates the object's "irreducible representation" in an uncertain universe.[3]

The observer only understands the QMO's state as an event in a system in a state of constant change. The ability for an observer to conceive of the QMO's identity requires one to keep distinct each of its states in its unending succession. The separation of the QMO and its system requires some covert employment of the concept of nothingness in the gap between the object's endless changes. Heisenberg's uncertainty theory, for example, formalizes the QMO's discontinuous disclosure. I understand Heisenberg's concept of discontinuity as just such an abstract reworking of the idea of nothingness. Heisenberg states that "the discontinuous change in the probability function takes place with the act of registration because it is the discontinuous change of our knowledge in the instance of registration that has its image in the discontinuous change of the probability function."[4] [2] This process finds its "holographic" equivalent in the imagination's ability to form images from memory's uneven and periodic representation of visual perceptions. Conscious memory partially bridges and equally reflects being's essential permanence in a universe that is inescapably free and uncertainty.[5] Seen as intractably linked, the unknown is fully revealed only in tandem with the subjective consciousness. The intelligibility of our universe is commensurate with its inescapable randomness, characteristic probable nature, and the relative and limited scope of our sensual grasp. The relative truth that follows from the imprecise nature of our measure and observation of the real is not a shortcoming—at least not from the perspective of free and rational beings that have any commitment to personal existence and a communal life. For "there can be no community without a certain balanced tempered, and average truth. This averageness is not only a natural datum based on the participation and communication of all subjects in a common human nature. Rather, it is always the product of a common renunciation in which all persons forgo the exclusivity of

[2] Ibid., 165.

[3] See Tegmark, *The Mathematical Universe*, 8.

[4] Heisenberg, *Physics and Philosophy* (New York, NY: Harper and Row, 1962), 55.

[5] See Lothar Schäfer, "Non-empirical Reality: Transcending the Physical and Spiritual in the Order of the One," *Zygon* 43, no. 2 (May 2008): 343.

their personal truth for the sake of the average truth that community demands."[6] [32] Scientific theories have a staggeringly impressive level of accuracy given the constraints that probability and indeterminacy place on all idealistic, positivist, and empirical accounts of the universe.

The infinite reach of the QMO reveals itself in its identity as a wave. The quantum wave amplitude will ultimately elongate to mirror the universe's spatiotemporal limit. One cannot isolate all the initial properties of a quantum state, as Heisenberg's uncertainty principle mathematically reminds us. Neither an object's physical appearance nor the sum of its causal relations can be captured in a single moment of surveillance, for "'if we know the present exactly we can calculate what will happen in the future,' [then] it is not the conclusion but the premise which is false. We cannot, even in principle, know every detail of the present."[7] [4] cited in [5].

By means of observational and instrumental measures, one attempts to define the QMO's precise identity. As Heisenberg asserts, "the discontinuous change in the probability function takes place with the act of registration; because it is the discontinuous change of our knowledge in the instance of registration that has its image in the discontinuous change of the probability function."[8] [2] In other words, every image of the QMO is relative to arrival in an oncoming relational investment and theoretical transference in its future disclosure. Thus, QMT helps us to understand divine action in the case of miracles in general by providing a plausible emplacement of divine and human action as relationally free at the conscious and subatomic levels. [6] The material structure and symbolic order of a being exist in dialectical opposition to all negative forces. This meeting takes place in the *espace de placement*, the "space of placement."[9] Theologians have found a similar theme to resonate in scripture's teaching that the universe's supernatural conclusion remains a mystery, teaching such as "you know neither the day nor the hour when the Son of man comes" (Matthew 25:13). Pure Being's eternal self-measure acts as an insurmountable boundary for all finite truth.

[6] *TL* 1, 192.

[7] Werner Heisenberg, "Uber den anschaulichen Inhalt der quantentheoretischen Kinematik und Mechanik," *Z. Physik* 43: 172–198 (1927), 197, Translated in John A. Wheeler and W. H. Zurek, *Quantum Theory and Measurement* (Princeton, NJ: Princeton University Press, 1983). See also Gordon Belot, *Primitive Ontologies*, 73–74. https://doi.org/10.1007S13194-011-0024-8.

[8] Heisenberg, *Physics and Philosophy* (New York, NY: Harper and Row, 1962), 55.

[9] On the poetic and philosophical origins of Badiou's thought on this issue see *Theory of Subject*, 76–77 and 184.

Theologically ordered, the universe is viewed as a subtractive or kenotic actualization of the Transcendent Creator—a perspective that allies individual evidence to a collection of irreversible physical laws. The QMO's distinctiveness is represented by the dual states of $|0\rangle$ and $|1\rangle$. With $|0\rangle$, we have an image of nothingness, both as a conceptual construct for universal potential and a QMO's specific locality given this placement dependence on the notion of the void or null dimensions.[10] [7] These considerations remain a thought experiment, as there is no way to finally determine the QMO as $|0\rangle$ qua $|0\rangle$ in a controlled environment—even in a vacuum. [9] Reeh-Schlieder's theorem outlines the mathematical proof for this instance of Nature's abhorrence of the empty and nothingness. [8] This theorem states that it is never possible to guarantee the observation of "0" in a given system, although its observation and approximate value is.[11] Reality's inchoate noise and randomness make a final measure of the QMO impossible.[12] The concept of nothingness is hinted at in the Fourth Lateran Council teachings on the asymmetrical nature (*maior dissimilitudo*) that exists between God and finite creatures. Where discontinuity is necessary to understand physical descriptions, nothingness introduces an obligatory theological divide for those who hold that a Creator and created exist. Ironically, it is this very nothingness that makes room for finite free action to express ontological differing in history.

Individuals only partially reveal themselves to other finite beings given their holo-somatic identity as a reflection of the Transcendent. The hidden remains in limited beings because the Transcendent ironically confirms one's existence against the impossibly imagined divine, for "the more closely a creature approaches God, the more it possesses *esse*...However, since a creature approaches God only in thus far as it participates in a finite *esse*, yet this distance from God is always infinite, it is said to have more *non-esse* than *esse*."[13] [10] Scripture testifies to this disjunctive relationship as a universal principle and philosophically minded thinkers lean upon it in their methodological formulations of analogy, for "The relationship between God and the world can be understood in various ways: as a

[10] On the quantum duality of vacuum polarization or the so-called Lamb shift see Silvan S. Schweber, *QED and the Men Who Made It: Dyson, Feynman, Schwinger, and Tomonaga* (Princeton, NJ: Princeton University Press, 1994), 86–87.

[11] See *Void*, 111–113, 125–127 and Peter. W. Milonni, *The Quantum Vacuum: An Introduction to Quantum Electrodynamics* (San Diego, CA: Academic Press, 1994).

[12] See *Void*, 112–117.

[13] *De ver.*, 2, 3, resp. 16.

relationship of knowledge or being, between God and the world or between the world and God."[14] [11] Every statement concerning God infers greater mystery for "it is easier to articulate what God is not (*non quid non est*), than what the Divine is (*quia est*)."[15] [12]

The Divine creates from nothing and then guides beings toward their ever-greater "actuality" through moral action. As perfect, the Divine intends Edenic perfection throughout the universe. Every subject-object relation is defined by a finite limitation that testifies to Transcendent infinity. Nyssa states: "the essence of the Mind as an object of thought only, since it is the 'image' of an Existence which is such; but it will not pronounce this image to be identical with the prototype. Just, then, as we have no doubts, owing to the display of a Divine Wisdom in the universe...a Divine Power existing in it all which it secures its continuance."[16] This follows as the Transcendent imprints Nothingness with an analogy of its own eternal or first principle of motion.[17] [13] A creature's perfectibility follows from the fact that its ground state is non-subsistent "zero" (*alpha privativum*). That is, its informational state sets parameters for nonbeing. Similarly, the universe premises its reflexive origins from Nothingness' "noetic" denotations as the empty set, the infinite and "ontologically" as void and emptiness.

Reality is multileveled and endowed with the power to evolve creatively. This is not to say that science dictates theological dogma, as many pantheists argue in their theories of a world—soul god. Indeed, QMT rejects such a reductive reading of natural theology. Science is indispensable to this work, as it gives a better perspective of reality than traditional Aristotelian philosophy or Newtonian science; it does so by clarifying the shape and locus of divine action in regard to transubstantiation. Nature apophatically resembles the eternal presence, thought, and energy of the Transcendent.[18] The different or distinct ways in which a being reflexively

[14] Walter Kern, "God-World Relationship," in *Sacramentum Mundi*, online, General Editor Karl Rahner, S.J. Consulted online on 23 June 2023 <http:11dx.doi.org/10.1163-483X_smuo_COM_00168>. First published online: 2016.

[15] See *Augustine, "Ac per hoc ne ineabilis quidem dicendus est Deus quia et hoc cum dicitur, aliquid dicitur,"* See *On Christian Doctrine*, *NPNF2*, Book 1, 6.

[16] Nyssa, *On the Soul and Resurrection*, *NPNF2*, vol. 5, Translated and Introduced by Philip Schaff, D.D., LL.D., and Henry Wace, D.D., (Grand Rapids, MI: Wm. Eerdmans Publishing Company, 1983), 437. See also *ST.* I-II, 1, 8.

[17] See *ST.* I, 104, 1.

[18] See *Polarity*, 46 and Paul Davis, *The Mind of God* (New York, NY: Simon and Schuster, 1992) and George Ellis, *The Far-Fetched Universe* (Radnor, PA: Templeton Foundation, 2002).

and reflectively relates to the infinite provide a basis upon which to set out a hierarchic path from nothingness to the Transcendent. From the perspective of Christian metaphysics, it is only with the Spirit's post-resurrection presence that finite beings can grasp how the Divine's presence moves from Prime Matter throughout reality.[19]

Nothingness is a pure medium that can receive the influence of divine exemplarity or supernatural information because it is without the metaphoric "noise" introduced by physical beings and motion.[20] The concept of theological nothingness allows some measure of an existing system or state's relative "distance" to metaphysical "normality." Consequently, the negentropic does not imprison existing beings into a sterile state but rather makes possible the essential and existential within the created order by "contemplatively" reducing existing "noise." Reversing the chaos within a system reveals the simplification of the informational state of the existing beings. Therefore, "whereas entropy represents the loss of difference differences constitutive of organizational structure, negentropy designates the temporary reversal of the process, which occurs when differentiated structures emerge in the midst of disorder."[21] [14] Nothingness is a "shared" medium that distinguishes how information differs in hylomorphic and scientifically based holographic systems.

The Trinity promotes equipoise between nothingness as Creation's underpinning and its material, physical manifestation as being by "condescending grace" that "stirs us by lifting is up."[22] [15] Ward points out that the grace-nature relationship can be read as a reiteration of how kenosis cooperates with the "natural negentropic forces that analogously remove physical and moral chaos. Thus, the created order at the informational level resembles the nature of participation for this arrangement presupposes that there are no ruptures, no conflicting ontologies or phenomenologies, no dualism, only re-cognitions."[23] The material substance of bread and wine properly corresponds not only to our existential need for food but also suggests the goodness of Creation's causal origins and the redemptive agency of the groundless Spirit. Whereas the concept of nothingness impels our conceptualization of creation, the idea of the void is the "window" or

[19] See *Homo Abyssus*, 449.

[20] See *De ver.*, 10.2 and resp. 5 on memory's role in knowing.

[21] *After God*, 322.

[22] Pseudo-Dionysius, *The Celestial Hierarchy*, 1.1 in *The Complete Works* Translated by Colm Luibheid (New York, NY: Paulist Press, 1987).

[23] See Graham Ward, *The Beauty of God*, 50.

ontological aperture indispensable to our grasping the idea of finite identity. In this stratagem, the physical body acts as a topological "manifold," "skin," or "screen" upon which its various ordering principles and defining forces manifest. An individual being's interior identity finds its corollary in a matter's distinction from what it is not. The void is the mean point between Nothingness or its mathematical impression in zero and its absolute actuality. The void is the mean point between nothingness or its mathematical impression in zero and its absolute actuality where every finite being resides, for "every object possesses, among its elements, an inexistent."[24] [16] Such abstract grouping stipulates bridging material (e.g., ontological) and formal (e.g., noetic) descriptions of being through the act of ontological differing.[25] The ontological and logical relationship in Badiou's theory of mathematical ontology demands further investigation then can be provided in this study. Badiou can be seen to invert what we take to be Balthasar's understanding of Jesus' interior-exterior relations that authorize finite consciousness, freedom, and ultimately a fidelity to the Transcendent God.[26] [17]

Most postmodern supporters argue against the hylomorphic ordering of immaterial form to material being, as they regard being as "soulless" as purely material in nature. Of the sequestering of individual identity to a material calculus, Graham Harman states that all physical beings should work toward "receding from all relations, always having an existence that perception or sheer causation can never adequately measure… [but that leads to] a universe packed full of elusive substances stuffed into mutually exclusive vacuums."[27] [18] Most postmodern theorists believe that all systems that employ supervenience demand an essential differing force and a limiting periphery that they disavow. In many ways, postmodernity repackages ancient theories of hierarchical cosmos—theism by promoting the world as a pantheistic lateral transcendent being. However, their arguments, like those of their predecessors, warrant our attention and critique.

[24] *Logics of Worlds*, 323. Badiou can be seen to invert what we take to be Balthasar's understanding of Jesus' interior-exterior relations that authorize finite consciousness, freedom, and ultimately fidelity to the Transcendent God.

[25] The conflating of the ontological and logical, the mathematical and the logical in Badiou's theory of mathematical ontology demands further investigation beyond what can be provided in this study.

[26] See Alain Badiou, *Deleuze: The Clamor of Being*, Translated by Louise Burchill (Minneapolis, MN: The University of Minnesota Press, 2000), 52.

[27] Graham Harman, *Guerilla Metaphysics: Phenomenology and the Carpentry of Things* (Chicago, IL: Open Court Press, 2005), 75–76.

The argument that dual notions of physical embodiment are to be rejected because pure immanence alone promotes universal freedom is a serious point to consider. If all relations are object-orientated, this line of argument continues, and all are of equal value. Given the substantial nature of the Eucharist, horizontal three-dimensional phenomenal space is ultimately an inadequate descriptor for this hyper-dimensional entangled reality. It is in the infinitely expandable dimensions of Hilbert Space utilized by quantum theorists that a new perspective emerges on how the nonlocal transference of Transcendent into material elements at the moment of transubstantiation takes place simultaneously in compounded dimensions.

The person of Jesus Christ metaphysically embraces Transcendent One's thinking of its own eradication (e.g., non(ONE)) or death. In this event, a conceptual template for dialectical and self-reflective rationalizations of nothingness and the void is offered. Like most postmodern academics, François Laruelle would most likely reject the idea that the Transcendent's registering of absolute Nothingness could serve as a generative principle for finite being. In my reading, his work highlights how any act of self-reflection leads to a real and total foreclosure of being. If my interpretation is correct, the ONE's parabolic projection of its own non-being in the non(ONE)'s holographic embrace of death and Nothingness is not redemptive. Laruelle argues that the ONE's self-conscious differentiation "virtually" introduces a "unilaterality" that suspends or invalidates the non(ONE). He states that this relationship is "*a mode of the One's being-foreclosed, either real and not effectuated ('uni-laterality') or transcendental and effectuated by the occasion of philosophical* 'nothingness' *('unilaterality'). It bears witness to the primary of (real) foreclosure over (philosophical)* negations."[28] [19] By extension, I suggest that the One's thought of its own destruction models how finite beings understand their own filiation to nonbeing. We can discern a symmetrical alliance between the ONE's theoretical non-existence as non(ONE) and finite being's authentication by self-consciously appropriating the meaning of nonbeing, death, and the concept of nothingness.

Badiou thematically consolidates negative apparitions of finite being into his idea of the "void." The void ontologically disposes being to be aware that nonbeing is the logical antipode of one's existence. Simply put, the void is the engine that delineates being as multiple and actual because the ONE's imagined antithetical non(One) suggests that it is possible to

[28] See *Dictionary of Non-Philosophy*, 97.

quantify even universal reality. The void circumscribes a margin, restriction, or frontier for beings, which aids in their intellectual appropriation by the subject. Badiou and Laruelle's argument for the conceptualization and quantification of what I have conceived of as the dialectical ONE-non(ONE). This correspondence I see as a self-reflective pause around the themes of Nothingness and void where being finds a plausible dialog partner with my own interpretation of the informational quantification of Ur-Kenosis in the Passion and the Eucharist. Indeed, human reason can only comprehend the apparent incongruity of Christ's death and descent into Hell when this mystery becomes the parabolic meaning of one's life. The believer focuses on Christ's relationship to death and thereby shapes their own sense of self: "He is for every person, in every age, the One who is already dead; each of them possesses the reality of his death in him, For the Son has not merely carried each sinner's sins, in order to redeem him; he [has] also died each one's death, so that all the dead might share in his being."[29] [20] The Son's inseparable but infinitely free relationship with the Spirit bears a "transcendent" evaluation of the abyss of Hell's infinitely constricting contingency. Jesus opens up this metaphorically collapsing infinite set because "[h]is fullness does not stand in contradiction to any void, because it is not the fulfillment of a determined quantity or a particular power of comprehension. It is fullness without antithesis."[30] [20] Only the eternal freedom and simplicity of the Divine Persons in the immanent Trinity are able to underwrite Jesus' descent into death and Hell not as moments of bad infinity but as the impassibility of the *perpetuum mobile*.

Balthasar does not understand Hell to be some kind of negative "henology," as Christ consciously transcends all experiences of Hell.[31] The neologism "henology" here assumes Plotinus' view that the Transcendent remains beyond all finite conceptual or material confinement. Hell is a privation and inversion of Creation's goodness, but the Eternal Logos stands above these realities. For, on "the basis of supernatural revelation, we can say: Just as the hypostasis of being is ruptured in the exinanitio [kenosis] and resurrection of the incarnate Logos and thus being's movement of finitization is sublated and recapitulated, so too the Antichrist, in the hubris of the hypostasis of being and by means of the negation of the exinanitio, recapitulates ontological spatiotemporality into anti-space and

[29] *Colossians*, 35.

[30] *Colossians*, 40.

[31] The neologism "henology" here assumes Plotinus' view that the Transcendent remains beyond all finite conceptual or material confinement.

anti-time, in which he sets up his kingdom."[32] Christ descended to the "lower parts of the earth, to see with his eyes that part of creation" what was inactive and resting in Creation.[33] [21] Such descriptions of Hell are congruent with our own interpretation of evil as virtual or holographic. A holographic reading of iniquity patterns redemption, for "all men being lost according to the transgression of Adam, His flesh before all others was saved and liberated, as being the Word's Body; and henceforth we, being incorporate with It, are saved after It's pattern."[34] [22] Balthasar's view follows from his larger theological vision of the nature of Christian witness and, more particularly, his apprehension of Jesus' sensual and conscious image making in Hell.

Trinitarian kenosis and Jesus' sacrificial acts in Creation align finite being, from its inexplicable beginnings in *creatio ex nihilo* to evil's last fraught attempts in Hell to overthrow the Father's plans for Creation.[35] [23] Jesus Christ's experiences of destabilization, angst, evil, nonbeing, and nothingness are ordered to his person and mission and hence to the Divine. Through differing relationships, Jesus perfects metaphysical limits as part of his cerebration of his eternal identity.[36] [24] Jesus experiences the extreme distances between the Divine, Creation, and evil as an incarnational moment.[37] [25] Christ's experience of alienation from the Father in Hell provokes a new understanding of absence and silence. Christ's experiences of Hell are not banausic in nature but act as an interstice into the mystery of the beatific vision—the Son and the Spirit's eternal holosomatic conscious self-reflection with the Father.[38] Jesus is the gateway between pure Trinitarian love and Creation's innermost yearnings.

The Eucharist is the apogee of Creation's deepest purposes, for in this sacrament bread and wine are converted into the actual presence of the

[32] *Homo Abyssus*, 338.

[33] Irenaeus, *Against Heresies*, *Ante-Nicene Fathers*, vol. 1, Book 2, chapter 31, Edited by Rev. Alexander Roberts, D.D, and James Donaldson, LL.D., (Grand Rapids, MI: Wm. B. Eerdmans Publishing Company, 1981).

[34] See Athanasius, *Against the Arians*, *NPNF2*, vol. 4, Translated and Introduced by Philip Schaff, D.D., LL.D., and Henry Wace, D.D., (Grand Rapids, MI: Wm. Eerdmans Publishing Company, 1983), 381.

[35] On the a-temporal nature and totality of Christ's mission, see *MP*, 173 and 189.

[36] *The Pillar and Ground of the Truth*, 67.

[37] Lucy Gardner and David Moss, "Something Like Time; Something like the Sexes—and Essay in Reception," in *End of Modernity*, 109. On how the spatiotemporal relates to the Passion, see *ST.*, III, 47, 3 and *TL* 3, 266–273.

[38] The *visio mortis* never denies Jesus' beatific vision see *GL* 7, 209 *TD* 5, 146 and *MP*, 175.

Eternal Son. The Paraclete relates Creation to the Transcendent via Christ's role as a trans-spatiotemporal scapegoat, a self-conscious "sin-offering" of love for all. "God made him to be sin who knew no sin," Paul teaches, "so that in him we might become the righteousness of God" (2 Corinthians 5:21). Whereas disharmony among people metaphorically erodes or subtracts from personal and communal relationships to the Transcendent, properly ordered human relationships are defined by their receptivity to the other's needs and gifts and foster a generative love that is both intimate and the basis for social and national harmony (Matthew 5:23–24). The offer of Divine love is made in an infinite variety of ways but always circles around the finite yet transcendent revelations of the Divine infinity and eternity. The liturgy is a ceremonious retelling of past events and a living "memory," a communal celebration of Christ's redemption of reordering of the real to its original divine purposes.[39]

7.2 An Ontological Theo-Drama

Sacramental ecstasy follows from the Divine's physical presence in history (John 14:6). The Eucharist universally categorizes Creation, for it contains Christ's somatic reorientation of created being.[40] The Eucharist is super-essential (*epiousios*) food that frees us from spatiotemporal limitation, a divine gift "so that I may eat together with him, and he with me" (Revelation 3:20). The Eucharist is a sacrifice of thanksgiving that images the Ur-Kenotic manifestation of Jesus Christ's person as part of his identity as Eternal Word's to the Father's Love that is in "perfect form in the Eucharist": "behind the sacrifice of the Son (to the world) stands the consubstantial loving surrender of the Father as the source of the Eucharist; and not only this, but also the absolute self-surrender of every Divine Person to every Other—and nothing in the 'economic' sphere can intensify this surrender."[41] [26] The human person searches for God in a manner proportionate to her identity as a creature, and this includes taking responsibility for the gifting of one's existence from nonbeing. The hylomorphic and holo-somatic accord of being personalizes this abstract process. Through relational differing, the human subject positively defines their identity as an integral person. The drama of non-subsistence naturally follows when one takes

[39] See *Constitution of the Sacred Liturgy*, 56. See also *ST*. III, 73, 1, 3.
[40] See *ST*. III, 73, 3, 5.
[41] *TD* 5, 264. See also ibid., 265.

creation as a gift manifested *ex nihilio*. Nonbeing and the enlivening Spirit bracket personal existence. The Spirit brings to completion the interior holo-somatic essential truth of a being. This event describes the person's perfection in the resurrected state. Of this process of divinization, Norman Russell writes that we "can benefit from this exaltation of the flesh, not by following Christ as an example in an external manner, but by participating personally in the new life which he inaugurated. Our participation in God has a twofold aspect, an ontological one in which we are raised from non-existence to createdness, and a dynamic one in which we advance from nonexistence to transcendence."[42] [27]

The concepts of difference and the negativity holo-somatically understood provide us with a way to theorize how finite relationships correspond to the Divine via conscious apprehension of Creation's birth in the Eternal's manipulation of Nothingness. These investigations are not definitive, but when placed in dialog with our own theological glosses on Ur-Kenotic theory of Trinitarian identity, inspire a new interpretation of the metaphysics of QMT, one whose quantifiable description of being dramatically realizes the event of transubstantiation as the logic of a holo-somatic idea of personhood. Ur-Kenosis provides an exegesis of divine existence as the resplendent fullness of the transcendent mystery, intra-Trinitarian hypostatic life as uncreated and in the pure *actus essendi* of the historical person, Jesus Christ.[43] Jesus Christ gives voice to Creation's source in Nothingness in his unconditional acceptance of the Cross and its eschatological completion with his resurrection. The Divine is eternal and therefore can complete its own revelation with the infinite lesser potency of the physical world.[44]

The person engages a physical reality that is fundamentally non-subsistent given its essential identity. This coordination is dramatized in the fact that the human person's onto-theological identity is naturally and supernaturally ordered. The human person bears a likeness to the Transcendent, but this state presents in terms of humanity's greater unlikeness.[45] The paradox of Transcendent and immanent identity is perfectly resolved in the person of Jesus Christ (*diversas proportonis ad aliquid unum*). The similarity between the Transcendent and immanent is not

[42] Norman Russell, *The Doctrine of Deification in the Greek Patristic Tradition*, (New York, NY: Oxford University Press, 2006), 191.

[43] See *AE*, 73.

[44] See *TL* 1, 80.

[45] See *Fourth Lateran Council* (1213–1215).

"balanced" out by their dissimilarity. However, the Transcendent-immanent structure implicitly inheres in the created order with its historical ratification in the person of Jesus Christ and his explicit truth in the Eucharist. The eternal complex dimensions of supernatural existence holographically define the infinite, pure potentiality of the spatiotemporal realm. These dimensions are manifested at consecration.[46] The mystery of the Eucharist's holo-somatic ordering illumines how each person is tasked with deciding what is possible and what is actual in a material universe.[47] Spatiotemporal limitations (*ubi et nunc*) are bolstered, the Christian believes, through differing relationships and more specifically through those made possible by the Eucharist.

7.3 Why and How Matter, Matters

The doctrine of transubstantiation gives an account of the mystery of divine identity without displacing the natural purposes of the created order. Indeed, the event of consecration is a specialized example of what is held to occur universally throughout reality, namely, a being constantly reveals its essential meaning, for such, "*unveiledness* is, first of all, an absolute property inherent in being as such."[48] [3] The Logos' revelation is an act of divine concession that allows the Divine's assimilation of all Creation within the "non-dimensional spaces" of Ur-Kenotic monotheistic identity, as "Christ appears as a milieu, an atmosphere, a world where man and God, man and man, communicate and are united. He is 'the One who fills and fulfills [*remplit*] all things.'"[49] [28]

Human reason's boundless reach is reduced by a variety of self-imposed and external constraints. These sensual and conceptual limitations are evident when the disciples fail to recognize Jesus Christ after his resurrection. The true form of Jesus Christ outside of faith and love is beyond our range of vision.[50] After his resurrection, the material components of Jesus' corporeal body were spiritually transformed. Jesus' material and corporeal body is now holo-somatically illumined as the resurrected state that

[46] See *AE*, 155, on the "meta" definition of a being.

[47] See *AE*, 208, on the fundamental structure of a being.

[48] *TL* 1, 37–38.

[49] See Henri de Lubac S.J., *Catholicism: Christ and Common Destiny of Man*, Foreword by Christopher Butler and Translated by Lancelot C. Sheppard (London, UK: Richard Clay and Company, LTD for Burns & Oates Limited, 1962), 30.

[50] See *GL* 1, 29.

grounds each being's essential corporeity. The post-Easter stories describe how Jesus Christ is unimpeded by all spatiotemporal and material obstacles (walls, time, distance, etc.). In these stories, Jesus' physical state is comparable to the angelic nature discussed in previous chapters. Scripture describes how Jesus possesses three different existential states. These three modes of bodily existence are (1) historical existence or typical state (*status connaturalis mortalis*); (2) the resurrected or anti-typical state of Jesus post-Easter (*status connaturalis gloriousus*); and (3) the synthesis and extension of the previous two via the Spirit's symbolic or sacramental work (*status sacramentalis*).[51] The Son's love for the Father and Spirit in the Trinity, Ur-Kenotically unifies all three revelatory modes.

As created beings, we experience the Transcendent through created relationships. These correspondences remain partially foreign given our material bodies or incomplete given our intellectual and spiritual scope. For the Christian, these inadequacies are vicariously overcome when interpreted from the perspective of the crucified redeemer. The anthropological and transcendent measure of Christian existence is found in Christ's Passion, where we see that "the total otherness of the man Jesus with respect to all other human beings (whom he calls his brothers only on the day of Easter (John 20:17)... [is] a total otherness within a perfect equality of human nature... He reveals that he is wholly other precisely in his abasement, his humility, his service of all (Mark 10:45). In addition, this very inimitable quality is what those who grasp its meaning are to imitate."[52] On the Cross, Jesus Christ paradoxically announces the resurrected state of immortality in the holographic presentation of the Divine's death or in my schema, the holographic One—non(ONE) correspondence.

The two contradictory states are held simultaneously in the person of Jesus Christ because this identity is also holo-somatically that of the Eternal Word. The Spirit maintains this "impossible" unity between the Transcendent and immanent existence. The Spirit's presence holo-somatically projects as grace, "first as a being, and in this way, it must be a finite being, since it is in the soul of Christ, as in a subject, and Christ's soul is a creature having a finite capacity; hence the being of grace cannot be infinite, since it cannot exceed its subject. Secondly, it may be viewed in its specific nature of grace; and thus the grace of Christ can be termed infinite, since it has whatsoever can pertain to the nature of grace, and

[51] See *Mystery of the Eucharist*, 205.
[52] *TL* 2, 70.

what pertains to the nature of grace is not bestowed on Him in a fixed measure."[53] [29] The concept of essential holo-somatic identity reinterprets how materialistic hylomorphic theory envisages mathematical infinity and the void.

The Christian accepts an ecclesial mission and becomes "the one sent" (*hapax legomenon*). This mission derives from one's vocation and relationship with the full range of existential states lived by Jesus Christ. The Christian is "not only negatively 'redeemed' but positively endowed with missions ('charisms') that makes them persons of profile and quality within the prototypical mission of Jesus."[54] [30] In receiving a Eucharistic identity, one's freedom is sanctioned as an integral part of the created order. Intra-divine Trinitarian intention redounds in created reality as glory and therefore "proceeds in order that the infinite understanding might become infinite goodness of love or delight or blessedness or joy."[55] The human person witnesses to this when they perfect their lives in the context of ecclesial existence. The commitment to the "other" and "strangers" is lived out through spiritual practice, conscious prayer, sacramental life, and charitable actions. These communal actions rebound one's individual identity, for as a being "a thing remains in itself; as living, it opens itself through the operations of life toward others; and as known or knowing it returns from others to itself."[56] [31]

Physical properties and laws are incorporated and historically divulged in Christ's person and mission.[57] Jesus Christ exemplifies the full truth of the analogy of being because his person is perfectly spiritual. Transubstantiation ritualistically prompts one to incorporate one's personality more fully by fruitfully bringing together three elements of moral agency. One's participation in the Eucharist entails one's material body, one's free, rational and self-reflective nature. The latter movement is central to one's participation in the Transcendent's eternal life and is, consequently, a manifestation of one's nonlocal holo-somatic personhood. The theological person has "positive intent" and a "negative passivity," that is,

[53] *ST.* III, 7, 11.

[54] *TD* 3, 231.

[55] *GL* 5, 218.

[56] John Milbank and Catherine Pickstock, *Truth in Aquinas*, (London, UK and New York, NY: Routledge, 2001), 7.

[57] See Hans Urs von Balthasar, *A Theology of History*, reprint (San Francisco, CA: Ignatius Press/Communio Press, 1994), 17.

one inheres to an interior image that is "absolute poverty of spirit."[58] A life of Christian poverty requires one to die spiritually with Christ (2 Corinthians 5:15) as part of one's Eucharistic life (Luke 22:19). The Christian vocation (Romans 12:6–8) is a divine calling that paradoxically realizes one's deepest desires. It is a sacrifice to something higher than oneself, but one made without diminution of who we are—persons who holo-somatically and thus, freely are able to participate in our own theonification (*imago Dei*).

Divine love is the fundamental starting point for all theological endeavors. Florensky explains that "[t]he problem of the certitude of truth is reducible to the problem of finding a criterion. The entire demonstrative force of a system is focused, as it were, in the answer to this problem of finding a criterion."[59] [33] We turn to looking to the common denominator of all systems, the free and conscious human subject. Human love is a self-conscious and affective act that defines each person as an unrepeatable possessor of a unique vision and mission in the world.

Louis Lavelle argues that consciousness is a contextual act insofar as it is bound to the emergent or changing states of an object's appearance.[60] The human person invests something of his or her intrinsic nature in his or her actions in the world. Every human action shares something of the mystery of the actor's past and unknown future with others. In the quantum universe, it is important to critique how one's presuppositions and limitations are reflected in their actions. An experiment's quantification is never wholly an objective affair, nor is one's understanding of the nature of the Eucharist. All that separates us from the Transcendent becomes for the "Eucharistic person" an interior principle of one's identity. The paradoxical assimilation of these "subtractive" and "alienating" norms is possible as they holo-somatically interpret the Absolute Love of the Ur-Kenotic communion.

A holo-cryptic interpretation of transubstantiation is possible in light of QMT. It pushes the speculative and experiential frontiers into these mysteries for the person in the pew. Indeed, the holo-cryptic interpretation of transubstantiation suggests three immediate advantages. First, this

[58] See Angelo Scola, *Hans Urs von Balthasar: A Theological Style* (Grand Rapids, MI: Wm. B. Eerdmans Publishing Group, 1995), 10.

[59] *The Pillar and The Ground of the Truth*, 20.

[60] See Louis Lavelle, *La Présence Totale* (Paris, France: Aubier aux Éditions Montaigne, 1934), 11. See also *TL* 1, 180–185.

approach gives individuals a way to understand how liturgical prayer functions nonlocally. The Spirit receives individual prayers analogous to the way an entangled QMO relays mutual identity. The Spirit's own hypostatic transcendent nature conveys these prayers to the Trinitarian community. In this way, the finite creature has access to the nonlocal, Transcendent Trinity. Because the Spirit continues the mission of Jesus Christ, the human person is able to participate holo-somatically in the person and mission of Jesus Christ. Second, holo-cryptic metaphysics presents a way to answer Arius' rejection of Jesus Christ's divinity. The Eternal Logos and the human person Jesus Christ are eternally "entangled" at the holo-somatic or spiritual level. The Incarnation reveals in the spatiotemporal realm or history the Transcendent—immanent state in a manner analogous to quantum superposition. The theological and metaphysical revision of the holographic principle gives the faithful a way to participate in liturgy without falling into the heretical tendencies of Arianism, Docetism, etc. Third, conceived as a holo-somatic being, the Eucharist's substantial transformation at the consecration evinces the most fundamental properties of subatomic particles. More specifically, the hyper-dimensional existence of the QMO gives us insights into the event of transubstantiation as a historical event that project into the supernatural or hyper-dimensional realms. In conclusion, the doctrine of transubstantiation encourages us to reflect on how created beings exist in nonsubsistent relationships in the created order and, further, how these correspondences paradoxically include a physically concrete relationship with the person of Jesus Christ and the Ur-Kenotic communion.

References

1. Henry P. Stapp, *Mind, Matter and Quantum Mechanics*, 3rd Edition (Springer, verlag; Berlin, Heidelberg, 2009), 165.
2. *Physics and Philosophy* (New York, NY: Harper and Row, 1962), 55.
3. Hans Urs von Balthasar, *Truth of the World* Vol.1, Translated by Adrian J. Walker (San Francisco, CA: Ignatius Press, 2000), 37–38.
4. Werner Heisenberg, "Uber denan schaulichen Inhalt der quanten theoretischen Kinematik und Mechanik," *Z. Physik* 43: 172–98 (1927), 197.
5. John A. Wheeler and W. H. Zurek, *Quantum Theory and Measurement* (Princeton, NJ: Princeton University Press, 1983).
6. Alain Badiou, *Theory of Subject* Translated and Introduced by Bruno Bosteels, (London, UK: Continuum, 2009), 76–77 and 184.

7. Silvan and S. Schweber, *QED and the Men Who Made It: Dyson, Feynman, Schwinger, and Sin-Itiro Tomonaga* (Princeton, NJ: Princeton University Press, 1994), 86–87.
8. James Owen Weatherall, *Void: The Strange Physics of Nothing* (New Haven, CT and London, UK: Yale University Press and Templeton Press, 2016), 111–113 and 125–127.
9. Peter Milonni, *The Quantum Vacuum: An Introduction to Quantum Electrodynamics*, (San Diego, CA: Academic Press, 1994),
10. Thomas Aquinas, *Truth* Vol. 2: Questions X–XX. Translated by James V. McGlynn, S.J. (Chicago, IL: Henry Regnery Company, 1953), 2, 3, 16.
11. *Sacrammentum Mundi:* Vol. 1. Edited by Karl Rahner (London: Burns Oates, 1993).
12. Augustine, "On Christian Doctrine", *The Nicene and Post-Nicene Fathers*, vol. 2, Book 1, chapter 6.
13. *Summa Theologiae* I, 104, 1, (Garden City, NY: Doubleday Company, 1964–1976).
14. Mark C. Taylor, *After God*, (Chicago, IL: The University of Chicago Press, 2009), 322.
15. *The Celestial Hierarchy*, 1.1, Pseudo-Dionysius, *The Complete Works*, Translated by Colm Luibhéid (New York, NY: Paulist Press, 1987).
16. Alain Badiou, *Logics of Worlds: Being and Event 2*, Translated by Alberto Toscano (London, UK: Continuum, 2009), 323.
17. Alain Badiou, *Deleuze: The Clamor of Being*, Translated by Louise Burchill (Minneapolis, MN: The University of Minnesota Press, 2000), 52.
18. Graham Harman, *Guerrilla Metaphysics: Phenomenology and the Carpentry of Things* (Chicago, IL: Open Court Press, 2005), 75–76.
19. François Laruelle and collaborators, Tony Brachet, Gilbert Kieffer, Laurent Leroy, Daniel Nicolet, Anne Françoise Schmid and Serge Valdinoci, *Dictionary of Non-Philosophy*, Translated by Taylor Adkins. (Minneapolis, MN: Univocal, 2013), 97.
20. Adrienne von Speyr, *Colossians*, Translated by Michael J. Miller, (San Francisco, CA: Ignatius Press, 1998).
21. Irenaeus. *The Ante-Nicene Fathers Translations of the Writing of the Fathers down to A.D. 325*, Series 1, Vol. 1: The Apostolic Fathers with Justin Martyr and Irenaeus: "Against Heresies." Edited by Alexander Roberts, D.D. and James Donaldson, LL.D. Revised and chronologically arranged, with brief prefaces and occasional notes by A. Cleveland Coxe, D.D. Catholic Logos Edition (Buffalo, NY: The Christian Literature Company, 1885), 4, 22, 1.
22. Athanasius of Alexandria, *Select Writings and Letters of Athanasius, Bishop of Alexandria*, "Against the Arians", Edited by Archibald Robertson. *The Nicene and Post-Nicene Fathers, Second Series*, Vol. 4. Oxford/New York, 1894; reprinted Edinburgh: T&T Clark, Edinburgh, 1989, Catholic Logos Edition) 4, 2, 61.

23. Balthasar, *Mysterium Paschale: The Mystery of Easter*, Translated and Introduced by Aidan Nichols, O.P. (Grand Rapids, MI: Wm. B. Eerdmans Publishing Company, 1990), 173.
24. Pavel Florensky, *The Pillar and Ground of the Truth: An Essay in Orthodox Theodicy in Twelve Letters*, Translated by Boris Jakim and Introduced by Richard F. Gustafson (Princeton, NJ: Princeton University Press, 1997), 67.
25. Lucy Gardner and David Moss, "Something Like Time; Something like the Sexes—and Essay in Reception," in *The End of Modernity*, Forward by Fergus Kerr. Afterward by Rowan Williams, (Edinburgh: T&T Clark Ltd, 1999), 109.
26. Hans Urs von Balthasar, *Theo-drama: Last Act*: Vol. 5. Translated by Graham Harrison (San Francisco, CA: Ignatius, 1998), 265.
27. Norman Russell, *The Doctrine of Deification in the Greek Patristic Tradition* (New York, New York: Oxford University Press, 2006), 191.
28. Henri de Lubac S.J., *Catholicism: Christ and the Common Destiny of Man*, Forward by Christopher Butler and Translated by Lancelot C. Sheppard, (London, UK: Richard Clay and Co. LTD., for Burns and Oates LTD., 1962), 30.
29. *Summa Theologiae* III, 7, 11 (Garden City, NY: Doubleday Company, 1964–1976).
30. Hans Urs von Balthasar, *Theo-Drama: Theological Dramatic Theory, Vol. 3, The Drama Personae: The Person of Christ*, Translated by Graham Harrison (San Francisco, CA: Ignatius Press, 1992), 231.
31. John Milbank and Catherine Pickstock, *Truth in Aquinas* (London, UK: Routledge, 2001), 7.
32. Hans Urs von Balthasar, *Truth of the World* Vol.1, Translated by Adrian J. Walker (San Francisco, CA: Ignatius Press, 2000), 192.
33. Pavel Florensky, *The Pillar and Ground of the Truth: An Essay in Orthodox Theodicy in Twelve Letters*, Translated by Boris Jakim and Introduced by Richard F. Gustafson (Princeton, NJ: Princeton University Press, 1997), 20.

Appendix A: The Quantum Object, Vectors, Eigen States, and Eigenvalues

Mathematics, philosophy, and theology are esteemed for their ability to describe the mysterious and partially known. Although recent technological advances allow a glimpse into the physical appearance of the QMO, their interior structure or essence defies easy access or explanation. While various mathematical vectors do not fully account for the physical and metaphysical properties of the QMO, they set out the necessary fundamental principles that are compatible with more in-depth formulas. In other words, mathematical vectors, Eigen states, and eigenvalues are tools we use to represent the complex, higher-dimensional spaces inhabited by the QMO. The physical properties of the quantum system are indicated by the sum of the vectors found within. A mathematical vector represents an object's path or direction, as well as the distance it travels. An arrow symbolically represents a mathematical vector whose nock signifies its definitive point of origin: → (see *Smith*, 140–42).

The distance between a vector's point of origin and its terminus emblematically presents a sense of an object's momentum, without reckoning fully for its intrinsic metaphysical freedom. An eigenvector details how a QMO's spatial transformation can occur within a given frame of reference without fundamentally changing its "essential" identity. Concomitantly, a vertical bar followed by a Greek letter and a bracket—e.g., [$|\psi\rangle$ or $|X\rangle$] is used as shorthand for the two most fundamental

M. P. Fusco, *The Physics and Metaphysics of Transubstantiation*,
https://doi.org/10.1007/978-3-031-34640-8

states of the QMO. Quantum spin gives us a way to talk about the QMO's reflexive nature as one of four possible orthogonal combinations—00, 01, 11, or 10. In this way, a QMO's causal momentum can be distinguished from its interior movement as spin-up or spin-down. Graphing the quantum spin in relationship to the Cartesian plane or the *x*, *y*, and *z* plane, a further insight follows: namely, spin-up is denoted by the mathematical location (1, 0), whereas the quantum state of spin-down identifies with the point (0, 1). This convention specifies a coherent way to discuss the formalization of QMO in relation to its place within a dimensional system in light of its internal states. Thus, the following equation holds for classical spatial representation and complex spatial renderings: $(x, y)(x'y') = (xx' - yy', xy' + yx')$. Idealized in its simplest configuration and with its motion conceptually arrested, a quantum particle's interior identity necessarily prevails in the state of spin-up ($|\uparrow_z\rangle$) along the *y*-axis or spin-down ($|\downarrow_z\rangle$) aligned to the *x*-axis of the Cartesian coordinate system.

Mathematically, a QMO's momentum is calculated by multiplying a vector by a nonzero complex number. The various positions of state vectors (e.g., $|\psi\rangle$) are symbolically rendered by a letter ($\alpha_1, \alpha_2, \ldots \alpha_n$) and some complex coefficient i ($i = 1, 2, 3\ldots\ n$) representing nonzero numbers. Taking O as our starting point, we can draw a vector OU and multiply it by a positive real number, α. Taking an imagined "transcendent" reference point R on the line UV that we can multiply $|OV|$ by some value α. If α is in the same direction, the following results: $\alpha|OU| = |OR|$. The number α must be in the same direction in the case of both vectors. When α is a negative number ($-\alpha$) one reverses the direction of the resultant vector.

If multiplied by zero ($\alpha = 0$), a vector's resultant value is 0 (*Smith*, 133–34) and suggests location rather than momentum or interior movement. Thus, I see the 0 value as a means to metaphorically speak of the local observer's perception of the QMO or the realization of the Absolute Freedom of the Spirit at the moment of consecration.

If the scalar α is a positive number, then the spatial point of origin O will not be a point found on a line between R and V. Point O theoretically "subtracts" or "distances" itself from all points on line UV (as represented by the vector movement away from its point of origin) and concurrently secures the addition of all points making up UV (e.g., the "difference" traveled). This "subtractive" or differential relating allows us to set our projected reference point O in the context of the line UV and the vector—a move that is proportional to all geometric constructs within a system created by the scalar multiplication of the vectors. In other words, the total sum of the eigenvectors from a common point of origin can be represented as quantum rays. These rays define the area of the quantum space (Σ) being considered (*Smith*, 62). The mathematical differentiation of the QMO when seen as part of its "synthetic" whole depicts something of what in our speculative holo-cryptic parlance means by finite kenotic relationships. Because the probable distribution of the sum of all divisions of vector space equals 1, the unique identity of each vector can be understood in relationship to the system as a whole (see *Forrest* 1988: 19–20). Additionally, the multiplication of state vectors by the same nonzero complex number does not alter the preexisting proportions between vectors. These formal manipulations of vectors have great predicative powers, as the value of an Eigen state corresponds to an eigenvector. The multiplication of an eigenvector by a complex number, for example, translates a QMO within a system without eradicating its preexisting spin (*Peter J. Lewis* 2016: 10–11).

The inner product stands in for the essential information concerning the geometric grounding of the quantum object's spin state—a property that allows one to attribute a logical state to geometric constants. An eigenvector of a quantum system is always positive because the sum of squared vectors always results in a nonzero number—it is physically impossible to fully remove a QMO from the quantum system that houses it (e.g., field, probability cloud) or analogously, two beings cannot exist in the same spatiotemporal place. A dyadic correspondence exists between the QMO's observation and its conceptualization (*A Beautiful Question*, 311).

The entangled state is determined by combining two corresponding vectors such that the result of obtaining spin-up for both electrons is determined by projecting the entangled state onto the basic state. An electron is in superposition when it is, for example, orientated along the z-axis such that its Eigen state is simultaneously twofold—it evidences an up spin ($|1\uparrow\rangle$) and down spin ($|0\downarrow\rangle$). The complexity of a wave's probability is reduced here to an idealized summary of the wave's complete spatial topology. The measuring of a QMO metaphorically "subtracts" something from its entangled partner, as its identity depends on its pair. This condensation of identity through the subtraction of another particle is conceptualized variously in terms of a particle's "decay" or a wave's "collapse" (see *Forrest* 1988, 45). The Eigen state of an observed particle is uniquely identified, as each perceived electron is both exceptional and in a complementary relationship with other particles or waves.

Appendix B: Complex Space

We first understand the QMO in its relationship to classical n-dimensional Euclidean space. However, the three values possible for the QMO's position and its three corresponding values for its momentum require six dimensions. This space is adequate to include representing the QMO's unchangeable geometry and its probabilistic structures (see *David Deutsch*, quant-ph/9906015 (1999)). The inner product of quantum vectors in complex Hilbert space can be used to quantify the interior properties of two or more vectors. In an orthodox reading of QMT, Hilbert spaces are reduced or normalized to represent the physical topology and mathematical configuration of the quantum system. Symmetry holds between the QMO's mathematical explanation and its physical topology as a boundary condition or projection.

Configuration space gives us a way to speak about the QMO's holographic and mereological identity. This follows because, as noted previously, the primitive identity of the physical QMO in closed linear subspaces is symmetrical to its local and emergent manifestation of a quantum particle in higher dimensional spaces. The properties of identity and difference, part and whole, feature the most elementary levels of reality when translated into holographic information. Introducing a part-whole schema within Hilbert space is valid because the x, y, and z coordinate plane transcribes into imaginary or higher dimensions (see *Smith*, 139). The

M. P. Fusco, *The Physics and Metaphysics of Transubstantiation*, https://doi.org/10.1007/978-3-031-34640-8

complex domain of the "hyper-dimensional" "holographically" builds upon and implicitly preserves something of a being or number's original physical and configuring dimensions. Both the physical object and its configuration as essence or mathematical configuration space are reduced to its spatiotemporal existence.

Appendix C: Holographic Information

According to Claude Shannon, information is defined in entropy's relationship to a thermodynamic system—that is, as a measure of order. Consequently, information can also be understood as a measure of disorder or negentropy (*Gleick* 2012: 269–286). Information can be likened to light (energy), as both can be physically and symbolically quantified, repeated, developed, and transmitted. The entropic "de-evolution" from complex patterning to the stilted chaos of a purely uniform environment highlights the relationship between material bodies and their informational characteristics: (1) entropy moves from pattern and order to a state of disorder, (2) entropic systems reflect unpredictability, (3) entropy quantifies motion and is therefore related to thermodynamic descriptions, and (4) further, entropy is a quantifiable informational state with an equilibrium point or "identity." The nexus of these various factors gives us an analogy for what Aristotle advances in his principle of the "mean" or midpoint, classical physics defines as a thermodynamic equilibrium point, and QMT analogously references when speaking of geometric structures.

Shannon's theory of information can easily be adapted to a broad range of methodological approaches because it mathematically defines information in its most primitive forms. The QMO's identity and its informational state can be symbolized as 0 or 1. The QMO's binary identity (e.g., ($|0\rangle$ and ($|1\rangle$) makes it an ideal model for all information. The theories of

M. P. Fusco, *The Physics and Metaphysics of Transubstantiation*,
https://doi.org/10.1007/978-3-031-34640-8

entropy and thermodynamics postulate that information must (1) be defined within a closed system and (2) be quantifiable in terms of its motion. These two properties of information are evident in the structure of the QMO and systematically manifest in the holographic nature of a black hole. A parallel can be drawn between the holographic projection of a quantum Eigen state ($\varphi\alpha$) in Hilbert subspaces and its holographic presentation in the thermodynamic and dissipated motion of Black Hole's event horizon (S. A. *Hartnoll* et al. 2007). Thus, a mutual relation holds, as in both cases, a description is offered for how information functions as a part of a boundary condition for a black hole or that of the total dimensional register of Hilbert space. A distinction can be made between informational states and their point of intersection at the surface gravity of the event horizon and that of the QMO. Subspaces of Hilbert space are commensurate with the structure of the black hole's maximal entropic state or degrees of perturbation or freedom. Consequently, in a black hole, it is possible to derive the entropic reading of the informational state of the QMO as a temperature and the change in density of the black hole's event horizon's radius (*Hartnoll* et al. 2018: 61).

Information imprinting at the event horizon takes place in lower spatiotemporal dimensions. Thus, a material object's transformation into light is bounded by the event horizon's surface area (*Bousso* 2002: 859). Hawking radiation is so extreme that informational changes at the event horizon require us to understand how it affects classical scientific equations and post-Newtonian estimations (*Hartnoll* et al. 2018: 4, 61). Indeed, at the black hole's singularity, we have an example of how a physical object theoretically exists in a vortex where a point of infinite spatial density exists (*Hawking* 1988). The smallest possible length in the physical universe is Planck's length. This constant gives us a way to speak of the relative informational density at the Black Hole's event horizon—the so-called Bekenstein—Hawking entropy. Even though a physical object presents holographic information (Hawking radiation) in two dimensions at the event horizon, theoretically, these data can be used to retroactively determine an object's four-dimensional spatiotemporal identity prior to entering a black hole (*Thorne and Macdonald* 1986).

A being's holographic representation is another manner of presenting the essence or informational state of the QMO. The micro-states of the quantum wave's motion can be normalized and mathematically defined within Hilbert space as a whole. This develops the conventional description of the QMO in terms of its energy and motion. Quantum reflexivity gives us a way to see how a QMO's metaphysical and mathematical truth theoretically and physically coheres. Consequently, we can see how the in-formational properties of the object holographically distend into spatiotemporal, hyper-dimensional spaces or hierarchic configurations (*Tegmark* 2007: 8). The interior informational state of the QMO is revealed at the edge where this holographic manifestation meets the boundaries of its spatiotemporal and dimensional "home." The QMO's holographic identity is reflexive to its geometric, physical, and dimensional structures.

The identity of the quanta or any other being includes a measure of the totality of information states it contains. The totality of quantifiable information "stored" in a given being can be termed its "density." The maximum storage of information or density of a being is quantified or distinguished by differentiating it from the system to which it belongs. Concurrently, a formal distinction can be made in a system in regard to a particular volume of space or motion in relation to the maximum amount of the system as a whole. In sacramental theology, this relationship is implicit when one speaks of the interior essence or informational state of the Eucharist in relationship to the eternal presence or metaphoric "volume" of the Spirit.

Glossary

Absolute Groundlessness A name for the divine essence and the Transcendent that highlights the Absolute Freedom of the Spirit.

Apophasis Apophasis is a logical process that gives us a way to speak about the unspeakable and partially unknown by elucidating it through negation. Thus, apophasis uses "subtraction" or "denial" to define the identity of an inaccessible being under investigation. Theologically apophasis is closely aligned to Thomistic remotion and Ur-Kenosis.

Aporia The indeterminacy that results when one's understanding of objective reality imperfectly corresponds to the truth.

Condensation The "gravitating" of a rational being's essential identity in terms of the three centering relationships of being-in-itself, being-for-itself, and being-for-others. Analogously, the QMO's identity condenses or collapses from a wave to a particle.

Density Matter's physical density is changed in a black hole into its photic or holographic identity as radiation. Thus, the holographic principle gives us a way to understand how a being's identity as form and matter, like energy and mass, are convertible, as Einstein proved with his $E = mc^2$ equation. The thermodynamic properties of a being within a system anticipate this realization. We speculatively apply these findings to explain that the physical properties of bread and wine holographically display the holo-somatic relationship shared between Jesus Christ and the Eternal Word through transubstantiation at the Eucharist.

M. P. Fusco, *The Physics and Metaphysics of Transubstantiation*, https://doi.org/10.1007/978-3-031-34640-8

Difference Through differing relationships, the identity of individual beings reveals. Difference is a nomological covariance that helps to establish a being's essential and accidental properties.

Dimensions A term for a configuration space of any number of reference points.

Eternal A supra-temporal state that defines the Transcendent. The concept of infinity is an analogy for how the resurrection holo-somatically defines the eternal for finite creatures.

Godself A non-gender specific term used to indicate divine self-reflectivity.

Heno-triadic Identifies the 3-in-1 relational natures of finite beings. In holo-cryptic metaphysics beings are seen to primarily exist in three configuring relationships: (1) self-reflective relationships, (2) non-subsisting correspondences with other beings, and (3) in transcendental relationships to the Divine. These three relationships holo-somatically image the person and mission of Jesus Christ in history and thus metaphysically confirm the economic Trinity's own source in the Immanent Trinity's eternal sharing of a single divine essence.

Holo-cryptic A new approach to metaphysics based on a cross-disciplinary reading of philosophy, theology, quantum physics, information theory, and black hole entropy. A synthesis of these various sources is achieved by seeing the fundamental symmetrical relationship between energy and mass in the universe, finding its privileged expression in the holographic principle.

Holographic Principle An explanation of how a physical body can be manifested as a holograph. When a physical body is swallowed into a black hole, its material structure is transformed into energy that represents its informational structure at a lower dimension on the event horizon.

Holo-somatic Holo-somatic is a term that denotes how an object or being's identity manifests as a holographic image in various spatiotemporal and complex dimensions and ultimately in the resurrection state.

Hylomorphism Hylomorphism is a philosophical theory that argues that all beings are intrinsically composed of form (essence) and matter (Greek hylē, "matter"; *morphē*, "form").

Infinity An unending series. Infinity is taken to be an apophatic image of the Eternal. This image is understood to be a holo-somatic predication of the Eternal. The infinite analogously spatiotemporally actualizes in the conception of the Eternal Word in the person of Jesus Christ and again at the moment of consecration with this divine-human person holo-somatically becoming present in the Eucharist.

Information At the fundamental level, information asks us to consider the nature of order and disorder. Thus, information conveys conceptual ideas (e.g., forms) and displays as a principle of embodiment in physical beings. In holo-cryptic metaphysics, holograph reveals how information is simultaneously a physical phenomenon (e.g., photic energy) and a form or immaterial measure of order and dis-order. Just as hylomorphic theory argues that form-matter is intrinsically connected or in a state of superposition, holo-cryptic metaphysics argues that a holograph's dimensional projection is symmetrical to a physical being's spatiotemporal identity. In this way, information is entropic when understood as a configuration or phase state. Consequently, prior to information transmission or holograph observation, it *a priori* anticipates its dimensional or spatiotemporal translation. Given information's properties, holo-cryptic metaphysics argues that it allows us to translate past interpretation of the concept of essence in light of Post-Newtonian science.

Meonic A term used to refer to revealed nature of Nothingness or void from the Greek: *τό μὴ ὅν*—"that which is not."

Mereological essentialism The attempt to explain that there is a relation of the part to the whole with reference to the essence of being.

Negentropy A term introduced by Léon Brillouin (1889–1969). Entropy measures the uncertainty (e.g., what is countable and what is uncountable) of an event, whereas negentropy can be viewed as a measure of entropy.

(non)One A term that alludes to the postmodern rejection of the ONE or the Transcendent's existence. The (non)One represents the ONE or the Transcendent's antithesis or its perceived opposite as non(ONE).

Nothingness The term Nothingness refers to the conceptual framing of the divine exemplar prior to the Transcendent contenting it with a specific being, concept, or meaning. This development is seen to be congruent with scriptural and Patristic accounts of Creation. Furthermore, this theological interpretation of Nothingness can be placed into conversation with many scientific descriptions of the void, nothingness, as well as the properties of a black hole.

ONE Is taken to be a homologous philosophical term for the Transcendent in the Christian tradition. The term ONE references the Transcendent's perfect self-unity against all finite composite realities.

One The use of lower case highlights how many postmodern philosophers such as Badiou reject all traditional faith-based theories about the Transcendent. The word One implies those theories and systems that argue against belief in the Divine but nonetheless need a logical placeholder for its meaning in their writings.

Ontological Difference The mutual differing of a being's essential (essence) and existential (esse) identities.

QED An abbreviation for quantum electrodynamics.

QMO An abbreviation that designates a quantum mechanical object. Thus, a QMO exists as both a wave and particle prior to its observation.

Quantum entanglement The event where the measure of a quantum particle immediately changes the measure of its entangled partner. This interaction takes place on the formal or informational level and is therefore a physical phenomenon that is independent of the limitations of spatial proximity.

Quantum superposition The QMO simultaneously exists as a single particle. These two states exist in superposition. A deeper level of investigation shows that superposition holds at the formal level as any two quantum states can be added together to create a third valid quantum state. The physical measure or immaterial conscious appropriation of quantum superposition is problematic, as both appear to introduce decoherence into the quantum system.

Reality The totality of all natural and supernatural properties, characteristics, and dimensions of Creation and Being. The incalculable diversity and multiplicity of beings within this whole allows for its description as a unified extensible totality with non-subsisting relations and other differences that can be infinitely separated and subdivided. The communication of reality in relational differing allows for the description of beings as discrete quanta (particular beings) that are part of reality as a whole (e.g., the universal).

Relation The "bridge" that joins two or more beings or terms. For finite beings, such relations are accidental, non-inhering, and historical. The hypostatic relations within the Trinity are eternal, and as all the Transcendent's choices are part of the Divine Persons' "conscious" state, even their revelation in history is non-accidental—introducing a real relation.

Resipiscence Refers to a change of mind or heart often to a correct position. Used as a specialized philosophical term it refers to a being or object's ongoing disclosure of its essential or informational identity in physical appearance. Each new phenomenal appearance and its subsequent historical dissipation make room for perspectives of accidental change. Resipiscence is part of a relational being's preconscious appropriation of identity's reflexive and intentional (telic) nature. Resipiscence is a forerunner to our scientifically grounded reading of the holographic principle and thus our own holo-cryptic metaphysics. Thus, resipiscence and holo-somatic identity inaugurate the interior state of an individual being and its physical identity in reality via one's relationships and thus actualize a finite analog of eternal Ur-Kenosis.

Symmetry Symmetry is the property of an object that defines it as invariant identity under a transformation (e.g., reflection, rotation, translation). Symmetry is part of the anagogic revelation of the divine and human relationship.

Transnihilation A term adapted from the work of Ferdinand Ulrich's work, *Homo Abyssus: The Drama of the Question of Being*. In Ulrich' work the term speaks to how a being's finitude and subsistence is related to its essential nature and the transformation of its composite or analogous "hypostatic" nature. In this work transnihilation refers to my speculative reading of a finite being's holographic experience of the Spirit as Absolute Nothingness, that is, as eternal hypostatic Absolute Freedom or Pure Spirit in the Trinity and holographically as non-subsistent analogous "kenotic" or differing correspondences in the created order.

Ur-Kenosis Is a term that defines how the Transcendent's identity actualizes through the eternally mutual and giving hypostatic relations among the three Divine Persons. According to Ur-Kenotic theory, the Transcendent's simple identity follows from the Divine Person's heno-triadic correspondence—the Father's gift of self to the Son who returns all back to the Father through the Spirit. My reading develops upon the work of Hans Urs von Balthasar.

Persons Index[1]

[1] Note: Page numbers followed by 'n' refer to notes.

M. P. Fusco, *The Physics and Metaphysics of Transubstantiation*, https://doi.org/10.1007/978-3-031-34640-8

C

Scriptural Passages Index[1]

G

E

K

M

P

W

I

E

D

[1] Note: Page numbers followed by 'n' refer to notes.

M. P. Fusco, *The Physics and Metaphysics of Transubstantiation*,
https://doi.org/10.1007/978-3-031-34640-8

Subject Areas Index[1]

[1] Note: Page numbers followed by 'n' refer to notes.

M. P. Fusco, *The Physics and Metaphysics of Transubstantiation*, https://doi.org/10.1007/978-3-031-34640-8

S

T

U

V

W

Z

Printed in the USA
CPSIA information can be obtained
at www.ICGtesting.com
LVHW010744240823
756131LV00007B/189